Bioenergetics

Bioenergetics

Dr. Siya Ram

RANDOM

RANDOM PUBLICATIONS
NEW DELHI (INDIA)

Bioenergetics

ISBN 978-93-5111-955-5

Published in 2016 in India by

Reprint 2019

RANDOM PUBLICATIONS
4376-A/4B, Gali Murari Lal, Ansari Road
New Delhi-110 002
Phone : +9111-43580356, 011-23289044, 011-43142548
e-mail: sales@randompublications.com,
info@randompublications.com, randomexports@gmail.com

Type Setting by : Friends Media, Delhi-110089
Printed at : Mehra Printers, Delhi-110 092

Preface

Energy and matter are the building blocks of the universe as we know it. Energy which has been defined as the capacity to do work is the basis of all changes that has occurred over a period of time since the origin of universe.

Bioenergetics is the part of biochemistry concerned with the energy involved in making and breaking of chemical bonds in the molecules found in biological organisms. It can also be defined as the study of energy relationships and energy transformations in living organisms.

Growth, development and metabolism are some of the central phenomena in the study of biological organisms. The role of energy is fundamental to such biological processes. The ability to harness energy from a variety of metabolic pathways is a property of all living organisms. Life is dependent on energy transformations; living organisms survive because of exchange of energy within and without. In a living organism, chemical bonds are broken and made as part of the exchange and transformation of energy. Energy is available for work (such as mechanical work) or for other processes (such as chemical synthesis and anabolic processes in growth), when weak bonds are broken and stronger bonds are made. The production of stronger bonds allows release of usable energy.

Living organisms produce ATP from energy sources via oxidative phosphorylation. The terminal phosphate bonds of ATP are relatively weak compared with the stronger bonds formed when ATP is broken down to adenosine monophosphate and phosphate and then dissolved in water. Here it is the energy of hydration that results in energy release. An organism's stockpile of ATP is used as a battery to store energy in cells, for intermediate metabolism. Utilization of chemical energy from such molecular bond rearrangement powers biological processes in every biological organism.

The book is meant to be a text for students of life science, pharmacy, agriculture etc. at the graduate and post graduate level.

–Author

Contents

Preface .... *v-vi*

1. Energy Flow through Living Systems **1**

Energy and Work 1
The Direction of Spontaneous Processes 5
Free Energy and Chemical Potential 7
The Electrochemical Potential 13

2. Biomolecules **18**

Biological Molecules 18
Biological Molecules: Structure and Function 21
Nucleic Acids 23
Amino Acids 26
Nucleosides and Nucleotides 30
Structure of DNA 37
Polysaccharides 41

3. Biological Functions of Proteins **56**

Proteins 56
Proteins Are Linear Polymers of Amino Acids 57
Architecture of Protein Molecules 66
Proteins are the agents of Biological Function 71
Chemical Groups 77
Reactions of Peptides and Proteins 80
Purification of Protein Mixtures 80

4. Thermodynamic Aspects of Bioenergetics **84**

Thermodynamic Measurements 84
Thermodynamic Parameters of b-lactoglobulin 91
Thermodynamics and Interactions 101
Fundamentals of Enzyme Kinetics 118

5. The Cell **120**

The Specialized Cells 120

The Structure of the Cell Membrane 147
The Permeability of Cell Membranes 152

6. Enzymes 160

Enzyme Classifications 160
Role of Coenzymes 164
Enzyme nomenclature 179
Enzyme units 182
The Mechanism of Enzyme Catalysis 183
Simple Kinetics of Enzyme Action 184
Enzyme Inhibition 203

7. Photosynthetic Energy Transformation 212

Role of Photosynthesis 212
Classification of Photosynthetic Organisms 215
Principles of Photosynthetic Energy Transformation 216
Oxygenic Photosynthesis 219
Photosynthetic Electron Transport System 235

8. Respiration The Hormones 246

Plant-Hormones .....
Cytokinins .....
Ethylene .....
Auxins .....
Gibberellins .....
Hormone Weedkillers .....

9. Biological Fixation of Nitrogen The Mitochondria 260

Introduction 260
The Chemical Nature of Mitochondria 265

Bibliography 309

Index 311

1

Energy Flow through Living Systems

The flow of matter through individual organisms and biological communities is part of everyday experience; the flow of energy is not, even though it is central to the very existence of living things. What makes concepts such as energy, work, and order so elusive is their insubstantial nature: We find it far easier to visualize the dance of atoms and molecules than the forces and fluxes that determine the direction and extent of natural processes.

The branch of physical science that deals with such matters is thermodynamics, an abstract and demanding discipline that most biologists are content to skim over lightly. Yet bioenergetics is so shot through with concepts and quantitative relationships derived from thermodynamics that it is scarcely possible to discuss the sub- ject without frequent reference to free energy, potential, entropy, and the second law.

The purpose of this chapter is to collect and explain, as simply as possible, the fundamental thermodynamic concepts and relationships that recur throughout this book. Readers who prefer a more extensive treatment of the sub-ject should consult either the introductory texts by Klotz and by Nicholls and Ferguson or the advanced texts by Morowitz and by Edsall and Gutfreund.

Thermodynamics evolved during the nineteenth century out of efforts to understand how a steam engine works and why heat is produced when one bores a cannon. The very name "thermodynamics," and much of the language of this science, recall these historical roots, but it would be more appropriate to speak of energetics, for the principles involved are universal. Living plants, like all other natural phenomena, are constrained by the laws of thermodynamics. By the same token, thermodynamics supplies an indispensable framework for the quantitative description of biological vitality.

ENERGY AND WORK

Let us begin with the meanings of "energy" and "work." Energy is defined in elementary physics, as in daily life, as the capacity to do work. The meaning of work is harder to come by and more narrow. Work, in the mechanical sense, is the displacement of any body against an opposing force. The work done is the product of the force and the distance displaced, as expressed in the following equation:*

$$W = f\,\Delta l \quad (2.1)$$

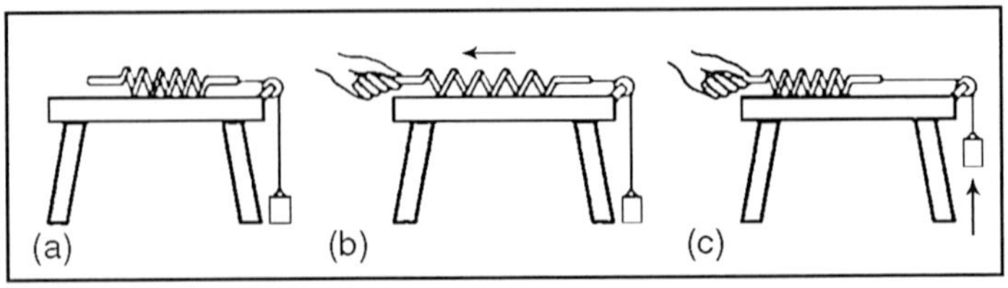

Fig. Energy and work in a Mechanical System.

(A) A Weight Restingon the Floor is Attached to a Spring via a String. (B) Pulling on the Spring Places the Spring Under Tension. (C) The Potential Energy Stored in the Extended Spring Performs the work of Raising the Weight when the Spring Contracts.

Mechanical work appears in chemistry because whenever the final volume of a reaction mixture exceeds the initial volume, work must be done against the pressure of the atmosphere; conversely, the atmosphere performs workwhen a system contracts. This work is calculated by the expression $P\Delta V$ (where P stands for pressure and V for volume), a term that appears frequently in thermodynamic formulas. In biology, work is employed in a broader sense to describe displacement against any of the forces that living things encounter or generate: mechanical, electric, osmotic, or even chemical potential.

A familiar mechanical illustration may help clarify the relationship of energy to work. The spring in Figure can be extended if force is applied to it over a particular distance—that is, if work is done on the spring. This work can be recovered by an appropriate arrangement of pulleys and used to lift a weight onto the table. The extended spring can thus be said to possess energy that is numerically equal to the work it can do on the weight (neglecting friction). The weight on the table, in turn, can be said to possess energy by virtue of its position in Earth's gravitational field, which can be utilized to do other work, such as turning a crank. The weight thus illustrates the concept of potential energy, a capacity to do work that arises from the position of an object in a field of force, and the sequence as a whole illustrates the conversion of one kind of energy into another, or energy transduction.

The First Law: The Total Energy Is Always Conserved

It is common experience that mechanical devices involve both the performance of work and the production or absorption of heat. We are at liberty to vary the amount of work done by the spring, up to a particular maximum, by using different weights, and the amount of heat produced will also vary. But much experimental work has shownthat, under ideal circumstances, the sum of the work done and of the heat evolved is constant and depends only on the initial and final extensions of the spring. We can thusenvisage a property, the internal energy of the spring, with the characteristic described by the following equation:

$$\Delta U = \Delta Q + \Delta W \ (2.2)$$

Here Q is the amount of heat absorbed by the system, and W is the amount of work done on the system. In Figure the work is mechanical, but it could just as well be electrical, chemical, or any other kind of work. Thus ΔU is the net amount of energy put into the system, either as heat or as work; conversely, both the performance of work and the evolution of heat entail a decrease in the internal energy. We cannot specify an absolute value for the energy content; only changes in internal energy can be measured. Note that Equation assumes that heat and work are equivalent; its purpose is to stress that, under ideal circumstances, ΔU depends only on the initial and final states of the system, not on how heat and work are partitioned.

Equation is a statement of the first law of thermodynamics, which is the principle of energy conservation. If a particular system exchanges no energy with its surroundings, its energy content remains constant; if energy is exchanged, the change in internal energy will be given by the difference between the energy gained from the surroundings and that lost to the surroundings.

The change in internal energy depends only on the initial and final states of the system, not on the pathway or mechanism of energy exchange. Energy and work are interconvertible; even heat is a measure of the kinetic energy of the molecular constituents of the system. To put it as simply as possible, Equation states that no machine, including the chemical machines that we recognize as living, can do work without an energy source.

An example of the application of the first law to a biological phenomenon is the energy budget of a leaf. Leaves absorb energy from their surroundings in two ways: as direct incident irradiation from the sun and as infrared irradiation from the surroundings. Some of the energy absorbed by the leaf is radiated back to the surroundings as infrared irradiation and heat, while a fraction of the absorbed energy is stored, as either photosynthetic products or leaf temperature changes. Thus we can write the following equation:

Total energy absorbed by leaf = energy emittedfrom leaf + energy stored by leaf.

Note that although the energy absorbed by the leaf has been transformed, the total energy remains the same, in accordance with the first law.

The Change in the Internal Energy of a System Represents the Maximum Work It Can Do

We must qualify the equivalence of energy and work by invoking "ideal conditions"—that is, by requiring that the process be carried out reversibly. The meaning of"reversible" in thermodynamics is a special one: The term describes conditions under which the opposing forces are so nearly balanced that an infinitesimal change in one or the other would reverse the direction of the process.Under these circumstances the process yields the maximum possible amount of work.

Reversibility in this sense does not often hold in nature, as in the example of the leaf. Ideal conditions differ so little from a state of equilibrium that any process or reaction would require infinite time and would therefore not take place at all. non-etheless, the concept of thermodynamic reversibility is useful: If we measure the change in internal energy that a process entails, we have an upper limit to the work that it can do; for any real process the maximum work will be less.

In the study of plant biology we encounter several sources of energy—notably light and chemical transformations—as well as a variety of work functions, including mechanical, osmotic, electrical, and chemical work. The meaning of the first law in biology stems from the certainty, painstakingly achieved by nineteenth-century physicists, that the various kinds of energy and work are measurable, equivalent, and, within limits, interconvertible.Energy is to biology what money is to economics: the means by which living things purchase useful goods and services.

Each Type of Energy Is Characterized by a Capacity Factor and a Potential Factor

The amount of work that can be done by a system, whether mechanical or chemical, is a function of the size of the system. Work can always be defined as the product of two factors—force and distance, for example. One is a potential or intensity factor, which is independent of the size of the system; the other is a capacity factor and is directly proportional to the size.

In biochemistry, energy and work have traditionally been expressed in calories; 1 calorie is the amount of heat required to raise the temperature of 1 g of water by 1°C, specifically, from 15.0 to 16.0°C. In principle, one can carry out the same process by doing the work mechanically with a paddle; such experiments led to the establishment of the mechanical equivalent of heat as 4.186 joules per calorie (J cal^{-1}).

We will also have occasion to use the equivalent electrical units, based on the volt: A volt is the potential difference between two points when 1 J of work is involved in the transfer of a coulomb of charge from one point to another. (A coulomb is the amount of charge carried by a current of 1 ampere [A] flowing for 1 s. Transfer of 1 mole [mol] of charge across a potential of 1 volt [V] involves 96,500 J of energy or work.) The difference between energy and work is often a matter of the sign. Work must be done to bring a positive charge closer to another positive charge, but the charges thereby acquire potential energy, which in turn can do work.

THE DIRECTION OF SPONTANEOUS PROCESSES

Left to themselves, events in the real world take a predictable course. The apple falls from the branch. A mixture of hydrogen and oxygen gases is converted into water. The fly trapped in a bottle is doomed to perish, the pyramids to crumble into sand; things fall apart. But there is nothing in the principle of energy conservation that forbids the apple to return to its branch with absorption of heat from the surroundings or that prevents water from dissociating into its constituent elements in a like manner. The search for the reason that neither of these things ever happens led to profound philosophical insights and generated useful quantitative statements about the energetics of chemical reactions and the amount of work that can be done by them. Since living things are in many respects chemical machines, we must examine these matters in some detail.

THE SECOND LAW: THE TOTAL ENTROPYALWAYS INCREASES

From daily experience with weights falling and warm bodies growing cold, one might expect spontaneous processes to proceed in the direction that lowers the internal energy—that is, the direction in which ΔU is negative. But there are too many exceptions for this to be a general rule. The melting of ice is one exception: An ice cube placed in water at 1°C will melt, yet measurements show that liquid water (at any temperature above 0°C) is in a state of higher energy than ice; evidently, some spontaneous processes are accompanied by an increase in internal energy. Our melting ice cube does not violate the first law, for heat is absorbed as it melts. This suggests that there is a relationship between the capacity for spontaneous heat absorption and the criterion determining the direction of spontaneous processes, and that is the case. The thermodynamic function we seek is called entropy, the amount of energy in a system not available for doing work, corresponding to the degree of randomness of a system. Mathematically, entropy is the capacity factor corresponding to temperature, Q/T. We may state the answer to our question, as well as the second law of thermodynamics, thus: The direction of all spontaneous processes is to increase the entropy of a system plus its surroundings.

Few concepts are so basic to a comprehension of the world we live in, yet so opaque, as entropy—presumably because entropy is not intuitively related to our sense perceptions, as mass and temperature are. The explanation given here follows the particularly lucid exposition by Atkinson (1977), who states the second law in a form bearing, at first sight, little resemblance to that given above:

We shall take [the second law] as the concept that any system not at absolute zero has an irreducible minimum amount of energy that is an inevitable property of that system at that temperature. That is, a system requires a certain amount of energy just to be at any specified temperature.

The molecular constitution of matter supplies a ready explanation: Some energy is stored in the thermal motions of the molecules and in the vibrations and oscillations of their constituent atoms. We can speak of it as isothermally unavailable energy, since the system cannot give up any of it without a drop in temperature (assuming that there is no physical or chemical change). The isothermally unavailable energy of any system increases with temperature, since the energy of molecular and atomic motions increases with temperature. Quantitatively, the isothermally unavailable energy for a particular system is given by ST, where T is the absolute temperature and S is the entropy.

But what is this thing, entropy? Reflection on the nature of the isothermally unavailable energy suggests that, for any particular temperature, the amount of such energy will be greater the more atoms and molecules are free to move and to vibrate—that is, the more chaotic is the system. By contrast, the orderly array of atoms in a crystal, with a place for each and each in its place, corresponds to a state of low entropy. At absolute zero, when all motion ceases, the entropy of a pure substance is likewise zero; this statement is sometimes called the third law of thermodynamics.

A large molecule, a protein for example, within which many kinds of motion can take place, will have considerable amounts of energy stored in this fashion—more than would, say, an amino acid molecule. But the entropy of the protein molecule will be less than that of the constituent amino acids into which it can dissociate, because of the constraints placed on the motions of those amino acids as long as they are part of the larger structure. Any process leading to the release of these constraints increases freedom of movement, and hence entropy.

This is the universal tendency of spontaneous processes as expressed in the second law; it is why the costly enzymes stored in the refrigerator tend to decay and why ice melts into water. The increase in entropy as ice melts into water is "paid for" by the absorption of heat from the surroundings. As long as the net change in entropy of the system plus its surroundings is positive, the process can take place spontaneously. That does not necessarily mean that the process will take place: The rate is usually deter-mined by kinetic factors separate from the entropy change. All the second law mandates is that the fate of the pyramids is to crumble into sand, while the sand will never reassemble itself into a pyramid; the law does not tell how quickly this must come about.

A Process Is Spontaneous If ΔS for the System and Its Surroundings Is PositiveThere is nothing mystical about entropy; it is a thermodynamic quantity like any other, measurable by experiment and expressed in entropy units. One method of quantifying it is through the heat capacity of a system, the amount of energy required to raise the temperature by 1°C. In some cases the entropy can even be calculated from theoretical principles, though only for simple molecules. For our purposes, what matters is the sign of the entropy change,

ΔS: A process can take place spontaneously when ΔS for the system and its surroundings is positive; a process for which ΔS is negative cannot take place spontaneously, but the opposite process can; and for a system at equilibrium, the entropy of the system plus its surroundings is maximal and ΔS is zero.

"Equilibrium" is another of those familiar words that is easier to use than to define. Its everyday meaning implies that the forces acting on a system are equally balanced, such that there is no net tendency to change; this is the sense in which the term "equilibrium" will be used here. A mixture of chemicals may be in the midst of rapid interconversion, but if the rates of the forward reaction and the backward reaction are equal, there will be no net change in composition, and equilibrium will prevail.

The second law has been stated in many versions. One version forbids perpetual-motion machines: Because ener-gy is, by the second law, perpetually degraded into heat and rendered isothermally unavailable ($\Delta S > 0$), continued motion requires an input of energy from the outside. The most celebrated yet perplexing version of the second law was provided by R. J. Clausius (1879): "The energy of the universe is constant; the entropy of the universe tends towards a maximum."

How can entropy increase forever, created out of nothing? The root of the difficulty is verbal, as Klotz (1967) neatly explains. Had Clausius defined entropy with the opposite sign (corresponding to order rather than to chaos), its universal tendency would be to diminish; it would then be obvious that spontaneous changes proceed in the direction that decreases the capacity for further spontaneous change.

Solutes diffuse from a region of higher concentration to one of lower concentration; heat flows from a warm body to a cold one. Sometimes these changes can be reversed by an outside agent to reduce the entropy of the system under consideration, but then that external agent must change in such a way as to reduce its own capacity for further change. In sum, "entropy is an index of exhaustion; the more a system has lost its capacity for spontaneous change, the more this capacity has been exhausted, the greater is the entropy".

Conversely, the farther a system is from equilibrium, the greater is its capacity for change and the less its entropy. Living things fall into the latter category: A cell is the epitome of a state that is remote from equilibrium.

FREE ENERGY AND CHEMICAL POTENTIAL

Many energy transactions that take place in living organisms are chemical; we therefore need a quantitative expression for the amount of work a chemical reaction can do. For this purpose, relationships that involve the entropy change in the system plus its surroundings are unsuitable.

We need a function that does not depend on the surroundings but that, like ΔS, attains a minimum under conditions of equilibrium and so can serve both as a criterion of the feasibility of a reaction and as a measure of the energy available from it for the performance of work. The function universally employed for this purpose is free energy, abbreviated G in honour of the nineteenth-century physical chemist J. Willard Gibbs, who first introduced it.

ΔG Is Negative for a Spontaneous Process at Constant Temperature and Pressure

Earlier we spoke of the isothermally unavailable energy, ST. Free energy is defined as the energy that is available under isothermal conditions, and by the following relationship:

$$\Delta H = \Delta G + T\Delta S \quad (2.3)$$

The term H, enthalpy or heat content, is not quite equivalent to U, the internal energy. To be exact, ΔH is a measure of the total energy change, including work that may result from changes in volume during the reaction, whereas ΔU excludes this work. (We will return to the concept of enthalpy a little later.) However, in the biological context we are usually concerned with reactions in solution, for which volume changes are negligible. For most purposes, then,

$$\Delta U \cong \Delta G + T\Delta S \quad (2.4)$$

and

$$\Delta G \cong \Delta U - T\Delta S \quad (2.5)$$

What makes this a useful relationship is the demonstration that for all spontaneous processes at constant temperature and pressure, ΔG is negative. The change in free energy is thus a criterion of feasibility. Any chemical reaction that proceeds with a negative ΔG can take place spontaneously; a process for which ΔG is positive cannot take place, but the reaction can go in the opposite direction; and a reaction for which ΔG is zero is at equilibrium, and no net change will occur. For a given temperature and pressure, ΔG depends only on the composition of the reaction mixture; hence the alternative term "chemical potential" is particularly apt. Again, nothing is said about rate, only about direction. Whether a reaction having a given ΔG will proceed, and at what rate, is determined by kinetic rather than thermodynamic factors.

There is a close and simple relationship between the change in free energy of a chemical reaction and the work that the reaction can do. Provided the reaction is carried out reversibly,

$$\Delta G = \Delta W_{max} \quad (2.6)$$

That is, for a reaction taking place at constant temperature and pressure, -ΔG is a measure of the maximum work the process can perform. More precisely, -ΔG is the maximum work possible, exclusive of pressure-volume work, and thus is a quantity of great importance in bioenergetics. Any process going towards equilibrium can, in principle, do work. We can therefore describe processes for which ΔG is negative as "energy-releasing," or exergonic. Conversely, for any process moving away from equilibrium, ΔG is positive, and we speak of an "energy-consuming," or endergonic, reaction. Of course, an endergonic reaction cannot occur: All real processes go towards equilibrium, with a negative ΔG.

The concept of endergonic reactions is nevertheless a useful abstraction, for many biological reactions appear to move away from equilibrium. A prime example is the synthesis of ATP during oxidative phosphorylation, whose apparent ΔG is as high as 67 kJ mol^{-1} (16 kcal mol^{-1}). Clearly, the cell must do work to render the reaction exergonic overall. The occurrence of an endergonic process in naturethus implies that it is coupled to a second, exergonic process. Much of cellular and molecular bioenergetics is concerned with the mechanisms by which energy coupling is effected.

THE STANDARD FREE-ENERGY CHANGE, ΔG^0, IS THE CHANGE IN FREE ENERGY WHEN THE CONCENTRATION OF REACTANTS AND PRODUCTS IS 1 M

Changes in free energy can be measured experimentally by calorimetric methods. They have been tabulated in two forms: as the free energy of formation of a compound from its elements, and as ΔG for a particular reaction. It is of theutmost importance to remember that, by convention, thenumerical values refer to a particular set of conditions. The standard free-energy change, ΔG^0, refers to conditions such that all reactants and products are present at a concentration of 1 M; in biochemistry it is more convenient to employ $\Delta G^0 2$, which is defined in the same way except that the pH is taken to be 7.

The conditions obtained in the real world are likely to be very different from these, particularly with respect to the concentrations of the participants. To take a familiar example, $\Delta G^0 2$ for the hydrolysis of ATP is about -33 kJ mol^{-1} (-8 kcal mol^{-1}). In the cytoplasm, however, the actual nucleotide concentrations are approximately 3 mM ATP, 1 mM ADP, and 10 mM P_i. As we will see, changes in free energy depend strongly on concentrations, and ΔG for ATP hydrolysis under physiological conditions thus is much more negative than $\Delta G^0 2$, about -50 to -65 kJ mol^{-1} (-12 to -15 kcal mol^{-1}). Thus, whereas values of $\Delta G^0 2$ for many reactions are easily accessible, they must not be used uncritically as guides to what happens in cells.

The Value of ΔG Is a Function of the Displacement of the Reaction from Equilibrium

The preceding discussion of free energy shows that there must be a relationship between ΔG and the equilibrium constant of a reaction: At equilibrium, ΔG is zero, and the farther a reaction is from equilibrium, the larger ΔG is and the more work the reaction can do. The quantitative statement of this relationship is

$$\Delta G^0 = -RT \ln K = -2.3RT \log K \quad (2.7)$$

where R is the gas constant, T the absolute temperature, and K the equilibrium constant of the reaction. This equation is one of the most useful links between thermodynamics and biochemistry and has a host of applications. For example, the equation is easily modified to allow computation of the change in free energy for concentrations other than the standard ones. For the reactions shown in the equation

$$A + B \Leftrightarrow C + D \quad (2.8)$$

the actual change in free energy, ΔG, is given by the equation

$$\Delta G = \Delta G + RTn1\frac{[C][D]}{[A][B]} \quad (2.9)$$

where the terms in brackets refer to the concentrations at the time of the reaction. Strictly speaking, oneshould use activities, but these are usually not known for cellular conditions, so concentrations must do.

Equation can be rewritten to make its import a little plainer. Let q stand for the mass:action ratio, [C][D]/[A][B]. Substitution of Equation into Equation, followed by rearrangement, then yields the following equation:

$$\Delta G = -2.3RT\log\frac{K}{a}$$

In other words, the value of ΔG is a function of the displacement of the reaction from equilibrium. In order to displace a system from equilibrium, work must be done on it and ΔG must be positive. Conversely, a system displaced from equilibrium can do work on another system, provided that the kinetic parameters allow the reaction to proceedand a mechanism exists that couples the two systems. Quantitatively, a reaction mixture at 25°C whose composition is one order of magnitude away from equilibrium (log K/q = 1) corresponds to a free-energy change of 5.7 kJ mol^{-1} (1.36 kcal mol^{-1}). The value of ΔG is negative if the actual mass:action ratio is less than the equilibrium ratio and positive if the mass:action ratio is greater.

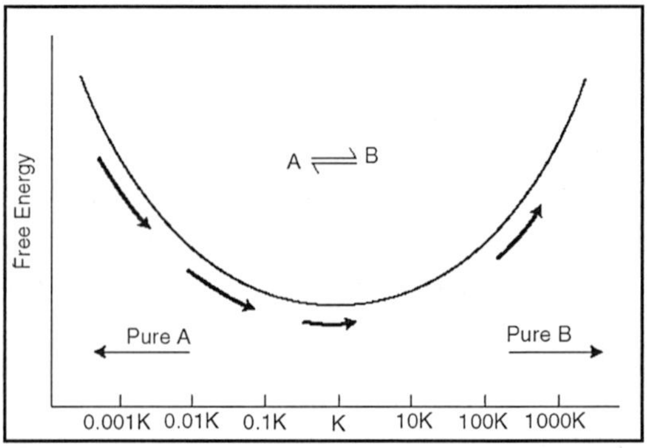

Fig. Free Energy of a Chemical Reaction as a Func- tion of Displacement from Equilibrium.

Imagine a closed system containing components A and B at concentrations [A] and [B]. The two components can be interconverted by the reaction A Δ B, which is at equilibrium when themass:action ratio, [B]/[A], equals unity. The curve shows qualitatively how the free energy, G, of the system varies when the total [A] + [B] is held constant but the mass:action ratio is displaced from equilibrium. The arrows represent schematically the change in free energy, ΔG, for a small con-version of [A] into [B] occurring at different mass:action ratios. The point that ΔG is a function of the displacement of a reaction (indeed, of any thermodynamic system) from equilibrium is central to an understanding of bioenergetics. Figure illustrates this relationship diagrammatical-ly for the chemical interconversion of substances A and B, and the relationship will reappear shortly in other guises.

The Enthalpy Change Measures the Energy Transferred as Heat

Chemical and physical processes are almost invariably accompanied by the generation or absorption of heat,which reflects the change in the internal energy of the system. The amount of heat transferred and the sign of the reaction are related to the change in free energy, as set out in Equation.

The energy absorbed or evolved as heat under conditions of constant pressure is designated as the change in heat content or enthalpy, ΔH. Processes that generate heat, such as combustion, are said to be exothermic; those in which heat is absorbed, such as melting or evaporation, are referred to as endothermic. The oxidation of glucose to CO_2 and water is an exergonic reaction (ΔG0 = –2858 kJ mol–1 [–686 kcal mol-1]); when this reaction takes place during respiration, part of the free energy is conserved through coupled reactions that generate ATP. The combustion of glucose dissipates the free energy of reaction, releasing most of it as heat

$$(\Delta H = -2804 \text{ kJ mol}^{-1} [-673 \text{ kcal mol}^{-1}]).$$

Bioenergetics is preoccupied with energy transduction and therefore gives pride of place to free-energy transac-tions, but at times heat transfer may also carry biological significance. For example, water has a high heat of vaporization, 44 kJ mol^{-1} (10.5 kcal mol^{-1}) at 25°C, which plays an important role in the regulation of leaf temperature. During the day, the evaporation of water from the leaf surface (transpiration) dissipates heat to the surroundings and helps cool the leaf. Conversely, the condensation of water vapour as dew heats the leaf, since water condensation is the reverse of evaporation, is exothermic. The abstract enthalpy function is a direct measure of the energy exchanged in the form of heat.

REDOX REACTIONS

Oxidation and reduction refer to the transfer of one or more electrons from a donor to an acceptor, usually to another chemical species; an example is the oxidation of ferrous iron by oxygen, which forms ferric iron and water. Reactions of this kind require special consideration, for they play a central role in both respiration and photosynthesis.

The Free-Energy Change of an Oxidation-Reduction Reaction Is Expressed as the Standard Redox Potential in Electrochemical Units

Redox reactions can be quite properly described in terms of their change in free energy. However, the participation of electrons makes it convenient to follow the course of the reaction with electrical instrumentation and encourages the use of an electrochemical notation. It also permits dissection of the chemical process into separate oxidative and reductive half-reactions.

For the oxidation of iron, we can write:

$$2Fe^{2+} \Leftrightarrow 2Fe^{3+} + 2e^{-}$$

$$\tfrac{1}{2}O_2 + 2H^{+} + 2E^{-} \Leftrightarrow H_2O$$

$$2Fe^{2+} + \tfrac{1}{2}O_2 + 2H^{+} \Leftrightarrow 2Fe^{3+} + H_2O$$

The tendency of a substance to donate electrons, its "electron pressure," is measured by its standard reduction (or redox) potential, E_0, with all components present at a concentration of 1 M. In biochemistry, it is more convenient to employ E2 $_0$, which is defined in the same way except that the pH is 7. By definition, then, E2 $_0$ is the electromotive force given by a half cell in which the reduced and oxidized species are both present at 1 M, 25°C, and pH 7, in equilibrium with an electrode that can reversibly accept electrons from the reduced species. By convention, the reaction is written as a reduction.

The standard reduction potential of the hydrogen electrode* serves as reference: at pH 7, it equals -0.42 V. The standard redox potential as defined here is often referred to in the bioenergetics literature as the midpoint potential, E_m.

A negative midpoint potential marks a good reducing agent; oxidants have positive mid-point potentials.

The redox potential for the reduction of oxygen to water is +0.82 V; for the reduction of Fe^{3+} to Fe^{2+} (the direction opposite to that of Equation), +0.77 V. We can therefore predict that, under standard conditions, the Fe^{2+}-Fe^{3+} couple will tend to reduce oxygen to water rather than the reverse. A mixture containing Fe^{2+}, Fe^{3+}, and oxygen will probably not be at equilibrium, and the extent of its displacement from equilibrium can be expressed in terms of either the change in free energy for Equation or the difference in redox potential, ΔE2 $_0$, between the oxidant and the reductant couples (+0.05 V in the case of iron oxidation). In general,

$$\Delta G^{0'} = -nF\,\Delta E'_0 \quad (2.14)$$

where n is the number of electrons transferred and F is Faraday's constant (23.06 kcal V^{-1} mol^{-1}). In other words, the standard redox potential is a measure, in electrochemical units, of the change in free energy of an oxidation-reduction process.

As with free-energy changes, the redox potential measured under conditions other than the standard ones depends on the concentrations of the oxidized and reduced species, according to the following equation (note the similarity in form to Equation):

$$Eh = E'_0 + \frac{23.RT}{nF} \log \frac{[\text{Oxidant}]}{[\text{Reductant}]}$$

Here E_h is the measured potential in volts, and the other symbols have their usual meanings. It follows that the redox potential under biological conditions may differ substantially from the standard reduction potential.

THE ELECTROCHEMICAL POTENTIAL

In the preceding section we introduced the concept that a mixture of substances whose composition diverges from the equilibrium state represents a potential source of free energy. Conversely, a similar amount of work must be done on an equilibrium mixture in order to displace its composition from equilibrium. In this section, we will examine the free-energy changes associated with another kind of displacement from equilibrium—namely, gradients of concentration and of electric potential.

TRANSPORT OF AN UNCHARGED SOLUTE AGAINST ITS CONCENTRATION GRADIENT

Consider a vessel divided by a membrane into two compartments that contain solutions of an uncharged solute at concentrations C_1 and C_2, respectively. The work required to transfer 1 mol of solute from the first compartment to the second is given by the following equation:

$$\Delta G = 23.RT \log \frac{C_2}{C_1}$$

This expression is analogous to the expression for a chemical reaction and has the same meaning. If C_2 is greater than C_1, ΔG is positive, and work must be done to transfer the solute. Again, the free-energy change for the transport of 1 mol of solute against a tenfold gradient of concentration is 5.7 kJ, or 1.36 kcal.

The reason that work must be done to move a substance from a region of lower concentration to one of higher concentration is that the process entails a change to a less probable state and therefore a decrease in the entropy of the system. Conversely, diffusion of the solute from the region of higher concentration to that of lower concentration takes place in the direction of greater probability; it results in an increase in the entropy of the system and can proceed spontaneously. The sign of ΔG becomes negative, and the process can do the amount of work specified by Equation, provided a mechanism exists that couples the exergonic diffusion process to the work function.

Movement an Ion from One Side to the Other By Membrane Potential

Matters become a little more complex if the solute in question bears an electric charge. Transfer of positively charged solute from compartment 1 to compartment 2 will then cause a difference in charge to develop across the membrane, the second compartment becoming electropositive relative to the first. Since like charges repel one another, the work done by the agent that moves the solute from compartment 1 to compartment 2 is a function of the charge difference; more precisely, it depends on the difference in electric potential across the membrane. This difference, called membrane potential for short, will appear again in later pages.

The membrane potential, ΔE,† is defined as the work that must be done by an agent to move a test charge from one side of the membrane to the other. When 1 J of work must be done to move 1 coulomb of charge, the potential difference is said to be 1 V. The absolute electric potential of any single phase cannot be measured, but the potentialdifference between two phases can be. By convention, the membrane potential is always given in reference to the movement of a positive charge. It states the intracellular potential relative to the extracellular one, which is defined as zero.

The work that must be done to move 1 mol of an ion against a membrane potential of ΔE volts is given by the following equation:

$$\Delta G = zF\ \Delta E \quad (2.17)$$

where z is the valence of the ion and F is Faraday's constant. The value of ΔG for the transfer of cations into a positive compartment is positive and so calls for work. Conversely, the value of ΔG is negative when cations move into the negative compartment, so work can be done. The electric potential is negative across the plasma membrane of the great majority of cells; therefore cations tend to leak in but have to be "pumped" out.

The Electrochemical-Potential Difference, Δ ~,Includes Both Concentration and Electric Potentials

In general, ions moving across a membrane are subject to gradients of both concentration and electric potential. Consider, for example, the situation depicted in Figure,which corresponds to a major event in energy transduction during photosynthesis. A cation of valence z moves from compartment 1 to compartment 2, against both a concentration gradient ($C_2 > C_1$) and a gradient of membrane elec-tric potential (compartment 2 is electropositive relative tocompartment 1).

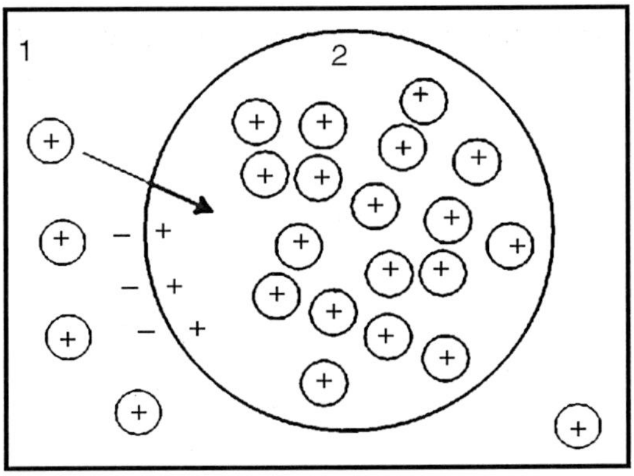

Fig. Transport against an Electrochemical-potential Gradient.

The agent that moves the charged solute (from compartment 1 to compartment 2) must do work to over- come both the electrochemical-potential gradient and the concentration gradient. As a result, cations in compartment 2 have been raised to a higher electrochemical potential than those in compartment 1. eutralizing anions have been omitted.

The free-energy change involved in this transfer is given by the following equation:

$$\Delta G = zF\,\Delta E + 23RT\log\frac{C_2}{C_1}$$

ΔG is positive, and the transfer can proceed only if coupled to a source of energy, in this instance the absorption of light. As a result of this transfer, cations in compartment 2 can be said to be at a higher electrochemical potential than the same ions in compartment 1.

The electrochemical potential for a particular ion is designated $\sim_{ion}$. Ions tend to flow from a region of high electrochemical potential to one of low potential and in so doing can in principle do work. The maximum amount of this work, neglecting friction, is given by the change in free energy of the ions that flow from compartment 2 to compartment 1 and is numerically equal to the electrochemical-potential difference, $\Delta\sim_{ion}$. This principle underlies much of biological energy transduction.

The electrochemical-potential difference, $\Delta\sim_{ion}$, is properly expressed in kilojoules per mole or kilocalories per mole. However, it is frequently convenient to express the driving force for ion movement in electrical terms, with the dimensions of volts or millivolts. To convert $\Delta\sim_{ion}$ into millivolts (mV), divide all the terms in Equation by F:

$$\frac{\Delta G\bar{\mu}_{ion}}{F} = z\Delta E + \frac{2.3RT}{F}\log\frac{C_2}{C_1}$$

Equations have proved to be of central importance in bioenergetics. First, they measure the amount of energy that must be expended on the active transport of ions and metabolites, a major function of bio-logical membranes.

Second, since the free energy of chemical reactions is often transduced into other forms via the intermediate generation of electrochemical-potential gradients, these gradients play a major role in descriptions of biological energy coupling.

It should be emphasized that the electrical and concentration terms may be either added, as in Equation, or subtracted, and that the application of the equations to particular cases requires careful attention to the sign of the gradients. We should also note that free-energy changes in chemical reactions are scalar, whereas transport reactions have direction; this is a subtle but critical aspect of the biological role of ion gradients.

Ion distribution at equilibrium is an important special case of the general electrochemical equation. Figure shows a membrane-bound vesicle (compartment 2) that contains a high concentration of the salt K_2SO_4, surrounded by a medium (compartment 1) containing a lower concentration of the same salt; the membrane is impermeable to anions but allows the free passage of cations. Potassium ions will therefore tend to diffuse out ofthe vesicle into the solution, whereas the sulfate anions are retained.

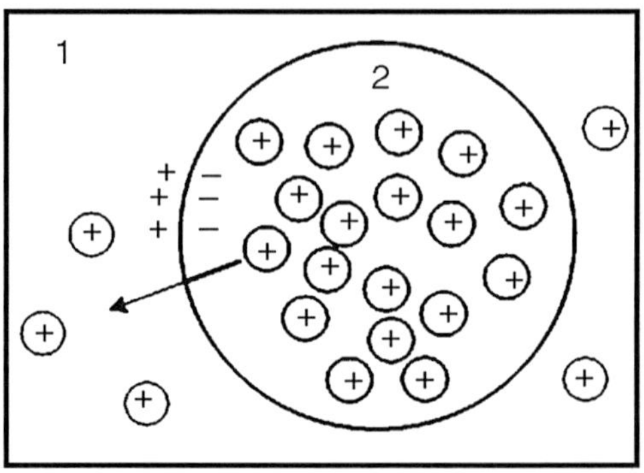

Fig. Generation of an Electric Potential by ion Diffusion.

Compartment 2 has a higher salt concentration than compartment 1 (anions are not shown). If the membrane is permeable to the cations but not to the anions, the cations will tend to diffuse out of compartment 2 into compartment 1, generating a membrane potential in which compartment 2 is negative.

Diffusion of the cations generates a membrane potential, with the vesicle interior negative, which restrains further diffusion. At equilibrium, ΔG and $\Delta\sim_{K}+$ equal zero (by definition). Equation can then be arranged to give the following equation:

$$\Delta E = \frac{-2.3RT}{zF}\log\frac{C_2}{C_1}$$

where C_2 and C_1 are the concentrations of K^+ ions in the two compartments; z, the valence, is unity; and ΔE is the membrane potential in equilibrium with the potassium concentration gradient.

This is one form of the celebrated Nernst equation. It states that at equilibrium, a permeant ion will be so distributed across the membrane that the chemical driving force (outward in this instance) will be balanced by the electric driving force (inward). For a univalent cation at 25°C, each tenfold increase in concentration factor corresponds to a membrane potential of 59 mV; for a divalent ion the value is 29.5 mV.

The preceding discussion of the energetic and electrical consequences of ion translocation illustrates a point that must be clearly understood—namely, that an electric potential across a membrane may arise by two distinct mechanisms. The first mechanism, illustrated in Figure, is the diffusion of charged particles down a pre-existing concentration gradient, an exergonic process.

A potential generated by such a process is described as a diffusion potential or as a Donnan potential. (Donnan potential is defined as the diffusion potential that occurs in the limiting case where the counterion is completely impermeant or fixed, as in Figure) Many ions are unequally distributed across biological membranes and differ widely in their rates of diffusion across the barrier; therefore diffusion potentials always contribute to the observed membrane potential.

But in most biological systems the measured electric potential differs from the value that would be expect-ed on the basis of passive ion diffusion. In these cases one must invoke electrogenic ion pumps, transport systems that carry out the exergonic process indicated in Figure at the expense of an external energy source. Transport systems of this kind transduce the free energy of a chemical reaction into the electrochemical potential of an ion gradient and play a leading role in biological energy coupling.

2

Biomolecules

BIOLOGICAL MOLECULES

ORGANIC COMPOUNDS

All living things are composed of organic compounds. Generally, organic compounds are distinguished from inorganic compounds by the presence of both carbon and hydrogen. We are referred to as carbon-based life forms, as is all life on earth. This means that carbon is a component of most of the chemical molecules that make up living organisms on this planet.

Why is carbon so ubiquitous in nature? Carbon's participation in organic chemistry is probably due to its relatively unique chemical bonding properties. These properties allow carbon to participate in the formation of an incredibly wide variety of chemical compounds and electron configurations. Carbon is also a relatively small, non-bulky atom so it can share electrons easily with other atoms.

Carbon, atomic number six, has six electrons. Two are in the first electron shell and four are in the second electron shell. Carbon must share four electrons with other atoms to fill its outermost electron shell and attain a stable configuration. Carbon atoms can share electrons with a wide variety of elements also commonly found in organic compounds, the most notable being other carbon atoms, hydrogen atoms and oxygen atoms. As a matter of fact, the simplest organic molecules are defined as being comprised of only carbon and hydrogen, and are therefore referred to as hydrocarbons.

Hydrocarbons are non-polar, hydrophobic compounds. The term hydrophobic comes from the Latin roots "hydro" (water) and "phobia" (fear). Literally speaking, a hydrophobic compound is afraid of water. The term non-polar is often used in conjunction with the term hydrophobic, because those compounds that have no charges on them will not mix with polar water. A non-polar compound repels a polar solvent and collects together in a film on top of the solvent, or forms a layer on the bottom of the container. Most oils, waxes, and fat compounds are hydrocarbons, as are the fossil fuels gasoline, kerosene, and similar compounds. Just think of how well these compounds mix with water and you'll easily grasp the concept of hydrophobicity.

Since we have introduced the concept of hydrophobicity its complement, hydrophilicity. Hydrophilic comes from the Latin roots "hydro" (water) and "philia" (love). Hydrophilic compounds love water, and mix with it easily. These compounds are polar, and so dissolve easily in the polar solvent water.

Methane CH_4	Ethane C_2H_6	Propane C_3H_8	Ethylene C_2H_4
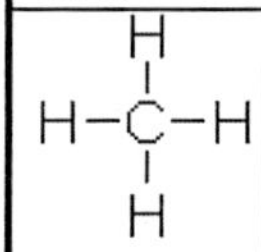	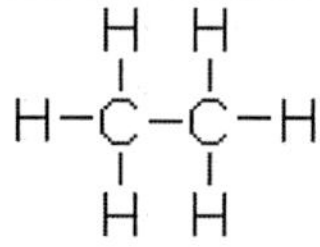	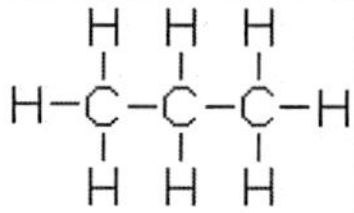	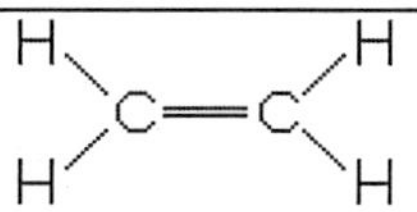

Fig. Structural Formulas of Some Simple Hydrocarbons

Now that you have seen the structures associated with the bonds around carbon. Carbon always needs to share four electrons to fill its outermost electron shell. Each straight line extending from the carbon in the diagrams represents one electron carbon is sharing with another atom. In turn, that atom shares one of its electrons with carbon, forming a covalent bond. Each atom involved in the covalent bond has essentially gained an electron. When carbon shares two electrons with another carbon or another atom a double covalent bond is formed. Two straight lines extending from the carbon to its electron-sharing partner represent a double bond. All structural formulas use the same characters to represent bonds. One line equals one shared electron from carbon, two lines equals two shared electrons, and three lines equals three shared electrons. These are called single bonds, double bonds, and triple bonds, respectively. Single and double bonds are the most common and single bonds are the most stable (least likely to react with other molecules). Carbon is not the only element that can form double bonds, as you will see when you review the structural formulas of other compounds.

Now that you have some information about carbon and simple organic compounds, we must consider some of the other elements essential to more complex organic compounds. While the simplest organic compounds consist of carbon and hydrogen, many organic compounds also include one or more of the following elements:

- Phosphorus
- Sulfur
- Oxygen
- Nitrogen

Not all complex organic compounds contain every one of these elements. For example, sulfur is a component of protein but is never a component of nucleic acid. Scientists have capitalized on this fact to differentiate between these two major biological molecules. Historically, the presence of phosphorus was used to determine that nucleic acid, and not protein, was the biological molecule of heredity!

The Major Classes of Biological Molecules

Now that we have introduced two groups of biological molecules all four major groups of biological molecules:

- Proteins
- Nucleic Acids
- Carbohydrates
- Lipids

These four types of biological molecules compose all life on earth.

Proteins have an incredible variety of functions in living organisms. The functions of proteins include but are not limited to:

- Acting as biological catalysts
- Forming structural parts of organisms
- Participating in cell signal and recognition factors
- Acting as molecules of immunity

Nucleic Acids: Consist of two distinct, but closely related chemical forms: deoxyribonucleic acid (DNA) and ribonucleic acid (RNA). The main functions of these biomolecules include the storage of all heritable information of all organisms on earth and the conversion of this information into proteins.

Carbohydrates: are the major source of food and the major form of energy for most living organisms. When carbohydrates combine together to form polymers they can function as long term food storage molecules, as protective coverings for cells and organisms, as the main structural support for land plants and constituents of many cells and their contents.

Lipids: are the major constituents of all membranes in all cells. They also serve as food storage molecules. This class of biological molecules includes the hydrophobic fats, oils and waxes.

Often times students have difficulty remembering the four major classes of biological molecules, so I have a way to help. To recall the four major classes of biological molecules I want you to think of lunch at a hamburger joint. No endorsements intended! Take a hamburger, fries, and a milkshake and you have all the components of the four major classes of biological molecules.

The hamburger is full of fat, and processors add fat to the hamburger bun. Mayonnaise is oil and egg yolk, which is mostly fat. Milk contains fat, and fat is added to nondairy shakes. As for the fries, we all know why they are called fries. The hamburger bun is made from wheat flour, which is full of starch, a carbohydrate. The sugar in your shake is the disaccharide sucrose, which is also a carbohydrate. As a matter of fact, the lettuce, onion and tomato on that burger all contain carbohydrates.Meat is composed of muscle tissue, and muscle tissue is a structural form of protein. Also most of the formerly living things that we eat have some form of protein in them: wheat, milk, potatoes, and so on. All cells contain DNA and RNA, except mature mammalian red blood cells, so DNA and RNA can be found in all organic matter. The lettuce, onion, and tomato contain nucleic acid, and the hamburger meat contains cow DNA.

BIOLOGICAL MOLECULES: STRUCTURE AND FUNCTION

The structural formula of a particular type of lipid which is the major component of cell membranes. Please notice the long hydrocarbon chains at one end of the molecule. This, as you should now realise, is a non-polar, hydrophobic, hydrocarbon structure. This part of the molecule is often referred to as the hydrophobic tail.

At the opposite end of the molecule you will see carbons attached to oxygen, nitrogen and phosphorus. These components make this end of the phospholipid hydrophilic and therefore water-soluble. This end of the phospholipid is called the hydrophilic head. The dual hydrophobic and hydrophilic nature of this phospholipid is essential to the structure of all cell membranes. Phospholipids will spontaneously form membranes in water because of their hydrophobic and hydrophilic ends. As hydrophobic elements try to avoid the polar water molecules they attract other hydrophobic molecules to them, excluding the water molecules. Hydrophobic forces are very important to both membrane and protein structure, and therefore are very important to the functions of these biomolecules. The force that comes from the hydrophobic attraction for other hydrophobic molecules and the repulsion of water is as essential to the processes of life as chemical bonds.

Imagine millions of these lipid molecules dumped into aqueous surroundings. Their hydrophobic nature causes them to aggregate with all the hydrophobic tails clustered together and all the hydrophilic heads oriented towards the water-based medium. If this phospholipid cluster occurs in a single layer then a structure called a micelle is formed. If this occurs in a double layer a bilipid layer membrane is formed. The bilipid layer membrane is the primary structure of all cells. Because the function of membranes is to either exclude or retain components solubilized in aqueous solutions.

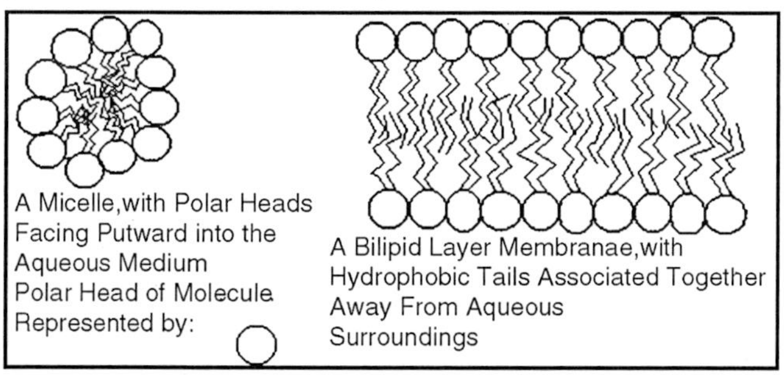

Fig. A Micelle and Bilipid Layer.

Carbohydrates are a class of biomolecules composed of carbon, hydrogen and oxygen. They can be referred to as hydrated skeletons of carbon. Basically, carbohydrates are carbons covalently bonded to each other and covalently bonded to oxygens and hydrogens. The best way to understand carbohydrate structure is to examine the linear and cyclic structures of glucose. Glucose ($C_6H_{12}O_6$) is the major food source for most cells.

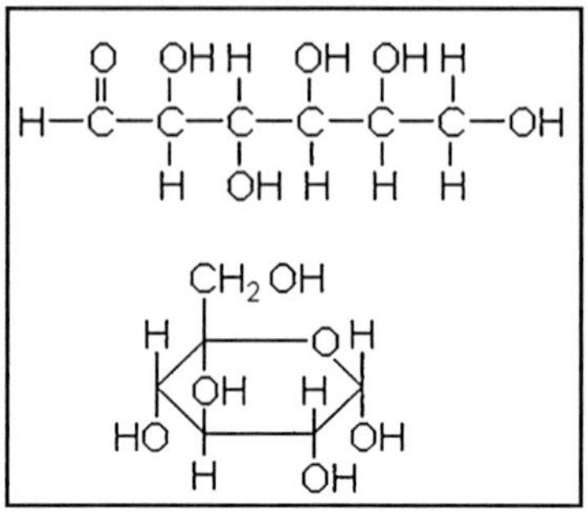

Fig. The Linear and Cyclic Structural Formulas of Glucose.

Not only is the glucose monomer the major source of energy for organisms, but it also has many polymeric forms. These include starch, the long-term glucose food storage molecule; and cellulose, the major structural support molecule in plants and chitin and a major component of animal exoskeletons. Monomer actually means "single unit," from the Latin "mono" (one) and "mer" (unit). Many chemical molecules exist as single units and may also chemically combine in a long string of repeating monomeric units called polymers. You've certainly heard of polyester - this is the polymer made of repeating ester units.

Just as artificial polymers are formed through the bonding of monomers, glucose monomers can covalently bond together to form naturally occurring biological polymers. Glucose can actually form several different polymers, as mentioned previously, which are quite different in their characteristics. For example, starch and cellulose, both polymers of glucose, have quite different properties.

Starch is easily broken down by organisms into the single glucose monomer and then taken up by cells for use as an energy source. Cellulose, however, resists digestion and breakdown by all organisms except a few unusual microbes. That is why the cellulose in plants is known as fibre, or the indigestible portion of plants. Fibre is still healthful to eat in moderate quantities because it facilitates the passage of materials through the digestive system.

Polymerization is an essential aspect of three of the four major biological molecules, so it is very important to understand this concept. Before we continue, lets review how monomers of biological molecules bond together to form polymers. In the cases of carbohydrate, protein, and nucleic acid, polymers are formed through the chemical reaction called a dehydration reaction. A dehydration reaction can take place when the oxygen and hydrogen on one end of a molecule combine with a hydrogen on another molecule. When these monomers come together under the appropriate conditions, an oxygen and two hydrogens are lost from the two molecules and a covalent bond is formed between the two monomers. The byproduct of this reaction is water, formed by the oxygen and two hydrogens that were removed from the monomeric units. In a sense, the monomers are now dehydrated, because they have lost water. As the bond formation is continued between several more monomers, a polymer or multiunit structure is formed.

The dehydration reaction is a very common reaction inside cells. The companion reaction to the dehydration reaction is the hydrolysis reaction. This is exactly opposite of what occurs in dehydration. In the hydrolysis reaction a molecule of water is added across the covalent bond, breaking the bond and adding the oxygen and hydrogens back on the ends of the molecules. Remember, what is built up in the cell must be able to be broken down eventually.

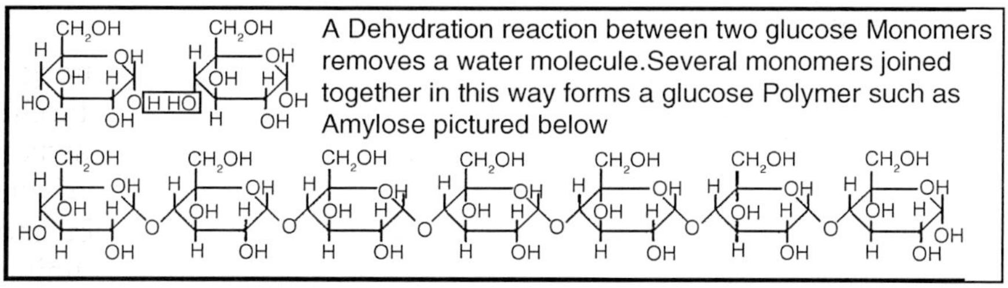

Fig. The Dehydration Reaction Between Glucose Monomers to form Amylose, a Simple form of Starch.

NUCLEIC ACIDS

Nucleic Acids are truly the biomolecules that define life. You may think that this is a rather startling statement when one considers the power of proteins; however, the information for the manufacture of all proteins - indeed for all biological molecules - resides in the nucleic acid molecule.

Nucleic acids are polymers as are proteins and carbohydrates. The monomeric units of nucleic acids are called nucleotides. Nucleotides are made up of three parts covalently bonded together:

- A ribose, which is a five carbon sugar
- A phosphate functional group comprised of one phosphorus and four oxygens
- One of four possible nitrogenous bases, which may be either a purine or pyrimidine

Each nucleotide is identical except for the specific nitrogenous base that is covalently bonded to the #1 carbon of ribose.

There are two main types of nucleic acid: deoxyribonucleic acid (DNA), and ribonucleic acid (RNA). These two types differ from each other in that the ribose of DNA lacks an oxygen on the #2 carbon, and the nitrogenous base thymine is replaced by uracil in RNA.

DNA is normally double-stranded, or consists of two polymers wound up together, while RNA is normally single-stranded.

RNA's function in the cell is to carry small pieces of information to various regions of the cell, while DNA functions as a storage for all genetic information.

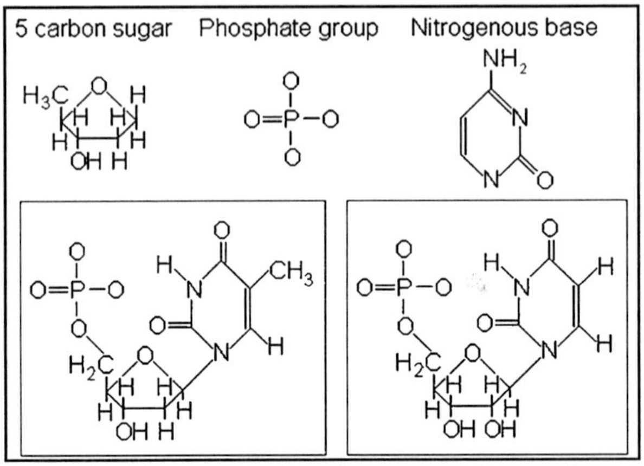

Fig. Structural Formula of one Complete Nucleotide Found in DNA and one Complete Nucleotide Found in RNA for Comparison.

When the nucleotide monomers of nucleic acids covalently bond together through a dehydration reaction, a special kind of bond, called a phosphodiester bond, is formed.

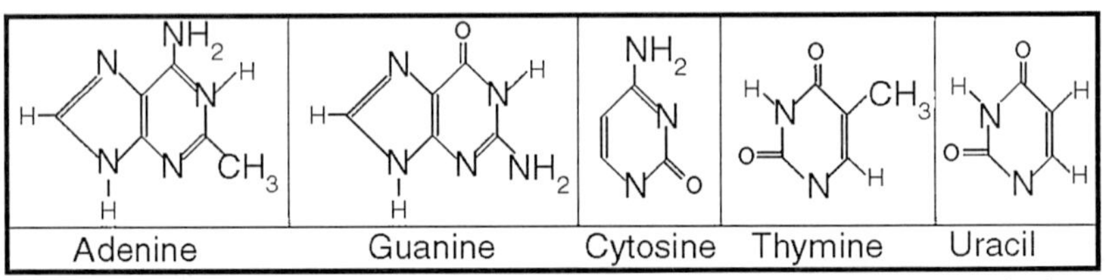

Fig. The Purines, Adenine and Guanine and the Pyrimidines Cytosine and Thymine, as well as Uracil.

The double-stranded structure of DNA is due to hydrogen bonds that form between the bases on one strand of DNA and the bases on a different strand.

The hydrogen bond, although one of the weakest chemical bonds, is absolutely essential to the structure and function of nucleic acids and proteins.

The potential for hydrogen bond formation arises when two atoms are bonded together that have very different electronegativities. In the case of the hydrogen bond, hydrogen is the electropositive element developing a partial positive charge. Oxygen or nitrogen are both electronegative elements, and develop a partial negative charge. In nucleic acids, a hydrogen on one nitrogenous base will form a hydrogen bond with either an oxygen or nitrogen on another nitrogenous base.

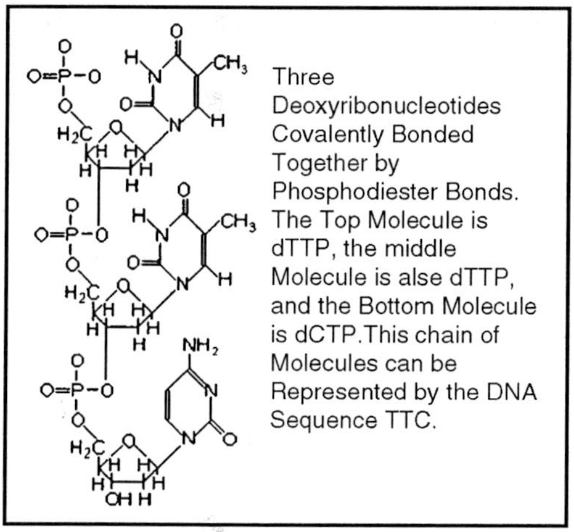

Fig. Three Nucleotides Covalently Bonded Together Through their Phosphodiester Bonds.

Hydrogen bonding occurs between bases on opposite strands. Adenine and Thymine (or Uracil) hydrogen bond together, as do Cytosine and Guanine. Hydrogen bonding cannot occur properly in any other combination of bases. The A and T bases share two hydrogen bonds, while C and G share three hydrogen bonds.

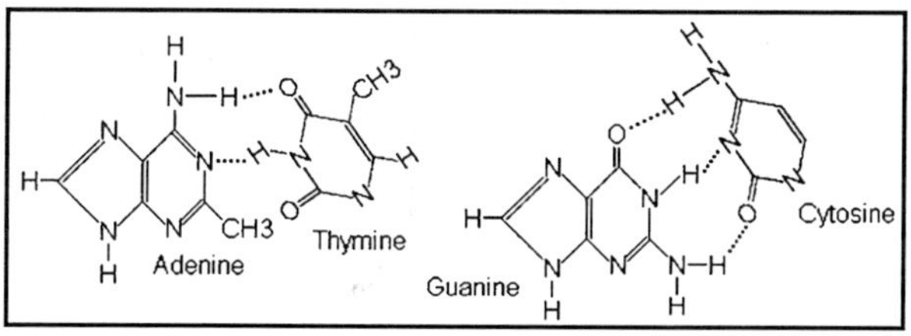

Fig. Hydrogen Bonds Formed between A and T or C and G.

The ability of the bases on one molecule of DNA to hydrogen bond to another molecule of DNA or RNA is called complementarity. This ability is the single most important feature of nucleic acid. Complementarity allows DNA to replicate or duplicate itself, and to be transcribed, or rewritten, in the RNA format. RNA travels through the cell to the organelles that translate its sequence of bases into the sequence of amino acids making up proteins. Complementarity also allows certain types of RNA molecules to fold up into 3-dimensional structures necessary for protein translation and a variety of cellular and regulatory processes.

Base Pairing Rules:

- *Adenine pairs with thymine*: A-T.
- *Guanine pairs with cytosine*: G-C.
- In RNA, where thymine has been replaced by the nitrogenous base uracil, the base-pairing rules do not change. Uracil also pairs with adenine.

The sequence of the amino acids in a protein is critically important to the structure and thus the function of a protein, and protein sequence is ultimately determined by DNA sequence. Therefore the linear sequence of bases in DNA determines the linear sequence of the amino acids in a protein. Change the base sequence of a DNA molecule and you will alter the amino acid sequence in the protein molecule obtained from that DNA sequence. The two critical features of nucleic acid function are the importance of the sequence of nucleotides, and complementarity, or the base pairing rules.

AMINO ACIDS

An AMINO ACID is by definition an organic compound containing an amine $-NH_2$ group and a carboxylic acid group in the same molecule. While there are many forms of amino acids, all of the important amino acids found in living organisms are alpha-amino acids. Alpha amino acids have the -COOH and -NH_2 groups both attached to the same carbon atom, called the alpha carbon atom. The simplest amino acid, which is the molecule glycine, H_2NCH_2COOH, contains no asymmetric carbon atoms. All of the other amino acids do contain such a carbon atom and are therefore optically active. The general structure of the alpha-amino acids is R-$CHNH_2$(alpha)-COOH, and optical activity for the alpha-amino acids more complex than glycine is found at the alpha carbon. The amino acids which have been found to be incorporated into the proteins of living organisms are listed in the Table below.

Table. Amino Acids Found to Be Incorporated Into Proteins

Name (Abb.)	**R Group**	**Class**	**Synth -esis**
Glycine (Gly)	H–	Hydrocarbon	Yes
Alanine (Ala)	CH_3–	Hydrocarbon	Yes
Valine (Val)	$(CH_3)_2CH$–	Hydrocarbon	No
Leucine (Leu)	$(CH_3)_2CH_2CH_2$–	Hydrocarbon	No
Isoleucine (Ile)	$CH_3CH_2CH(CH_3)$ –	Hydrocarbon	No
Aspartic Acid (Asp)	$HOOCCH_2$–	Acidic	Yes
Glutamic Acid (Glu)	$HOOCCH_2CH_2$–	Acidic	Yes
Lysine (Lys)	$H_2N(CH_2)_4$–	Basic	No
Arginine (Arg)	$HN{=}C(NH_2)NH(CH_2)_3$–	Basic	No
Histidine* (His)	Ring Structure	Basic	No
Phenylalanine (Phe)	$C_6H_5CH_2$–	Aromatic	No
Tyrosine* (Tyr)	$HOC_6H_5CH_2$–	Aromatic	Yes
Tryptophan* (Trp)	Ring Structure	Aromatic	No
Serine (Ser)	$HOCH_2$–	Alcohol	Yes
Threonine (Thr)	$HOCH(CH_3)$ –	Alcohol	No
Methionine (Met)	$H_3CSCH_2CH_2$–	Sulfur	No
Cysteine (Cys)	$HSCH_2$–	Sulfur	No
Asparagine (Asp)	H_2NCOCH_2–	Amide	Yes
Glutamine (Gln)	$H_2NCOCH_2CH_2$–	Amide	Yes
Proline* (Pro)	Heterocyclic Ring	Unique	Yes

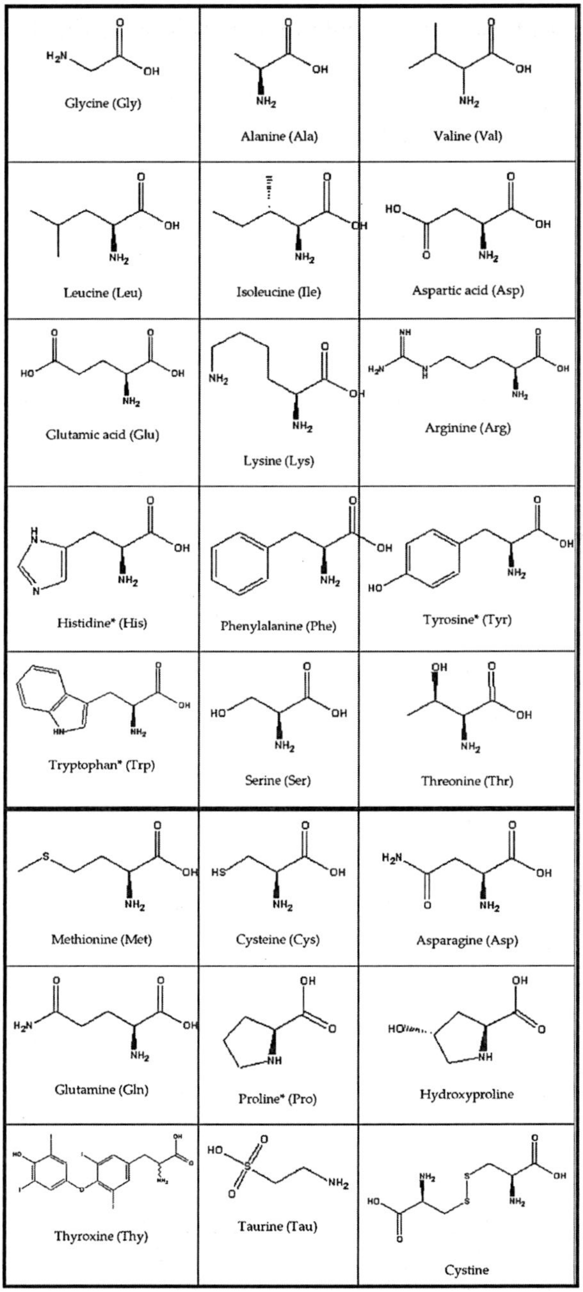

Amino Acids: have both an acidic group, in the carboxylic acid, and a basic group, in the amine. Under physiological aqueous conditions a proton transfer from the acid to the base occurs, forming a dipolar ion or zwitterion, because the carboxylic acid is a much stronger acid than is the ammonium ion. The actual structure of glycine in solution, for example, is $^{+}H_3NCH_2COO^{-}$ at pH 7 rather than H_2NCH_2COOH. At very low pH the acid group can be protonated and at very high pH the ammonium group can be deprotonated, but the forms of amino acids relevant to living organisms are the zwitterions. Each asymmetric carbon gives rise to two optical isomers which are traditionally distinguished by the letters D or L.

Only those amino acids which are the L forms (left-handed) at the alpha carbon are found in terrestrial life.

Amino Acids: are the "building Blocks" of the body. Besides building cells and repairing tissue, they form antibodies to combat invading bacteria and viruses; they are part of the enzyme and hormonal system; they build nucleoproteins (RNA and DNA); they carry oxygen throughout the body and participate in muscle activity. When protein is broken down by digestion the result is 22 known amino acids. Eight are essential (cannot be manufactured by the body) the rest are non-essential.

- *Tryptophan (Essential Amino Acid)*: A natural relaxant, helps alleviate insomnia by inducing normal sleep; reduces anxiety and depression; helps in the treatment of migraine headaches; helps the immune system; helps reduce the risk of artery and heart spasms; works with Lysine in reducing cholesterol levels.
- *Lysine (Essential Amino Acid)*: Insures the adequate absorption of calcium; helps form collagen (which makes up bone cartilage and connective tissues); aids in the production of antibodies, hormones and enzymes. Recent studies have shown that Lysine may be effective against herpes by improving the balance of nutrients that reduce viral growth. A deficiency may result in tiredness, inability to concentrate, irritability, bloodshot eyes, retarded growth, hair loss,anemia and reproductive problems.
- *Methionine (Essential Amino Acid)*: Is a principle supplier of sulfur which prevents disorders of the hair, skin and nails; helps lower cholesterol levels by increasing the liver's production of lecithin; reduces liver fat and protects the kidneys; a natural chelating agent for heavy metals; regulates the formation of ammonia and creates ammonia-free urine which reduces bladder irritation; influences hair follicles and promotes hair growth.
- *Phenylalaine (Essential Amino Acid)*: Used by the brain to produce Norepinephrine, a chemical that transmits signals between nerve cells and the brain; keeps you awake and alert; reduces hunger pains; functions as an antidepressant and helps improve memory.
- *Threonine (Essential Amino Acid)*: Is an important constituent of collagen, Elastin, and enamel protein; helps prevents fat build-up in the liver; helps the digestive and intestinal tracts function more smoothly; assists metabolism and assimilation.

- *Valine (Essential Amino Acid)*: Promotes mental vigour, muscle coordination and calm emotions.
- *Leucine and Isoleucine (Essential Amino Acid)*: They provide ingredients for the manufacturing of other essential biochemical components in the body, some of which are utilized for the production of energy, stimulants to the upper brain and helping you to be more alert.
- *Arginine (Non-Essential Amino Acid)*: Studies have shown that it has improved immune responses to bacteria, viruses and tumor cells; promotes wound healing and regeneration of the liver; causes the release of growth hormones; considered crucial for optimal muscle growth and tissue repair.
- *Tyrosine (Non-Essential Amino Acid)*: Transmits nerve impulses to the brain; helps overcome depression; Improves memory; increases mental alertness; promotes the healthy functioning of the thyroid, adrenal and pituitary glands.
- *Glycine (Non-Essential Amino Acid)*: Helps trigger the release of oxygen to the energy requiring cell-making process; Important in the manufacturing of hormones responsible for a strong immune system.
- *Serine (Non-Essential Amino Acid)*: A storage source of glucose by the liver and muscles; helps strengthen the immune system by providing antibodies; synthesizes fatty acid sheath around nerve fibres.
- *Glutamic Acid (Non-Essential Amino Acid)*: Considered to be nature's "Brain food" by improving mental capacities; helps speed the healing of ulcers; gives a "lift" from fatigue; helps control alcoholism, schizophrenia and the craving for sugar.
- *Aspartic Acid (Non-Essential Amino Acid)*: Aids in the expulsion of harmful ammonia from the body. When ammonia enters the circulatory system it acts as a highly toxic substance which can be harmful to the central nervous system. Recent studies have shown that Aspartic Acid may increase resistance to fatigue and increase endurance.
- *Taurine (Non-Essential Amino Acid)*: Helps stabilize the excitability of membranes which is very important in the control of epileptic seizures. Tau rine and sulfur are considered to be factors necessary for the control of many biochemical changes that take place in the aging process; aids in the clearing of free radical wastes.
- *Cystine (Non-Essential Amino Acid)*: Functions as an antioxidant and is a powerful aid to the body in protecting against radiation and pollution. It can help slow down the aging process, deactivate free radicals, neutralize toxins; aids in protein synthesis and presents cellular change. It is necessary for the formation of the skin, which aids in the recovery from burns and surgical operations. Hair and skin are made up 10-14 per cent Cystine.

- *Histidine (Non-Essential Amino Acid)*: Is found abundantly in hemoglobin; has been used in the treatment of rheumatoid arthritis, allergic diseases, ulcers and anemia. A deficiency can cause poor hearing.
- *Proline (Non-Essential Amino Acid):* Is extremely important for the proper functioning of joints and tendons; also helps maintain and strengthen heart muscles.
- *Alanine (Non-Essential Amino Acid)*: Is an important source of energy for muscle tissue, the brain and central nervous system; strengthens the immune system by producing antibodies; helps in the metabolism of sugars and organic acids.

POLYPEPTIDES TO PROTEINS

Amino acids form long chains called peptide chains or polypeptides by elimination of water in an amide formation reaction. The amide bonds formed between amino acids when they form polypeptides are often called peptide bonds. The general reaction of peptide bond formation is

$$^{+}H_3NCHRCOO^{-} + {}^{+}H_3NCHR'COO^{-} \longleftrightarrow$$
$$^{+}H_3NCHRCONHCHR'COO^{-} + H_2O.$$

Although this reaction is written as an equilibrium reaction, the equilibrium lies far to the left, on the side of hydrolysis rather than synthesis. The synthesis of polypeptides requires energy.

A polypeptide consists of an unbranched chain of amino acids. One end of the chain has a free amine group and the other end of the chain has a free carboxylate group. As a consequence, a polypeptide chain has a direction; the two end groups are not identical. By convention, the free amino group is taken to be the beginning of the chain. The structure of the tripeptide Glu-Thy-Met, for example, is not identical to the structure of Met-Thy-Glu. Polypeptide chains are read and written from left to right beginning with the free amino group.

NUCLEOSIDES AND NUCLEOTIDES

Two pentose sugars, ribose and 2-deoxyribose, react with certain nitrogenous bases to form the biologically significant nucleosides. A nucleoside molecule contains one nitrogenous base and one pentose sugar. The only two different kinds of nitrogenous bases commonly found in nucleosides are the pyrimidines and the purines. There are only five of them, three pyrimidines and two purines, which are commonly found in living organisms.

These purine and pyrimidine bases can attach to sugars through the amine hydrogens at the bottom of the structures shown. Pyrimidine bases have a single six-membered ring while purine bases have the structures with two fused rings shown. Their names and abbreviations are adenine (A), guanine (G), thymine (T), cytosine (C), and uracil (U). The nucleosides take their names from the names of the bases and sugars which comprise them. The four major ribose nucleosides are adenosine, guanosine, cytidine, and uridine; the four major 2-deoxyribose nucleosides are 2-deoxya-denosine, 2-deoxyguanosine, 2-deoxycytidine, and 2-deoxythymine.

Mononucleotides

A nucleoside can react with phosphoric acid at one of the sugar ring hydroxyl groups to lose water. When it does so it is said to form a mononucleotide such as adenosine monophosphate (AMP). Adenosine monophosphate can react with additional inorganic phosphate to form high-energy phosphate bonds in adenosine diphosphate (ADP) and adenosine triphosphate (ATP). Hydrolysis of these phosphate bonds returns the energy used to form them. Living organisms use ATP as an energy carrier, since movement of the ATP and ADP molecules is equivalent to the movement of energy. The most ubiquitous and perhaps the most important energy carrier in living organisms is the nucleotide adenosine triphosphate, normally abbreviated ATP. The name adenosine triphosphate denotes adenine coupled to one of the ribose sugars, with three phosphate groups attached in a chain structure. The actual high-energy bonds in ATP are the bonds linking the phosphate groups to each other. Thus the hydrolysis of such a bond releases energy while the formation of it requires energy, and it is in the form of ATP that energy is moved around the organism to the point at which it is required.

While ATP and, to a much lesser extent, other similar molecules serve as energy transport molecules, the amount of ATP present in an organism at any one time is comparatively small. Thus sustained muscular work, for example, could not be provided by a handy supply of ATP stored in the muscle. Muscular work is a complex process in which ATP gives up its energy and is thus converted into adenosine diphosphate (ADP) and inorganic phosphate (H_3PO_4 or, more correctly, simply Pi since it is not in acid form):

$$\text{ATP} \longleftrightarrow \text{ADP} + \text{Pi} + \text{energy}.$$

In higher animals, such as humans, muscles all contain some stored energy in the form of creatine phosphate. Lower animals, such as invertebrates, use other compounds. The storage is based upon a reaction whose equilibrium constant is about 0.6 catalyzed by creatine kinase:

$$\text{creatine phosphate} + \text{ADP} \longleftrightarrow \text{creatine} + \text{ATP}.$$

This reaction is catalyzed by the enzyme creatine kinase, one of the many enzymes present in the human body; its concentration in human blood serum is about 10^{-7} mol/litre.

Living organisms use, through the intermediate of ATP, energy for a wide variety of things. Some of it must be expended to produce more energy, as in the oxidation of glucose. Some of it is used in muscular work. Some is used to produce light, as in the firefly; much of it appears as waste heat. Some of it is used in the interesting area of osmotic work.

POLYNUCLEOTIDE

A polymer of mononucleotides is called a polynucleotide. In polynucleotides, only one phosphoric acid is present on each ribose sugar so hydrolysis of polynucleotides produces equimolar solutions of nitrogenous base, ribose sugar, and phosphate. The phosphoric acid component of polynucleotides readily loses a proton and so polynucleotides are also called nucleic acids. Nucleic acids include either ribose sugars (ribonucleic acids, often abbreviated RNA), or 2'-deoxyribose sugars (deoxyribonucleic acids, often abbreviated DNA). Only one type of sugar is found to be included in each polymer. Polynucleotides, both DNA and RNA, are the information carriers of living organisms and play the central role in reproduction. The simplest form of life may be the virus. A virus is a packet of infectious polynucleotides, which may be either RNA or DNA and either single-stranded or double-stranded, surrounded by a protective protein coat. One of the most well-understood small viruses is the tobacco mosaic virus (TMV virus, which infects the leaves of tobacco plants. The tobacco mosaic virus particle, or virion, contains one single-stranded RNA molecule which is about 6400 nucleotides long. Its coat is a protein which contains 336,540 amino acid residues, made up of 2130 identical subunits each 158 amino acid residues long.

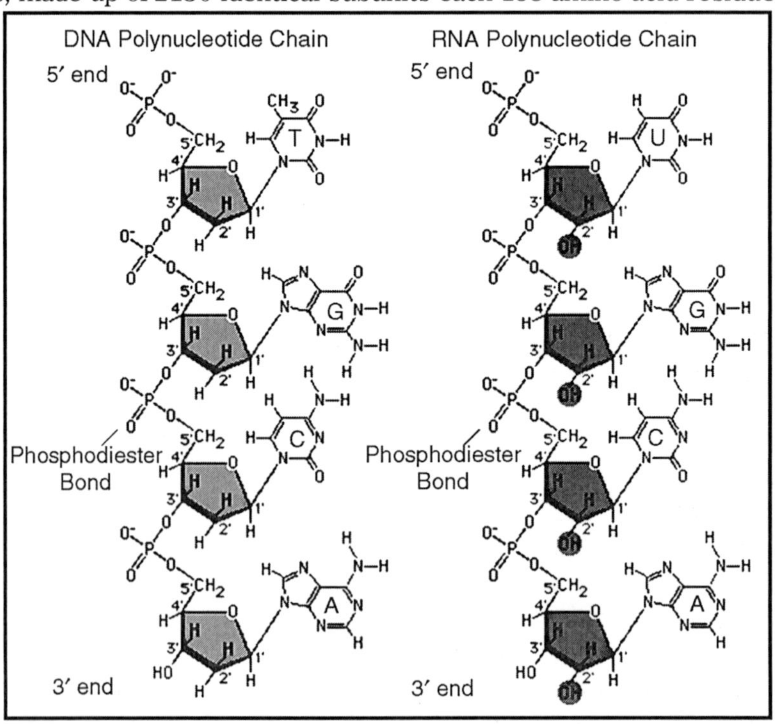

Fig. Polynucleotide

The structure of the TMV virion is very stable and the virions can remain infective for decades, although in the absence of the protein coat the TMV RNA is easily attacked by enzymes. In 1955, Heinz Fraenkel-Conrat and Robley Williams showed that under suitable conditions of pH and salt content the separate coat subunits and RNA of TMV spontaneously self-assemble into virions which are indistinguishable in structure and infectivity from the original TMV which contained them. This stable and infective form of the virion has a protein coat which is just sufficient to cover the RNA. Each virion is a rod about 30 micrometers long and 1.8 micrometers in diameter which has a mass of about 30 million daltons.

The segment of a DNA or RNA molecule which carries the genetic information required for the synthesis of a single polypeptide chain is called a gene. Each TMV virion is believed to contain six genes. More complex viruses, and all higher forms of life, contain far more genes.

The T4 bacteriophage virus, which attacks the common Escherichia coli human intestinal bacteria, is a much more complex virus than the tobacco mosaic virus. Its genetic information, about 165 genes, is contained in a double-stranded DNA molecule. The T4 bacteriophage virion consists of this DNA molecule inside a complex structured protein coat. The protein coat consists of a hollow head which contains the DNA molecule, a tail, and a baseplate with six short spikes and six long fibres. The tail can contract, and the entire assembly serves much as a hypodermic syringe to inject the DNA of the virion into the host bacterial cell. Viruses, whether simple or complex, cannot generate metabolic energy nor can they synthesize proteins. They reproduce the proteins necessary to make up their protective coats by taking over control of the protein-synthesizing mechanism of the cells they infect, and use the metabolic energy generated by the host cell. The effect of this is usually to destroy the host cell. For example, the T4 bacteriophage virion ruptures the bacterial cell wall, destroying the cell, when the assembly of its daughter virions is complete. The entire process from infection to release of about two hundred new virions takes about twenty minutes.

Higher organisms are made up of cells which are surrounded by a cell wall or cell membrane. The simplest cells are the prokaryotic cells or prokaryotes which are found as some of the simpler one-celled algae and bacteria. The common human intestinal bacterium Escherichia coli is a typical eucaryote; its cells are about 200 micrometers in diameter. Prokaryotic cells contain only one chromosome, consisting of a single molecule of closely coiled double-stranded DNA. Different lengths of this chromosome each correspond to a different gene of the procaryotic cell. A bacterium may contain several thousand genes. Human cells, like those of all higher animals and plants, are eukaryotic cells. Eucaryotic cells are 100 to 10,000 times as large as prokaryotes. Eukaryotic cells contain a nucleus, which is separated from the rest of the cell contents by a nuclear membrane. This nuclear membrane is in addition to the cell membrane. The nucleus contains the genetic material of the cell, which consists of several to many chromosomes. Each chromosome consists of closely coiled double-stranded DNA. Human cells contain 21 chromosomes, and like other eukaryotic cells contain millions of different genes.

The transcription of genetic information into molecular and ultimately cellular structure is the subject of molecular genetics. The central canon of molecular genetics is that information flows from DNA to RNA to proteins. This means that the order of bases in DNA, which is the actual genetic information transferred in reproduction in more complex living organisms, is first expressed in the order of bases within smaller RNA polynucleotides. The sequence or order of bases within the RNA molecules then is expressed as the sequence of the amino acids which form a protein.

The order of amino acids in a protein is its primary structure. Most proteins will under physiological conditions spontaneously assume the most stable configurations, forming sulfide bonds and hydrogen bonds, and so establish unique secondary and tertiary structures. The information coded in the base sequence of DNA is thus expressed in the actual physical structure and nature of the protein. The mechanism by which this transfer of information occurs, and the actual assembly of proteins by living cells, is probably the single most active area of research in biochemistry and molecular genetics.

NUCLEIC ACIDS

Nucleic acids are colourless, complex, amorphous compounds made up of three units : purine and pyrimidine bases, sugar and phosphoric acid. The nucleic acids are of two types DNA and RNA. It is very important to know the structure of nucleic acids and their components.

Pyrimidine Bases

Structure of Pyrimidine Bases

Pyrimidine bases are derived from the parent compound pyrimidine. The pyrimidine bases present in nucleotides are cytosine, uracil and thymine. It is 6 membered heterocyclic compound contains carbon, nitrogen, hydrogen and oxygen atoms.

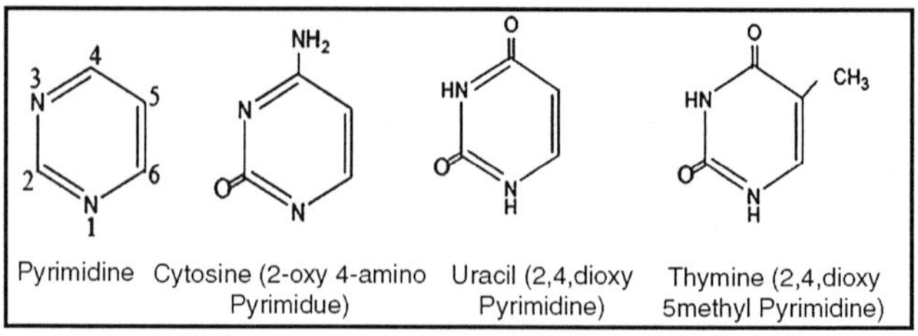

Fig. Structure of Pyrimidine Bases

Properties of Pyrimidine

- Pyrimidine bases are soluble in water.
- They absorb UV light at 260 nm. This property is used to detect and estimate pyrimidine nucleotides.
- They are capable of forming hydrogen bonds with other purine bases
- They exhibit keto-enol tautormerism.

Purine Bases

Structure of Purine Bases

Purine bases are derived from the parent compound purine. Purine contains the heterocyclic ring system. Fusion of the pyrimidine ring with imidazole yields the purine ring.

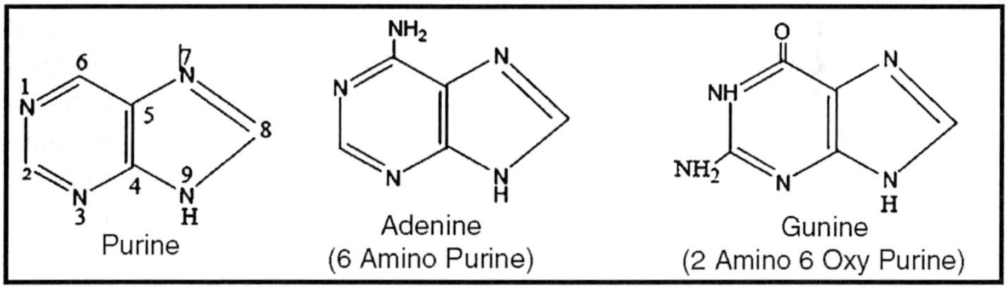

Fig. Structure of Purine Bases

The purine bases present in nucleic acids are adenine and guanine. Other purine bases are hypoxanthine and xanthine. They are intermediates in the formation of adenine and guanine nucleotides.

Properties of Purine Bases

- Purine bases are sparingly soluble in water.
- They absorb light in UV region at 260 nm. This property is used for the detection of and the quantification of purine nucleotides.
- They are capable of forming hydrogen bonds.
- They exhibit keto-enol tautomerism at body pH.

NUCLEIC ACIDS

Two types of nucleic acids are present in all mammalian cells. They are DNA - deoxy ribonucleic acid and RNA- ribonucleic acid. DNA is present in the nucleus and mitochondria. RNA is present in the nucleus, risosome and cytoplasm. Nucleic acids are acidic sustances containing nitrogenous bases, pentose sugar and phosphoric acid. Both DNA and RNA are polynucleotides. They are polymers of mononucleotides. In nucleic acids, nucleotides are joined together by phosphodiester linkages.

Nucleosides

A nucleoside is composed of purine or pyrimidine base and a pentose sugar. Two types of pentose sugar are present in nucleoside, they are ribose and deoxy ribose. In the case of purine nucleosides, the sugar is attached to N-9 of the purine ring, whereas in pyrimidine nucleosides, the sugar is attached to N-1 of the pyrimidine ring. The type of linkage is N-glycosidic linkage.

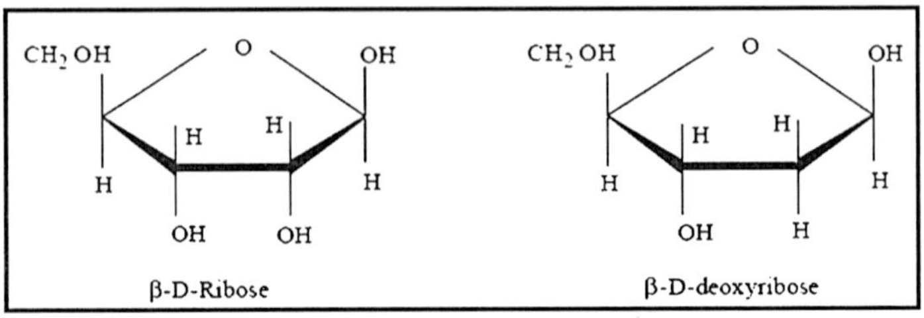

Fig. Structure of Sugars in Nucleic Acids

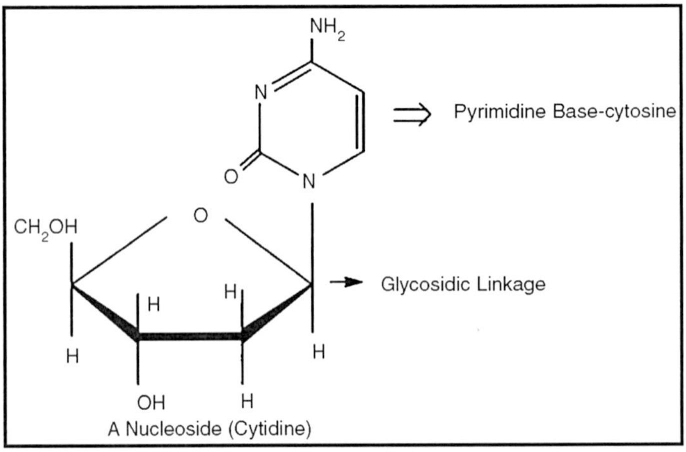

Fig. Structure of a Nucleoside

Nucleotides

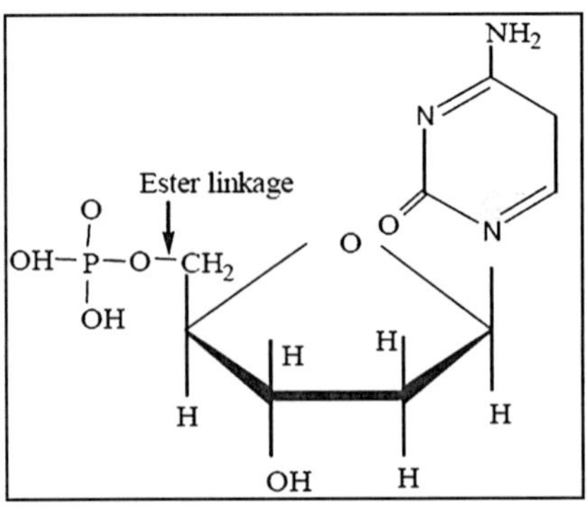

Fig. Structure of a Nucleotide (Cytidylic Acid)

Nucleotides are phosphorylated nucleosides usually one or two of hydroxyl groups of ribose (or) deoxyribose are phosphorylated. Thus a nucleotide has three structural components. They are nitrogenous base, sugar and phosphate. Phosphate is attached to ribose (or) deoxy ribose through an ester linkage.

STRUCTURE OF DNA

PRIMARY STRUCTURE

Nucleotide sequence of a nucleic acid is known as its primary structure which confers individuality to the polynucleotide chain. Polynucleotide chain has direction. They are represented in 5'---> 3' and 3'----> directions. Each polynucleotide chain has 2 ends. The 5' end carrying a phosphate group and 3' end carrying an unreacted hydroxyl group.

In 1953, J.D. Watson and F.H.C. Crick proposed a precise three dimensional model of DNA structure based on model building studies, base composition and X-ray diffraction studies. This model is popularly known as the DNA double helix. The purine bases present in DNA are adenine and guanine and the pyrimidine bases present are thymine and cytosine. The purine and pyrimidine bases of DNA carry genetic information where as the sugar and phosphate groups perform the structural role.

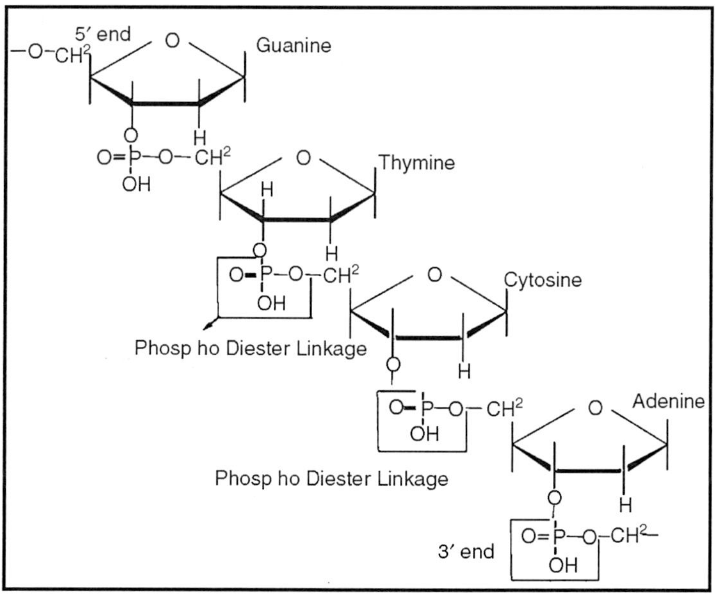

Fig. Primary Structure of DNA

Salient Features of Double Helix

- Two polynuleotide chains are coiled around a central axis in the form of a right handed double helix.

- Each polynucleotide chain is made up of 4 types of nucleotides. They are adenylate, guanidylate, thymidylate and cytidinilate.
- Each polynucleotide chain has direction or polarity. Further each polynucleotide chain has 5' phosphorylated and 3' hydroxyl ends.
- The backbone of each strand consists of alternating sugar and phosphate. The bases project inwards and they are perpendicular to the central axis.
- The 2 strands run in opposite direction (ie.) they are antiparallel.
- The strands are complementary to each other. Base composition of one strand is complementary to the opposite strand. If adenine appears in one strand, thymine is found in the opposite strand and vice versa. When guanine is found in one strand, cytosine is present in the opposite strand and vice versa.
- Bases of opposite strands are involved in pairing. Pairing occurs through hydrogen bonding and it is specific. Adenine pairs with thymine through two hydrogen bonds. Guanine pairs with cytosine with three hydrogen bonds.
- Major and minor grooves are present on the double helix. They arise because glycosidic linkages of base pairs are not opposite to each other. Protein interact with DNA through the minor and major grooves without disrupting the DNA strands.
- According to Chargaff's observation, the number of adenine base is equal to thymine base and the number of quanine base is equal to number of cytocine base ie. $A = T$ and $G = C$. Also $A + T = G + C$ and the ratio of $A+T/G+C$ = nearly 1.0. The total number of purine bases = the total number of pyrimidine bases.

Functions of DNA

- DNA is the genetic material of living organisms. It is the greatest super chip ever made by man.
- DNA contain all the information required for the information of an individual organism.
- The genetic information in DNA is converted to characteristic features of living organisms like colour of the skin and eye, height, intelligence, ability to metabolise particular substance, ability to withstand stress, susceptibility to desease and ability to produce or synthesise certain substances.
- DNA is the source of information for the synthesis of all cellular proteins. The segment of DNA that contain information for a protein is known as gene.
- DNA is transmitted from parents to offsprings and hence transmit genetic information from one generation to another.

- The amount of DNA in any given species or cell is constant and is not affected by nutritional and metabolic states.

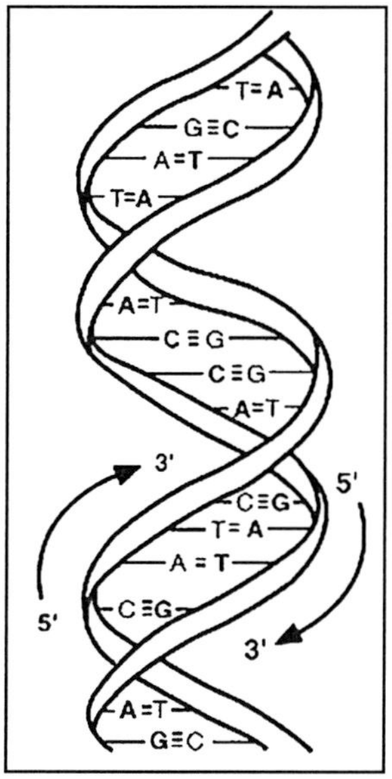

Fig. Double Helical Structure of DNA

- Avery Macleod and Mc Carty in 1944 first demonstrated that DNA contained the genetic information and they referred DNA as 'transforming factor'.
- The nucleotides present in DNA are deoxy adenylic acid, deoxy quanidylic acid, deoxy cytidylic acid and deoxy thymidylic acid.

Structure of RNA

RNAs are present in the nucleus, ribosomes and cytoplasm of eukaryolic cells. They are involved in the transfer and expression of genetic information. They act as primer for DNA formation. Some act as enzymes and as coenzymes. RNA also function as genetic material for viruses. RNAs are also polynucleotides. In RNA polymer, purine and pyrimidine nucleotides are linked together through phosphodiester linkages. The sugar present is ribose. The nitrogenous bases present in RNA are adenine and guanine (purine bases), uracil and cytosine (pyrimidine bases). The nucleotides present in RNA are adenylic acid, quanidylic acid, cytidylic acid and uridylic acid.

Types of RNA

There are mainly three types of RNAs in all prokaryotic and eukaryotic cells. They are (1) Messenger RNA (mRNA) 2) Transfer RNA (tRNA) 3) Ribosomal RNA (rRNA). They differ from each other by size, formation and stability.

Messenger RNA

It accounts for 1-5 per cent of cellular RNA. They have a primary structure. They are single standed linear molecules. They consist of 1000-10,000 nucleotides. They have a free or phosphorylated 3' and 5' end. They have different life span ranging from few minutes to days. mRNA molecules are capped at 5' end by methylated guanine triphosphate. Capping protects mRNA from nuclease attack. At 3' end a polymer of adenylate (poly A) is found as the tail. Poly A tail protects mRNA from nuclease attack. Intrastrand base pairing among complementary bases allows folding of the linear molecule. As a result, haripin or loop like secondary structure is formed.

Functions

- mRNA is a direct carrier of genetic information from the nucleus to the cytoplasm.
- It contain information required for the synthesis of protein molecules.

TRANSFER RNA

It accounts for 10-15 per cent of total cell RNA. They are the smallest of all the RNAs. Usually they consist of 50-100 nucleotides. They are single standard molecules.

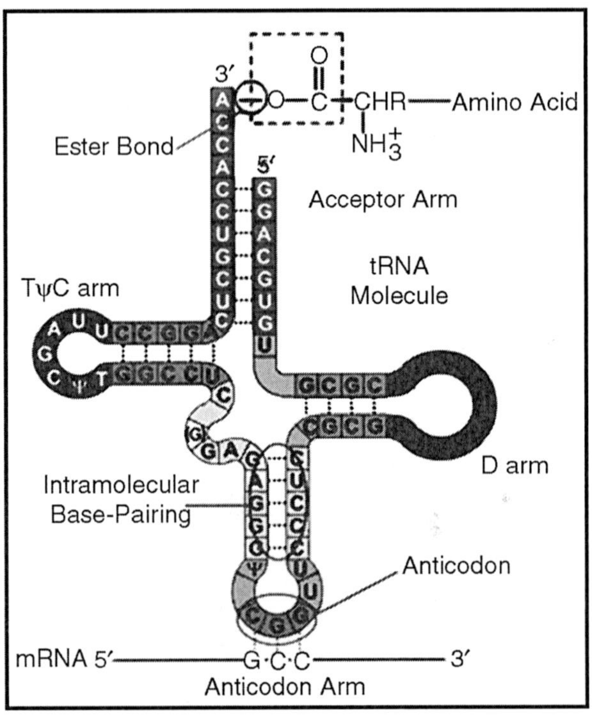

Fig. Secondary Structure of t-RNA

They contain unusual bases such as methylated adenine, guanine, cytosine and thymine,dihydrouracil and pseudouridine. These unusual bases are important for binding 6-RNA to intra chain base pairing. Further some bases are not involved in base pairing, resulting in loops and arms formation in tRNA. These folding in the primary structure generates a secondary structure.

Secondary structure of t-RNA is in the form of a clover leaf. The important feature of the clover-leaf structure are, 1. An acceptor arm with base sequence "CCA" 3'–OH of adenosine moiety of t-RNA.

Functions

- It is the carrier of amino acids to the site of protein synthesis.
- There is at least one t-RNA molecule to each of 20 amino acids required for protein synthesis.

RIBOSOMAL RNA

This accounts for 80 per cent of the total cellular RNA. It is present in ribosomes. In ribosomes, r-RNA is found in combination with protein. It is known as ribonucleoprotein. The length of rRNA ranges from 100-600 nucleotides. rRNA molecules have a secondary structure. Intra strand base pairing between complementary bases generate double helical segments or loops.

Functions

- They are required for the formation of risosomes
- They are involved in the initiation of protein synthesis.

DIFFERENCES BETWEEN DNA AND RNA

	DNA	RNA
•	Sugar moiety is deoxy ribose	Sugar moeity is ribose
•	Uracil is absent	Thymine is usually absent
•	Double stranded molecules	Single stranded molecules
•	Sum of purine and pyrimidine bases are equal G+C = A+T equal.G+C = A+T	Sum of purine and pyrimindine bases are not
•	Bases are not modified	Bases are modified
•	Resistant to alkali hydrolysis hydrolysis.	Highly susceptible to alkali
•	NO catalytic activity	Some are catalytically active
•	Mostly DNA is present in nucleus and also in mitochondria	Present in nucleolus and cytoplasm.

POLYSACCHARIDES

Polysaccharides are the complex carbohydrates. They are made up of chains of monosaccharides (the sugars) which are linked together by glycosidic bonds, which are formed by the condensation reaction. The linkage of monosaccharides into chains creates chains of greatly varying length, ranging from chains of just two monosaccharides, which makes a disaccharide to the polysaccharides, which consists of many thousands of the sugars.As a group, the polysaccharides play diverse and important roles within the biology of life processes; however, they can be divided into two broad groups.

STRUCTURAL POLYSACCHARIDES

Cellulose

Cellulose is a major component of plant cell walls. It is an unbranched polymer with about ten thousand glucose units per chain. Hydroxyl groups (–OH) project out from each chain, forming hydrogen bonds with neighbouring chains which creates a rigid cross-linking between the chains, making cellulose the strong support material that it is. Despite the combined strength of cellulose, it is fully permeable to water and solutes which makes it ideal for allowing water and solutes into and out of the cell.It is the most abundant organic substance in the living world and it has been estimated that more than half the total organic carbon on the planet is in cellulose.

Chitin

Chitin is closely related in structure to cellulose, also being an unbranched polysaccharide. However, instead of the hydroxyl groups (–OH), the chains have the following structure $-NH.CO.CH_3$ replacing it. Large amounts of chitin is found in the cuticles of arthropods, with smaller amounts being found in sponges, molluscs and annelids. Chitin can also be found in the cell walls of most fungi and in some green algae.

POLYSACCHARIDES USED AS ENERGY STORES

Polysaccharides make ideal storage molecules for energy for a number of reasons; a) they are large, this makes them insoluble in water and therefore they exert no osmotic or chemical effect on the cell; b) they fold into compact shapes; c) they are easily converted into the required sugars when needed.

Glycogen

Glycogen is a branched polysaccharide found in nearly all animal cells and in certain protozoa and algae. In humans and other vertebrates it is principally stored in the liver and muscles and is the main form of stored carbohydrate in the body, acting as a reservoir of glucose for when the body is being starved of food. In an animal that has been fed well, as much as 10 per cent of its liver weight may be glycogen. The liver's glycogen store can be almost completely depleted during a 24 hour fast or during muscular exertion but glycogen is quickly re-synthesised from newly ingested carbohydrate.

Starch

Starch is similar to glycogen, however it is found in plant cells, protists and certain bacteria. The starch granules can be seen in the cell through a light microscope. The starch granules are made up of two polysaccharides, amylose and amylopectin. Amylose is an unbranched molecule made up of several thousand glucose units, coiled helically into a more compact shape. Amylopectin is also compact but has a branched structure and is made up of twice as many glucose units as amylose.

COMBINING POLYSACCHARIDES WITH PROTEINS AND LIPIDS

Polysaccharides can be combined with proteins or lipids to form other biological molecules. For example, peptidoglycans, which are a combination of protein and polysaccharide and are found in the cell wall of certain bacteria. Glycolipids, a combination of polysaccharides and lipids are found in the cell membrane.

Bacterial Capsule Polysaccharides

Pathogenic bacteria commonly produce a thick, mucous-like, layer of polysaccharide. This "capsule" cloaks antigenic proteins on the bacterial surface that would otherwise provoke an immune response and thereby lead to the destruction of the bacteria. Capsular polysaccharides are water soluble, commonly acidic, and have molecular weights on the order of 100-1000 kDa. They are linear and consist of regularly repeating subunits of one ~ six monosaccharides. There is enormous structural diversity; nearly two hundred different polysaccharides are produced by E. coli alone. Mixtures of capsular polysaccharides, either conjugated or native are used as vaccines. Bacteria and many other microbes, including fungi and algae, often secrete polysaccharides as an evolutionary adaptation to help them adhere to surfaces and to prevent them from drying out. Humans have developed some of these polysaccharides into useful products, including xanthan gum, dextran, gellan gum, and pullulan.

Cell-surface polysaccharides play diverse roles in bacterial ecology and physiology. They serve as a barrier between the cell wall and the environment, mediate host-pathogen interactions, and form structural components of biofilms. These polysaccharides are synthesized from nucleotide-activated precursors (called nucleotide sugars) and, in most cases, all the enzymes necessary for biosynthesis, assembly and transport of the completed polymer are encoded by genes organized in dedicated clusters within the genome of the organism. Lipopolysaccharide is one of the most important cell-surface polysaccharides, as it plays a key structural role in outer membrane integrity, as well as being an important mediator of host-pathogen interactions.

The enzymes that make the *A-band* (homopolymeric) and *B-band* (heteropolymeric) O-antigens have been identified and the metabolic pathways defined. The exopolysaccharide alginate is a linear copolymer of β-1,4-linked D-mannuronic acid and L-guluronic acid residues, and is responsible for the mucoid phenotype of late-stage cystic fibrosis disease. The *pel* and *psl* loci are two recently discovered gene clusters that also encode exopolysaccharides found to be important for biofilm formation. Rhamnolipid is a biosurfactant whose production is tightly regulated at the transcriptional level, but the precise role that it plays in disease is not well understood at present. Protein glycosylation, particularly of pilin and flagellin, is a recent focus of research by several groups and it has been shown to be important for adhesion and invasion during bacterial infection.

PROTEIN SEQUENCING STRATEGY

The usual strategy for determining the amino acid sequence of a protein involves eight basic steps:

1. If the protein contains more than one polypeptide chain, the chains are separated and purified.
2. Intrachain S-S (disulfide) cross-bridges between cysteine residues in the polypeptide chain are cleaved. (If these disulfides are interchain linkages, then step 2 precedes step 1.)
3. The amino acid composition of each polypeptide chain is determined.
4. The N-terminal and C-terminal residues are identified.
5. Each polypeptide chain is cleaved into smaller fragments, and the amino acid composition and sequence of each fragment are determined.
6. Step 5 is repeated, using a different cleavage procedure to generate a different and therefore overlapping set of peptide fragments.
7. The overall amino acid sequence of the protein is reconstructed from the sequences in overlapping fragments.
8. The positions of S-S cross-bridges formed between cysteine residues are located.

Step 1. Separation of Polypeptide Chains

If the protein of interest is a heteromultimer (composed of more than one type of polypeptide chain), then the protein must be dissociated and its component polypeptide subunits must be separated from one another and sequenced individually. Subunit associations in multimeric proteins are typically maintained solely by noncovalent forces, and therefore most multimeric proteins can usually be dissociated by exposure to pH extremes, 8 *M* urea, 6 *M* guanidinium hydrochloride, or high salt concentrations. (All of these treatments disrupt polar interactions such as hydrogen bonds both within the protein molecule and between the protein and the aqueous solvent.) Once dissociated, the individual polypeptides can be isolated from one another on the basis of differences in size and/or charge.

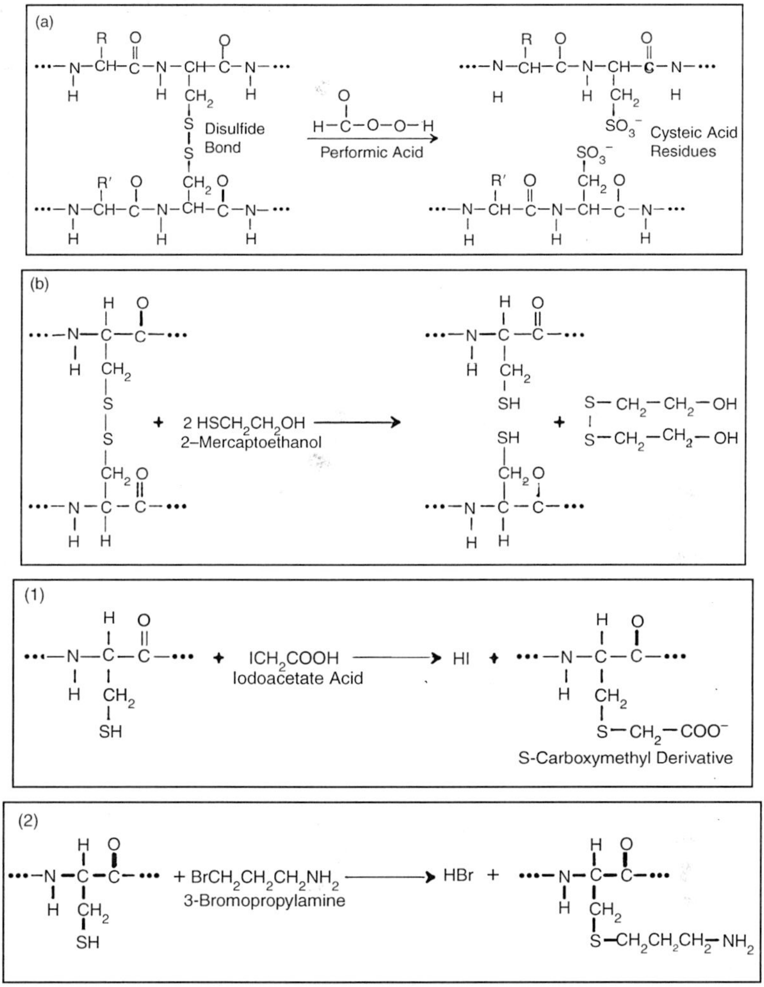

Fig. Methods for Cleavage of Disulfide bonds in Proteins. (a) Oxidative Cleavage by Reaction with Performic Acid. (b) Re-ductive Cleavage with Sulfhydryl compounds. Disulfide Bridges can be broken by Reduction of the S–S link with sulfhydryl agents such as 2-mercaptoethanol or Dithiothreitol. Because reaction between the newly reduced –SH groups to re-establish Disulfide bonds is a Likelihood, SOS Reduction must be followed by –SH modification: (1) Alkylation with Iodoacetate (ICH_2COOH) or (2) Modification with 3-bromopropylamine ($Br–(CH_2)_3–NH_2$).

Occasionally, heteromultimers are linked together by interchain S-S bridges. In such instances, these cross-links must be cleaved prior to dissociation and isolation of the individual chains. The methods described under step 2 are applicable for this purpose.

Step 2. Cleavage of Disulfide Bridges

A number of methods exist for cleaving disulfides (Figure). An important consideration is to carry out these cleavages so that the original or even new S-

S links do not form. Oxidation of a disulfide by performic acid results in the formation of two equivalents of cysteic acid (Figure a). Because these cysteic acid side chains are ionized SO_3^- groups, electrostatic repulsion (as well as altered chemistry) prevents S-S recombination. Alternatively, sulfhydryl compounds such as 2-mercaptoethanol readily reduce S-S bridges to regenerate two cysteine-SH side chains (Figure b).

However, these SH groups recombine to re-form either the original disulfide link or, if other free Cys-SHs are available, new disulfide links. To prevent this, S-S reduction must be followed by treatment with alkylating agents such as iodoacetate or 3-bromopropylamine, which modify the SH groups and block disulfide bridge formation (Figure a).

Step 3. Analysis of Amino Acid Composition

The standard protocol for analysis of the amino acid composition of proteins. Results of such analyses allow the researcher to anticipate which methods of polypeptide fragmentation might be useful for the protein.

Step 4. Identification of the N- and C-Terminal Residues

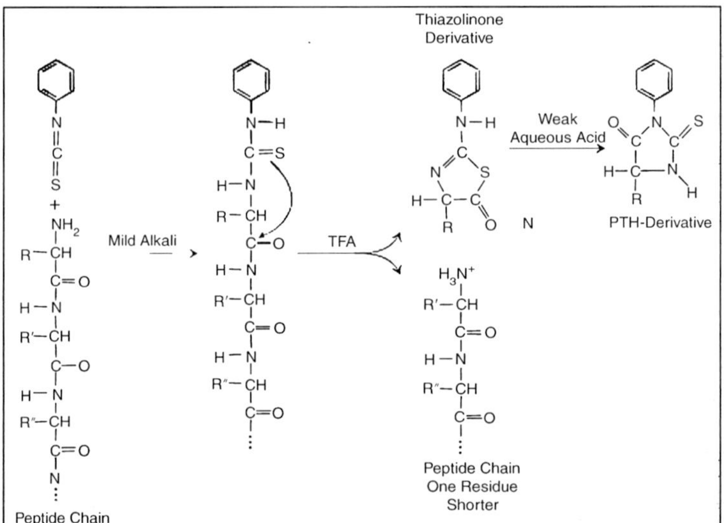

Fig. N-Terminal Analysis Using Edman's Reagent, Phenylisothiocyanate. Phenylisothiocyanate Combines with the N-terminus of a Peptide under Mildly Alkaline conditions to form a Phenylthiocarbamoyl Substitution. Upon treatment with TFA (Trifluoroacetic acid), this Cyclizes to release the N-terminal amino acid Residue as a Thiazolinone Derivative, but the other Peptide bonds are not Hydrolyzed. Organic Extraction and Treatment with Aqueous Acid Yield the N-terminal Amino Acid as a Phenylthiohydantoin (PTH) Derivative.

End-group analysis reveals several things. First, it identifies the N- and C-terminal residues in the polypeptide chain. Second, it can be a clue to the number of ends in the protein. That is, if the protein consists of two or more different polypeptide chains, then more than one end group may be discovered, alerting the investigator to the presence of multiple polypeptides.

A. N-Terminal Analysis

The amino acid residing at the N-terminal end of a protein can be identified in a number of ways; one method, Edman degradation, has become the procedure of choice. This method is preferable because it allows the sequential identification of a series of residues beginning at the N-terminus (Figure). In weakly basic solutions, phenyliso-thiocyanate, or Edman's reagent (phenyl-N=C=S), combines with the free amino terminus of a protein, which can be excised from the end of the polypeptide chain and recovered as a phenylthiohydantoin (PTH) derivative. This PTH derivative can be identified by chromatographic methods. Importantly, in this procedure, the rest of the polypeptide chain remains intact and can be subjected to further rounds of Edman degradation to identify successive amino acid residues in the chain. Often, the carboxyl terminus of the polypeptide under analysis is coupled to an insoluble matrix, allowing the polypeptide to be easily recovered by filtration following each round of Edman reaction.

Thus, Edman reaction not only identifies the N-terminus of proteins but can also reveal additional information regarding sequence. Automated instruments (so-called *Edman sequenators*) have been designed to carry out the reaction cycle of the Edman procedure. In practical terms, as many as 50 cycles of reaction can be accomplished on 50 pmol (about 0.1 mg) of a polypeptide 100 to 200 residues long, generating the sequential order of the first 50 amino acid residues in the protein. The efficiency with larger proteins is less; a typical 2000-amino acid protein provides only 10 to 20 cycles of reaction.

B. C-Terminal Analysis

For the identification of the C-terminal residue of polypeptides, an enzymatic approach is commonly used.

Enzymatic Analysis with Carboxypeptidases

Carboxypeptidases are enzymes that cleave amino acid residues from the C-termini of polypeptides in a successive fashion. Four carboxypeptidases are in general use: A, B, C, and Y. *Carboxypeptidase* A (from bovine pancreas) works well in hydrolyzing the C-terminal peptide bond of all residues except proline, arginine, and lysine. The analogous enzyme from hog pancreas, *carboxypeptidase* B, is effective only when Arg or Lys are the C-terminal residues. Thus, a mixture of carboxypeptidases A and B liberates any C-terminal amino acid except proline. *Carboxypeptidase* C from citrus leaves and *carboxypeptidase* Y from yeast act on any C-terminal residue. Because the nature of the amino acid residue at the end often determines the rate at which it is cleaved and because these enzymes remove residues successively, care must be taken in interpreting results. Carboxypeptidase Y cleavage has been adapted to an automated protocol analogous to that used in Edman sequenators.

Steps 5 and 6. Fragmentation of the Polypeptide Chain

The aim at this step is to produce fragments useful for sequence analysis. The cleavage methods employed are usually enzymatic, but proteins can also be fragmented by specific or non-specific chemical means (such as partial acid hydrolysis). Proteolytic enzymes offer an advantage in that they may hydrolyze only specific peptide bonds, and this specificity immediately gives information about the peptide products. As a first approximation, fragments produced upon cleavage should be small enough to yield their sequences through end-group analysis and Edman degradation, yet not so small that an over-abundance of products must be resolved before analysis. However, the determination of total sequences for proteins predates the Edman procedure, and alternative approaches obviously exist.

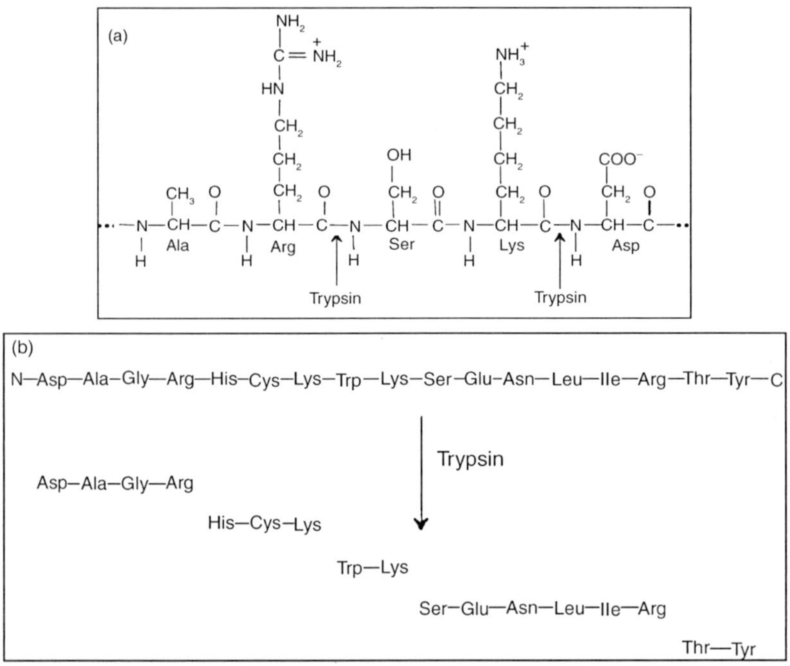

Fig. Trypsin is a Proteolytic Enzyme, or *Protease,* that Specifically cleaves only those Peptide bonds in which Arginine or Lysine Contributes the Carbonyl Function. The products of the Reaction are a Mixture of Peptide Fragments with C-terminal Arg or Lys Residues *and* a single Peptide Derived from the Polypeptide's C-terminal end.

Trypsin

The digestive enzyme *trypsin* is the most commonly used reagent for specific proteolysis. Trypsin is specific in hydrolyzing only peptide bonds in which the carbonyl function is contributed by an arginine or a lysine residue. That is, trypsin cleaves on the C-side of Arg or Lys, generating a set of peptide fragments having Arg or Lys at their C-termini.

The number of smaller peptides resulting from trypsin action is equal to the total number of Arg and Lys residues in the protein *plus* one—the protein's C-terminal peptide fragment.

Chymotrypsin

Chymotrypsin shows a strong preference for hydrolyzing peptide bonds formed by the carboxyl groups of the aromatic amino acids, phenylalanine, tyrosine, and tryptophan.

However, over time chymotrypsin also hydrolyzes amide bonds involving amino acids other than Phe, Tyr, or Trp. Peptide bonds having leucine-donated carboxyls become particularly susceptible. Thus, the specificity of chymotrypsin is only relative. Because chymotrypsin produces a very different set of products than trypsin, treatment of separate samples of a protein with these two enzymes generates fragments whose sequences overlap. Resolution of the order of amino acid residues in the fragments yields the amino acid sequence in the original protein.

Other Endopeptidases: A number of other *endopeptidases* (proteases that cleave peptide bonds within the interior of a polypeptide chain) are also used in sequence investigations. These include *clostripain*, which acts only at Arg residues, *endopeptidase Lys*-C, which cleaves only at Lys residues, and *staphylococcal protease*, which acts at the acidic residues, Asp and Glu. Other, relatively non-specific endopeptidases are handy for digesting large tryptic or chymotryptic fragments.

Pepsin, papain, subtilisin, thermolysin, and *elastase* are some examples. Papain is the active ingredient in meat tenderizer and in soft contact lens cleaner as well as in some laundry detergents. The abundance of papain in papaya, and a similar protease (bromelain) in pineapple, causes the hydrolysis of gelatin and prevents the preparation of Jell-O® containing either of these fresh fruits. Cooking these fruits thermally denatures their proteolytic enzymes so that they can be used in gelatin desserts.

Cyanogen Bromide

Several highly specific chemical methods of proteolysis are available, the most widely used being *cyanogen bromide* (*CNBr*) cleavage. CNBr acts upon methio-nine residues (Figure). The nucleophilic sulfur atom of Met reacts with CNBr, yielding a sulfonium ion that undergoes a rapid intramolecular rearrangement to form a cyclic iminolactone.

Water readily hydrolyzes this iminolactone, cleaving the polypeptide and generating peptide fragments having C-terminal homoserine lactone residues at the former Met positions.

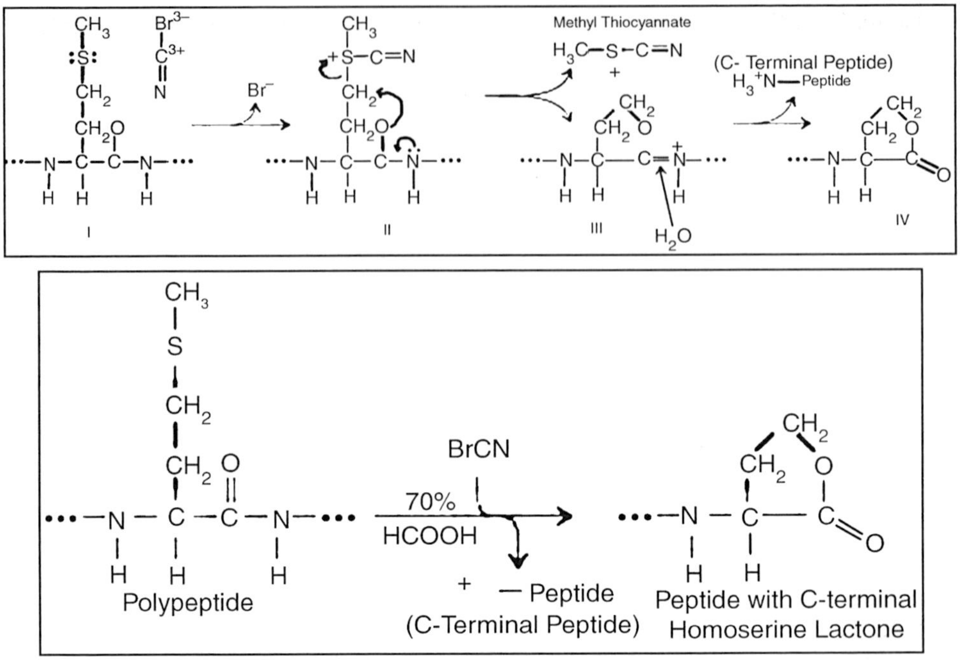

Fig. Cyanogen Bromide (CNBr) is a Highly Selective Reagent for Cleavage of Peptides only at Methionine Residues. (I) The Reaction Occurs in 70 per cent Formic Acid via Nucleophilic Attack of the Met S Atom on the –CºN Carbon Atom, with Displacement of Br. (II) The Cyano Intermediate Undergoes Nucleophilic Attack by the Met Carbonyl Oxygen Atom on the R Group, (III) Resulting in Formation of the Cyclic Derivative, which is Unstable in Aqueous Solution. (IV) Hydrolysis Ensues, Producing Cleavage of the Met Peptide Bond and Release of Peptide Fragments, with C-terminal Homoserine Lactone Residues where Met Residues once were. One Peptide does not have a C-terminal Homoserine Lactone: The Original C-terminal end of the Polypeptide.

Other Methods of Fragmentation

A number of other chemical methods give specific fragmentation of polypeptides, including cleavage atasparagine-glycine bonds by hydroxylamine (NH_2OH) at pH 9 and selective hydrolysis at aspartyl-prolyl bonds under mildly acidic conditions. Table summarizes the various procedures described here for polypeptide cleavage. These methods are only a partial list of the arsenal of reactions available to protein chemists.

Cleavage products generated by these procedures must be isolated and individually analysed with respect to amino acid composition, end-group identity, and amino acid sequence to accumulate the information necessary to reconstruct the protein's total amino acid sequence. In the past, sequence was often deduced from exhaustive study of the amino acid composition and end-group analysis of small, overlapping peptides.

Peptide sequencing today is most commonly done by Edman degradation of relatively large peptides.

Table. Specificity of Representative Polypeptide Cleavage Procedures Used in Sequence Analysis

Method	Peptide Bond on Carboxyl (C) or Amino (N) Side of Susceptible Residue	Susceptible Residue(s)
Proteolytic enzymes		Arg or Lys
Trypsin	C	Phe, Trp, or
Chymotrypsin	C	Tyr; Leu
Clostripain	C	Arg
Staphylococcal	C	Asp or Glu
protease		
Chemical methods	C	Met
Cyanogen bromide	Asn-Gly bonds	
NH_2OH	Asp-Pro bonds	
pH 2.5, 40°C		

Sequence Determination by Mass Spectrometry (MS)

Mass spectrometers exploit the difference in the mass-to-charge (*m/z*) ratio of ionized atoms or molecules to separate them from each other. The *m/z* ratio of a molecule is also a highly characteristic property that can be used for determining chemical and structural information. Further, molecules can be fragmented in distinctive ways in mass spectrometers, and the fragments that arise also provide quite specific structural information about the molecule. The basic operation of a mass spectrometer is to (1) evaporate and ionize molecules in a vacuum, creating gas-phase ions, (2) separate the ions in space and/or time based on their *m/z* ratios, and (3) measure the amount of ions with specific *m/z*ratios.

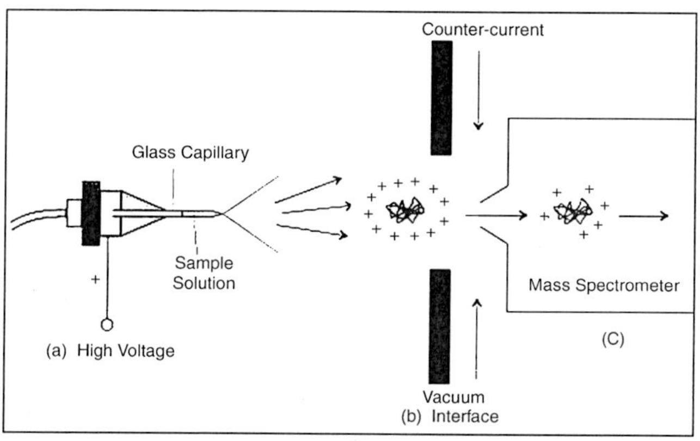

Fig. The Three Principal Steps Inelectrospray Mass Spectrometry (ES-MS). (a) Small, Highly Charged Droplets are Formed by Electrostatic Dispersion of a Protein Solution through a Glass Capillary Subjected to a High Electric Field; (b) Protein Ions are de-sorbed from the Droplets into the Gas Phase (Assisted by Evaporation of the Droplets in a Stream of hot N_2 gas; and (c) Analysis of the Protein Ions in a mass Spectrometer.

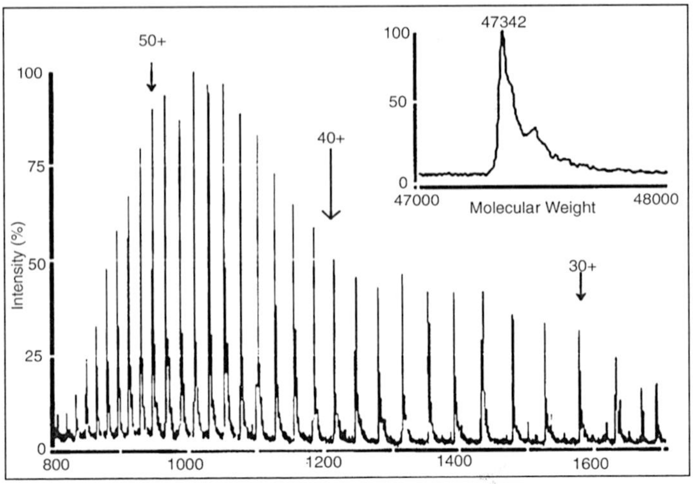

Fig. Electrospray Mass Spectrum of the Protein, Aerolysin K.

The attachment of many protons per protein molecule (from less than 30 to more than 50 here) leads to a series of *m/z* peaks for this single protein. The inset shows a computer analysis of the data from this series of peaks that generates a single peak at the correct molecular mass of the protein.

Figure illustrates the basic features of electrospray mass spectrometry (ES-MS). In this technique, proteins pick up, on average, about one positive charge (proton) per kilodalton, leading to the spectrum of m/z ratios for a single protein species (Figure). Computer algorithms can convert these data into a single spectrum having a peak at the correct protein mass.

Sequencing by Tandem Mass Spectrometry

Tandem MS (or MS/MS) allows sequencing of proteins by hooking two mass spectrometers in tandem. The first mass spectrometer is used to separate oligopeptides from a protein digest and then to select in turn each of these oligopeptides for further analysis.

A selected ionized oligopeptide is directed towards the second mass spectrometer; on the way, this oligopeptide is fragmented by collision with helium or argon gas mole-cules, and the collection of fragments is analysed by the second mass spectrometer (Figure). Fragmentation occurs primarily in the peptide bonds linking successive amino acids in the oligopeptide, so the fragments

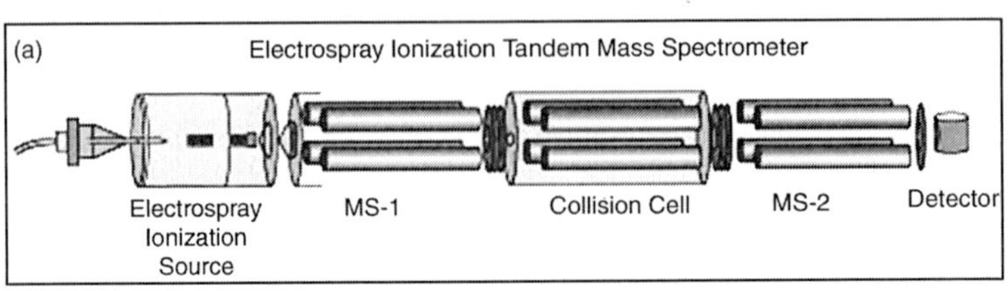

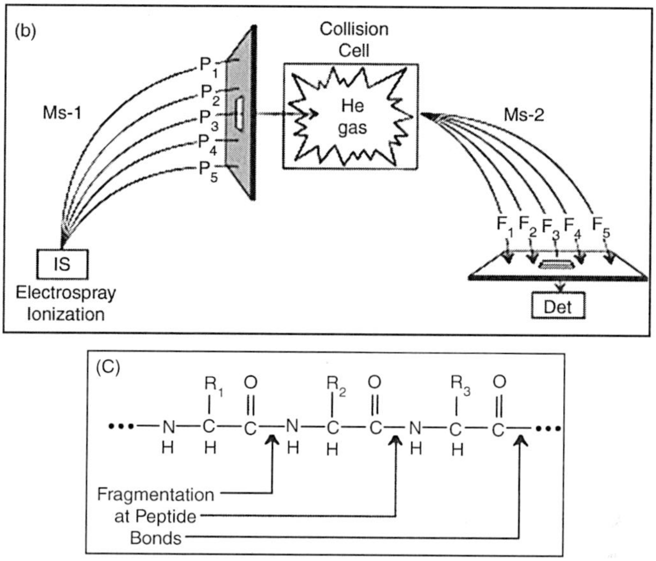

Fig. Tandem Mass Spectrometry. (a) Configuration used in Tandem MS. (b) Schematic Description of Tandem MS: Tandem MS Involves Electrospray Ionization of a Protein Digest (IS in this figure), Followed by Selection of a Single Peptide ion Mass for Collision with Inert Gas Molecules (He) and Mass Analysis of the Fragment Ions Resulting from the Collisions. (c) Fragmentation usually Occurs at Peptide bonds, as indicated.

The fragments differ in mass by 56 atomic mass units (the mass of the peptide backbone atoms (NH-CH-CO)) *plus*the mass of the R group at each position, which ranges from 1 atomic mass unit (Gly) to 130 (Trp). MS sequencing has the advantages of very high sensitivity, fast sample processing, and the ability to work with mixtures of proteins. Subpicomoles (less than 10^{-12} moles) of peptide can be analysed.

However, in practice, tandem MS is limited to rather short sequences (no longer than 15 or so amino acid residues). Nevertheless, capillary HPLC-separated peptide mixtures from trypsin digests of proteins can be directly loaded into the tandem MS spectrometer. Further, separation of a complex mixture of proteins from a whole-cell extract by two-dimensional gel electrophoresis, followed by trypsin digest of a specific protein spot on the gel and injection of the digest into the HPLC/tandem MS, gives sequence information that can be used to identify specific proteins. Often, by comparing the mass of tryptic peptides from a protein digest with a database of all possible masses for tryptic peptides (based on all known protein and DNA sequences), a protein of interest can be identified without actually sequencing it.

Step 7. Reconstruction of the Overall Amino Acid Sequence

The sequences obtained for the sets of fragments derived from two or more cleavage procedures are now compared, with the objective being to find overlaps that establish continuity of the overall amino acid sequence of the polypeptide chain. Peptides generated from specific hydrolysis of the polypeptide can be aligned to reveal the overall amino acid sequence. Such comparisons are also useful in eliminating errors and validating the accuracy of the sequences determined for the individual fragments.

Step 8. Location of Disulfide Cross-Bridges

Strictly speaking, the disulfide bonds formed between cysteine residues in a protein are not a part of its primary structure. Nevertheless, information about their location can be obtained by procedures used in sequencing, provided the disulfides are not broken prior to cleaving the polypeptide chain.

Because these covalent bonds are stable under most conditions used in the cleavage of polypeptides, intact disulfides link the peptide fragments containing their specific cysteinyl residues and thus these linked fragments can be isolated and identified within the protein digest.

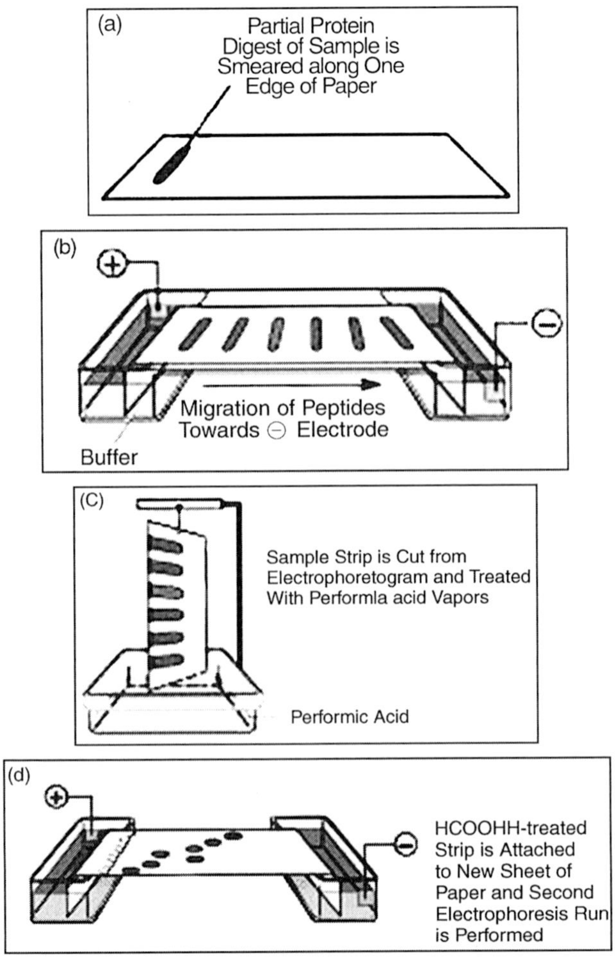

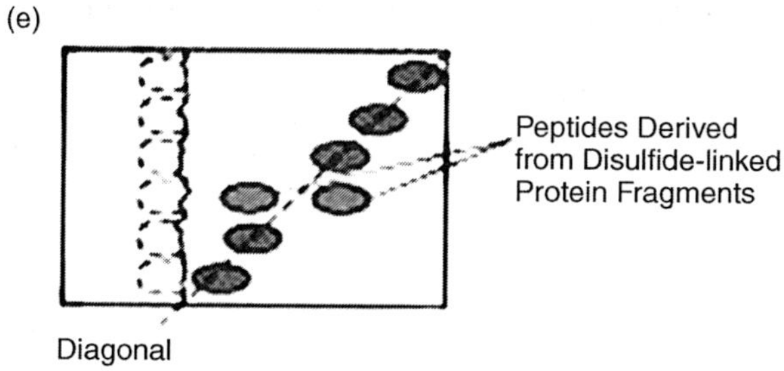

Fig. Disulfide Bridges Typically are Cleaved Prior to Determining the Primary Structure of a Polypeptide.

Consequently, the positions of disulfide links are not obvious from the sequence data. To determine their location, a sample of the polypeptide with intact SOS bonds can be fragmented and the sites of any disulfides can be elucidated from fragments that remain linked. Diagonal electrophoresis is a technique for identifying such fragments.

(a) A protein digest in which any disulfide bonds remain intact and link their respective Cys-containing peptides is streaked along the edge of a filter paper and
(b) Subjected to electrophoresis.
(c) A strip cut from the edge of the paper is then exposed to performic acid fumes to oxidize any disulfide bridges.
(d) Then the paper strip is attached to a new filter paper so that a second electrophoresis can be run in a direction perpendicular to the first.
(e) Peptides devoid of disulfides experience no mobility change, and thus their pattern of migration defines a diagonal.

Peptides that had disulfides migrate off this diagonal and can be easily identified, isolated, and sequenced to reveal the location of cysteic acid residues formerly involved in disulfide bridges.

An effective way to isolate these fragments is throughdiagonal electrophoresis. Peptides that were originally linked by disulfides now migrate as distinct species following disulfide cleavage and are obvious by their location off the diagonal. These cysteic acid-containing peptides are then isolated from the paper and sequenced. From this information, the positions of the disulfides in the protein can be stipulated.

3

Biological Functions of Proteins

PROTEINS

Proteins are composed of repeating monomeric units called amino acids. These monomers are named for the presence of the functional groups common to all amino acids, the amine group at one end of the molecule and the carboxylic acid group at the other.

The amine functional group is NH_2 and the carboxylic acid functional group is COOH. Although there are 20 different amino acids that compose all proteins, all amino acids have the amine and carboxylic acid ends.

The ends of the amino acid monomer are essential in the formation of bonds between amino acids. The bond that links amino acids together is called the peptide bond. Because of the peptide bond formation between all amino acids, the polymeric form of amino acids is called a polypeptide. Most people use the terms polypeptide and protein interchangeably; however, the term protein implies a functional, three-dimensional molecule while a polypeptide is simply a string of amino acids bonded together.

The 20 essential amino acids are quite different from each other. Biochemists usually categorize groups of amino acids together according to the types of functional characteristics they exhibit. These groups include the charged, or ionizable, amino acids; the non-polar amino acids; the polar, uncharged, amino acids; and those with unique structural properties. The amino acids can be categorized in many different ways but the grouping itself is insignificant. The real significance in the structure of amino acids resides in the ability of the chemical groups on different amino acids to impart very important structural and functional information into the protein that they form.

All of the proteins in living systems perform essential functions for organisms, and the functions of proteins are defined by the characteristics of the amino acids that compose them. In particular, the sequence of amino acids in a protein determines the function of that protein. This explains how there can be seemingly limitless types of proteins from a mere 20 different amino acids.

Compare two polypeptides: one composed of a string of 99 glycine amino acids and a final asparagine amino acid on the end. The other composed of a string of 98 glycine amino acids, with one asparagine amino acid in the next position, and then a final glycine on the end. Are these two polypeptides different? Absolutely! One might make a functional protein and the other may not. Change the amino acid sequence of any polypeptide or protein and the polypeptide or protein structure changes.

For example, consider the genetic condition called sickle cell anemia. People who have this condition make a hemoglobin protein (the oxygen carrying protein molecule of red blood cells) that has one different amino acid from normal hemoglobin. One charged aspartic acid amino acid is replaced with an uncharged glycine amino acid. This amino acid replacement results in a drastic alteration in the structure and function of the hemoglobin protein. The sickle cell hemoglobin protein actually precipitates out of solution under low oxygen conditions and causes the red blood cell to become sickle shaped. This, as you may imagine, has devastating effects upon individuals who have only this type of hemoglobin and usually results in a greatly shortened life span. Never underestimate the power of the sequence of amino acids, for it determines the structure and function of all proteins.

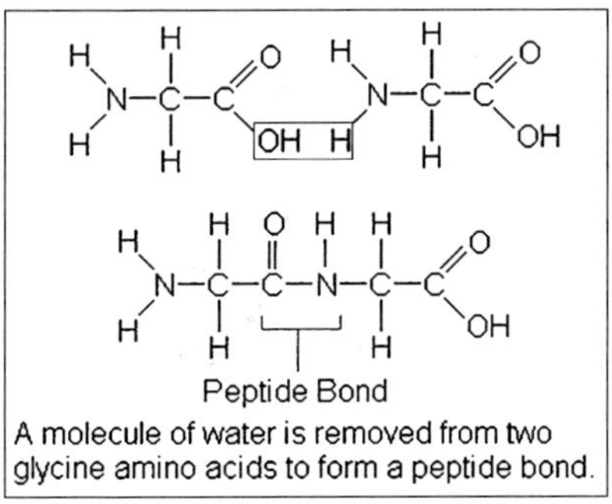

Fig. Two Glycine Amino Acids Combining Through the Formation of a Peptide Bond to form a Dipeptide.

PROTEINS ARE LINEAR POLYMERS OF AMINO ACIDS

Chemically, proteins are unbranched polymers of amino acids linked head to tail, from carboxyl group to amino group, through formation of covalent peptide bonds, a type of amide linkage (Figure).

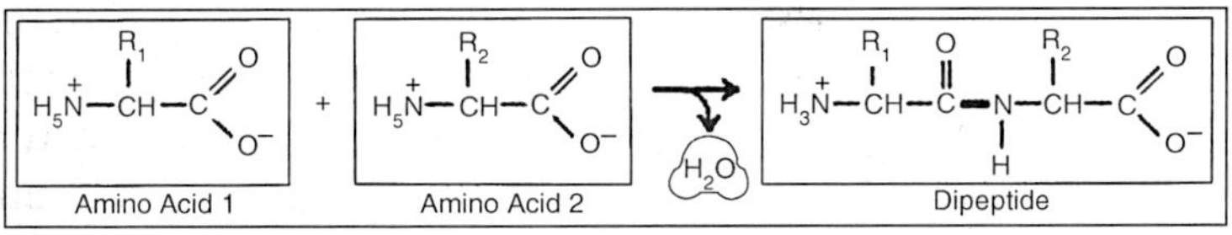

Fig. Peptide Formation is the Creation of an Amide Bond between the Carboxyl group of one Amino Acid and the Amino group of another Amino acid. R_1 and R_2 Represent the R groups of two Different Amino Acids.

Peptide bond formation results in the release of H_2O. The peptide "backbone" of a protein consists of the repeated sequence -N-C_a-C-, where the N represents the amide nitrogen, the C_a is thea-carbon atom of an amino acid in the polymer chain, and the final C is the carbonyl carbon of the amino acid, which in turn is linked to the amide N of the next amino acid down the line. The geometry of the peptide backbone is shown in Figure 5.2. Note that the carbonyl oxygen and the amide hydrogen are *trans* to each other in this figure. This conformation is favoured energetically because it results in less steric hindrance between nonbonded atoms in neighbouring amino acids. Because the a-carbon atom of the amino acid is a chiral center (in all amino acids except glycine), the polypeptide chain is inherently asymmetric. Only $_L$-amino acids are found in proteins.

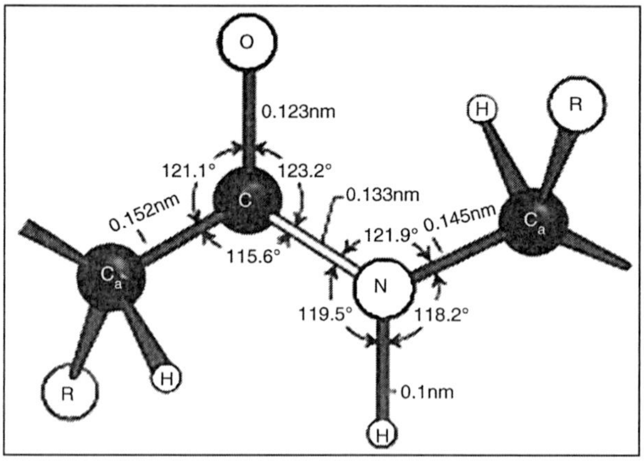

Fig. The Peptide bond is Shown in its usual *Trans* Conformation of carbonyl O and Amide H.

The C_α atoms are the α-carbons of two Adjacent Amino Acids Joined in Peptide Linkage. The Dimensions and Angles are the Average Values Observed by Crystallographic Analysis of Amino acids and small peptides.

THE PEPTIDE BOND HAS PARTIAL DOUBLE BOND CHARACTER

The peptide linkage is usually portrayed by a single bond between the carbonyl carbon and the amide nitrogen (Figure a). Therefore, in principle, rotation may occur about any covalent bond in the polypeptide backbone because all three kinds of bonds (N-C_α, C_α-C_o, and the C_o-N peptide bond) are single bonds.

In this representation, the C and N atoms of the peptide grouping are both in planar sp^2hybridization and the C and O atoms are linked by a p bond, leaving the nitrogen with a lone pair of electrons in a $2p$ orbital. However, another resonance form for the peptide bond is feasible in which the C and N atoms participate in a p bond, leaving a lone e^- pair on the oxygen (Figure b).

(A)

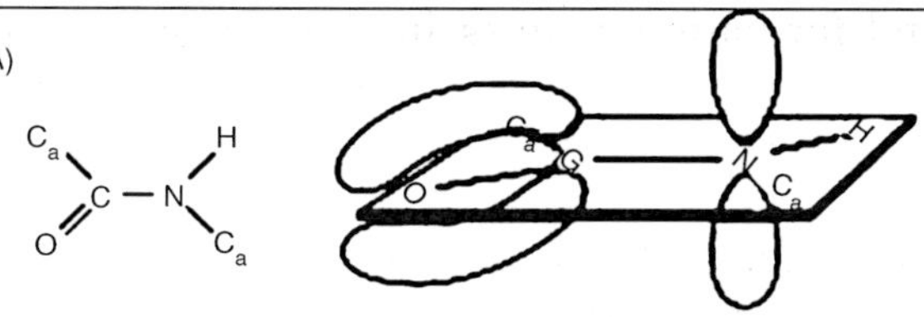

A Pure double bond between C and O would permit free rotation around the C–N bon.

(B)

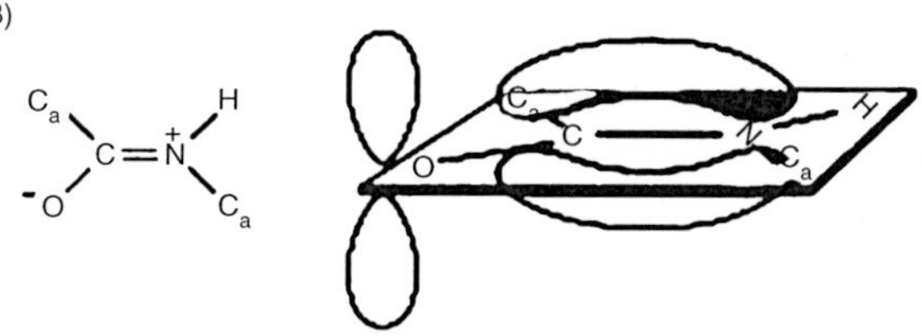

The other extreme would prohibit C–N bond rotation but would place too great a charge on O and N.

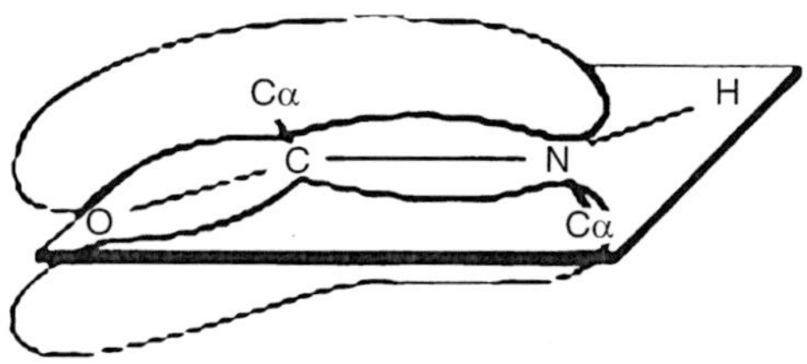

(c) The true Electron Density is Intermediate, The Barrier to C – N Bond Rotation of about 88 Kj/mol is enough to Keep the Amide group Planner.

Fig. The Partial Double bond Character of the Peptide Bond. Resonance Interactions among the Carbon, Oxygen, and Nitrogen Atoms of the Peptide group can be Represented by two Resonance Extremes (a and b). (a) The usual way the Peptide atoms are drawn. (b) In an equally feasible form, the Peptide bond is now a Double bond; the Amide N bears a positive Charge and the Carbonyl O has a Negative Charge. (c) The Actual Peptide Bond is best Described as a Resonance Hybrid of the forms in (a) and (b). Significantly, all of the Atoms Associated with the Peptide group are Coplanar, Rotation about C_0–N is Restricted, and the Peptide is Distinctly polar.

This structure prevents free rotation about the C_0–peptide bond because it becomes a double bond. The real nature of the peptide bond lies somewhere between these extremes; that is, it has partial double bond character, as represented by the intermediate form shown in Figure c.

Peptide bond resonance has several important consequences. First, it restricts free rotation around the peptide bond and leaves the peptide backbone with only two degrees of freedom per amino acid group: rotation around the N– C_α bond and rotation around the C_α–C_0 bond. Second, the six atoms composing the peptide bond group tend to be coplanar, forming the so-called amide plane of the polypeptide backbone (Figure 5.4). Third, the C_o–N bond length is 0.133 nm, which is shorter than normal C–N bond lengths (for example, the C_α–N bond of 0.145 nm) but longer than typical C=N bonds (0.125 nm). The peptide bond is estimated to have 40 per cent double-bond character.

The angle of ratation about the N–C_α bond is designated φ, phi, whereas the C_α–C_o angle of roation is designated ψ, psi.

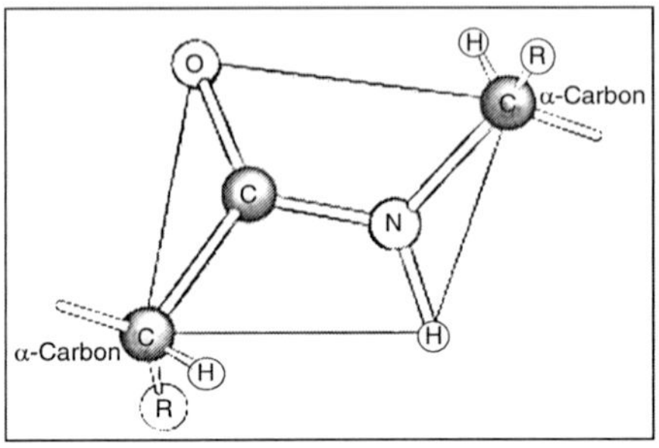

Fig. The Coplanar Relationship of the Atoms in the Amide Group is Highlighted as an Imaginary Shaded Plane Lying Between two Successive α-Carbon atoms in the Peptide Backbone.

THE POLYPEPTIDE BACKBONE IS RELATIVELY POLAR

Peptide bond resonance also causes the peptide backbone to be relatively polar. As shown in Figure b, the amide nitrogen represents a protonated or positively charged form, and the carbonyl oxygen becomes a negatively charged atom in the double-bonded resonance state. In actuality, the hybrid state of the partially double-bonded peptide arrangement gives a net positive charge of 0.28 on the amide N and an equivalent net negative charge of 0.28 on the carbonyl O. The presence of these partial charges means that the peptide bond has a permanent dipole. Nevertheless, the peptide backbone is relatively unreactive chemically, and protons are gained or lost by the peptide groups only at extreme pH conditions.

PEPTIDE CLASSIFICATION

Peptide is the name assigned to short polymers of amino acids. Peptides are classified by the number of amino acid units in the chain. Each unit is called an amino acid residue, the word *residue* denoting what is left after the release of H_2O when an amino acid forms a peptide link upon joining the peptide chain. Dipeptides have two amino acid residues, tripeptides have three, tetrapeptides four, and so on. After about 12 residues, this terminology becomes cumbersome, so peptide chains of more than 12 and less than about 20 amino acid residues are usually referred to as oligopeptides, and, when the chain exceeds several dozen amino acids in length, the term polypeptide is used. The distinctions in this terminology are not precise.

PROTEINS ARE COMPOSED OF ONE OR MORE POLYPEPTIDE CHAINS

The terms *polypeptide* and *protein* are used interchangeably in discussing single polypeptide chains. The term protein broadly defines molecules composed of one or more polypeptide chains. Proteins having only one polypeptide chain are monomeric proteins. Proteins composed of more than one polypeptide chain are multimeric proteins.

Table. Size of Protein Molecules

Protein	M_r	Number of Residues Per Chain	Subunit Organization
Insulin (bovine)	5,733	21 (A)	
		30 (B)	
Cytochrome *c* (equine)	12,500	104	α_1
Ribonuclease A (bovine pancreas)	12,640	124	α_1
Lysozyme (egg white)	13,930	129	α_1
Myoglobin (horse)	16,980	153	α_1
Chymotrypsin (bovine pancreas)	22,600	13 (α) 132 (α) 97 (γ)	αβγ
Hemoglobin(human)	64,500	141 (α) 146 (β)	$\alpha_2\beta_2$
Serum albumin (human)	68,500	550	α_1
Hexokinase (yeast)	96,000	200	α_4
γ-Globulin (horse)	149,900	214 (α) 446 (β)	$\alpha_2\beta_2$
Glutamate dehydrogenase (liver)	332,694	500	α_6
Myosin (rabbit)	470,000	1800 (heavy, *h*) 190 (α) 149 (α) 160 (β)	$h_2\alpha_1\alpha'_2\beta_2$
Ribulose bisphosphate carboxylase (spinach)	560,000	475 (α) 123 (β)	$\alpha_8\beta_8$
Glutamine synthetase (E.*coli*)	600,000	468	α_{12}

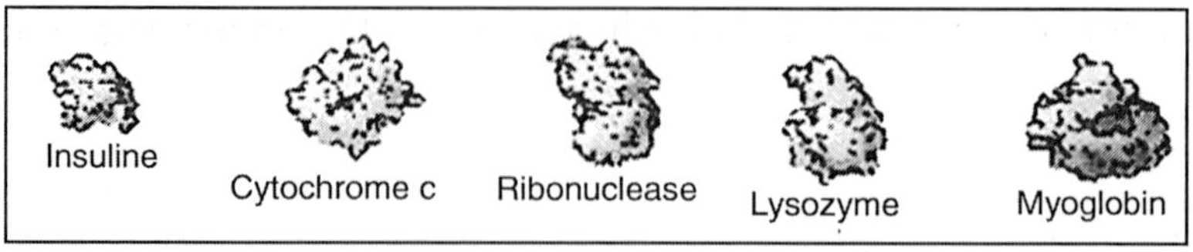

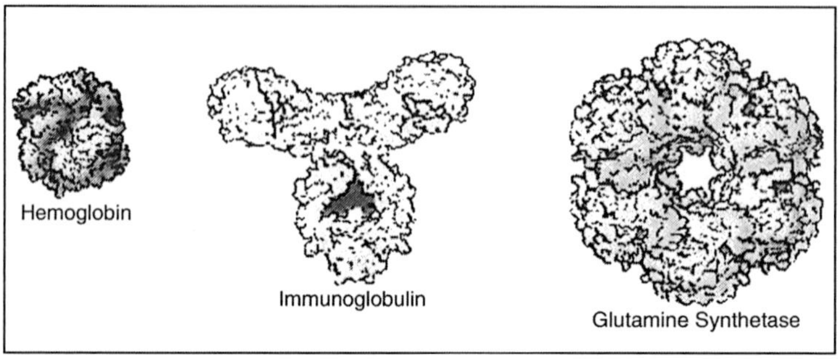

Multimeric proteins may contain only one kind of polypeptide, in which case they are homomultimeric, or they may be composed of several different kinds of polypeptide chains, in which instance they are heteromultimeric. Greek letters and subscripts are used to denote the polypeptide composition of multimeric proteins. Thus, an α_2-type protein is a dimer of identical polypeptide subunits, or a homodimer. Hemoglobin (Table) consists of four polypeptides of two different kinds; it is an $\alpha_2 \beta_2$ hetero-multimer.

Polypeptide chains of proteins range in length from about 100 amino acids to 1800, the number found in each of the two polypeptide chains of myosin, the contractile protein of muscle. However, titin, another muscle protein, has nearly 27,000 amino acid residues and a molecular weight of 2.8×10^6.

The average molecular weight of polypeptide chains in eukaryotic cells is about 31,700, corresponding to about 270 amino acid residues. The molecular weights (M_r) of proteins can be estimated by a number of physicochemical methods such as polyacrylamide gel electrophoresis or ultracent-rifugation. Precise determinations of protein molecular masses are best obtained by simple calculations based on knowledge of their amino acid sequence.

No simple generalizations correlate the size of proteins with their functions. For instance, the same function may be fulfilled in different cells by proteins of different molecular weight. The *Escherichia coli* enzyme responsible for glutamine synthesis (a protein known as *glutamine synthetase*) has a molecular weight of 600,000, whereas the analogous enzyme in brain tissue has a molecular weight of just 380,000.

ACID HYDROLYSIS OF PROTEINS

Peptide bonds of proteins are hydrolyzed by either strong acid or strong base. Because acid hydrolysis proceeds without racemization and with less destruction of certain amino acids (Ser, Thr, Arg, and Cys) than alkaline treatment, it is the method of choice in analysis of the amino acid composition of proteins and polypeptides. Typically, samples of a protein are hydrolyzed with 6 N HCl at 110°C for 24, 48, and 72 hr in sealed glass vials.

Tryptophan is destroyed by acid and must be estimated by other means to determine its contribution to the total amino acid composition. The OH-containing amino acids serine and threonine are slowly destroyed, but the data obtained for the three time points allow extrapolation to zero time to estimate the original Ser and Thr content (Figure).

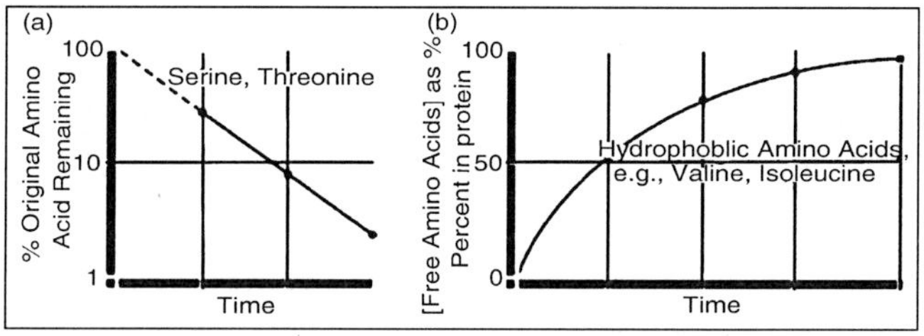

Fig. (a) The Hydroxy Amino Acids Serine and Threonine are slowly destroyed during the course of Protein Hydrolysis for amino acid Composition Analysis. Extrapolation of the data back to time zero allows an Accurate estimation of the amount of these amino acids originally Present in the protein sample. (b) Peptide bonds involving Hydrophobic Amino Acid Residues such as valine and isoleucine Resist Hydrolysis by HCl. With time, these Amino Acids are Released and their free concentrations approach a limiting value that can be Approximated with reliability.

In contrast, peptide bonds involving hydrophobic residues such as valine and isoleucine are only slowly hydrolyzed in acid. Another complication arises because the b- and g-amide linkages in asparagine (Asn) and glutamine (Gln) are acid labile. The amino nitrogen is released as free ammonium, and all of the Asn and Gln residues of the protein become aspartic acid (Asp) and glutamic acid (Glu), respectively. The amount of ammonium released during acid hydrolysis gives an estimate of the total number of Asn and Gln residues in the original protein, but not the amounts of either. Accordingly, the concentrations of Asp and Glu determined in amino acid analysis are expressed as Asx and Glx, respectively. Because the relative contributions of [Asn + Asp] or [Gln + Glu] cannot be derived from the data, this information must be obtained by alternative means.

AMINO ACID ANALYSIS OF PROTEINS

The complex amino acid mixture in the hydrolysate obtained after digestion of a protein in 6 *N* HCl can be separated into the component amino acids by either ion exchange chromatography or by reversed-phase high-pressure liquid chromatography (HPLC). The amount of each amino acid can then be determined. In ion exchange chrom-atography, the amino acids are separated and then quantified following reaction with ninhydrin (so-called postcolumn derivatization). In HPLC, the amino acids are converted to phenylthiohydantoin (PTH) derivatives via reaction with Edman's reagent prior to chromatography (precolumn derivatization). Both of these methods of separation and analysis are fully automated in instruments called amino acid *analys*ers. Analysis of the amino acid composition of a 30-kD protein by these methods requires less than 1 hour and only 6 mg (0.2 nmol) of the protein.

Table gives the amino acid composition of several selected proteins: ribonuclease A,alcohol dehydrogenase, myoglobin, histone H3, and collagen. Each of the 20 naturally occurring amino acids is usually represented at least once in a polypeptide chain. However, some small proteins may not have a representative of every amino acid. Note that ribonuclease (12.6 kD, 124 amino acid residues) does not contain any tryptophan. Amino acids almost never occur in equimolar ratios in proteins, indicating that proteins are not composed of repeating arrays of amino acids. There are a few exceptions to this rule. Collagen, for example, contains large proportions of glycine and proline, and much of its structure is composed of (Gl-*x*-Pro) repeating units, where x is any amino acid. Other proteins show unusual abundances of various amino acids. For example, histones are rich in positively charged amino acids such as arginine and lysine. Histones are a class of proteins found associated with the anionic phosphate groups of eukaryotic DNA.

Table. Amino Acid Composition of Some Selected Proteins
Values expressed are per cent representation of each amino acid.

		Proteins*			
Amino Acid	**RNase**	**ADH**	**Mb**	**Histone H3**	**Collagen**
Ala	6.9	7.5	9.8	13.3	11.7
Arg	3.7	3.2	1.7	13.3	4.9
Asn	7.6	2.1	2.0	0.7	1.0
Asp	4.1	4.5	5.0	3.0	3.0
Cys	6.7	3.7	0	1.5	0
Gln	6.5	2.1	3.5	5.9	2.6
Glu	4.2	5.6	8.7	5.2	4.5
Gly	3.7	10.2	9.0	5.2	32.7
His	3.7	1.9	7.0	1.5	0.3
Ile	3.1	6.4	5.1	5.2	0.8
Leu	1.7	6.7	11.6	8.9	2.1
Lys	7.7	8.0	13.0	9.6	3.6
Met	3.7	2.4	1.5	1.5	0.7
Phe	2.4	4.8	4.6	3.0	1.2
Pro	4.5	5.3	2.5	4.4	22.5
Ser	12.2	7.0	3.9	3.7	3.8
Thr	6.7	6.4	3.5	7.4	1.5
Trp	0	0.5	1.3	0	0
Tyr	4.0	1.1	1.3	2.2	0.5
Val	7.1	10.4	4.8	4.4	1.7
Acidic	8.4	10.2	13.7	8.1	7.5
Basic	15.0	13.1	21.8	24.4	8.8

Aromatic	6.4	6.4	7.2	5.2	1.7
Hydrophobic	18.0	30.7	27.6	23.0	6.5

*Proteins are as follows:

- *RNase*: Bovine ribonuclease A, an enzyme; 124 amino acid residues. Note that RNase lacks tryptophan.
- *ADH*: Horse liver alcohol dehydrogenase, an enzyme; dimer of identical 374 amino acid polypeptide chains. The amino acid composition of ADH is reasonably representative of the norm for water-soluble proteins.
- *Mb*: Sperm whale myoglobin, an oxygen-binding protein; 153 amino acid residues. Note that Mb lacks cysteine.
- *Histone H3*: Histones are DNA-binding proteins found in chromosomes; 135 amino acid residues. Note the very basic nature of this protein due to its abundance of Arg and Lys residues. It also lacks tryptophan.
- *Collagen*: Collagen is an extracellular structural protein; 1052 amino acid residues. Collagen has an unusual amino acid composition; it is about one-third glycine and is rich in proline. Note that it also lacks Cys and Trp and is deficient in aromatic amino acid residues in general.

Amino acid analysis itself does not directly give the number of residues of each amino acid in a polypeptide, but it does give amounts from which the percentages or ratios of the various amino acids can be obtained (Table). If the molecular weight *and* the exact amount of the protein analysed are known (or the number of amino acid residues per molecule is known), the molar ratios of amino acids in the protein can be calculated.

Amino acid analysis provides no information on the order or sequence of amino acid residues in the polypeptide chain. Because the polypeptide chain is unbranched, it has only two ends, an amino-terminal or N-terminal end and a carboxyl-terminal or C-terminal end.

THE SEQUENCE OF AMINO ACIDS IN PROTEINS

The unique characteristic of each protein is the distinctive sequence of amino acid residues in its polypeptide chain(s). Indeed, it is the amino acid sequence of proteins that is encoded by the nucleotide sequence of DNA. This amino acid sequence, then, is a form of genetic information. By convention, the amino acid sequence is read from the N-terminal end of the polypeptide chain through to the C-terminal end.

As an example, every molecule of ribonuclease A from bovine pancreas has the same amino acid sequence, beginning with N-terminal lysine at position 1 and ending with C-terminal valine at position 124 (Figure). Given the possibility of any of the 20 amino acids at each position, the number of unique amino acid sequences is astronomically large. The astounding sequence variation possible within polypeptide chains provides a key insight into the incredible functional diversity of protein molecules in biological systems, which is discussed shortly.

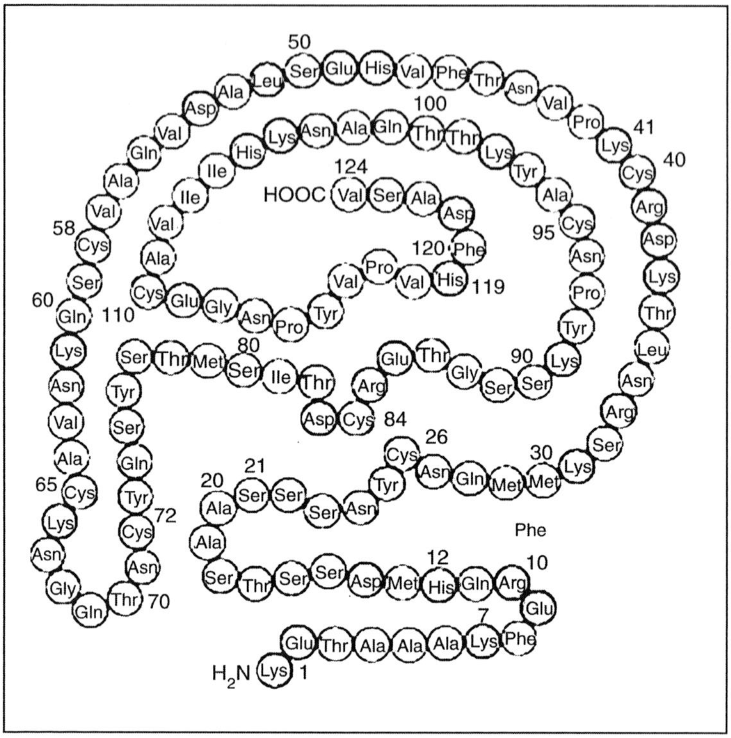

Fig. Bovine Pancreatic Ribonuclease A contains 124 Amino Acid Residues, none of which are Tryptophan. Four Intrachain Disulfide Bridges (S–S) form cross-links in this Polypeptide between Cys^{26} and Cys^{84}, Cys^{40} and Cys^{95}, Cys^{58} and Cys^{110}, and Cys^{65} and Cys^{72}.

ARCHITECTURE OF PROTEIN MOLECULES

PROTEIN SHAPE

As a first approximation, proteins can be assigned to one of three global classes on the basis of shape and solubility: fibrous, globular, or membrane (Figure). Fibrous proteins tend to have relatively simple, regular linear structures. These proteins often serve structural roles in cells. Typically, they are insoluble in water or in dilute salt solutions.

In contrast, globular proteins are roughly spherical in shape. The polypeptide chain is compactly folded so that hydrophobic amino acid side chains are in the interior of the molecule and the hydrophilic side chains are on the outside exposed to the solvent, water. Consequently, globular proteins are usually very soluble in aqueous solutions. Most soluble proteins of the cell, such as the cytosolic enzymes, are globular in shape.

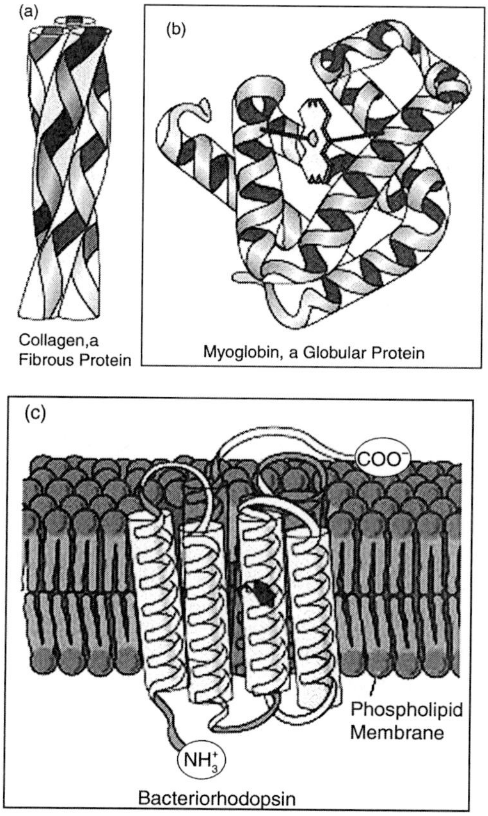

Fig. (a) Proteins having Structural Roles in Cells are Typically Fibrous and often water insoluble. Collagen is a good example. Collagen is Composed of three Polypeptide chains that Intertwine. (b) Soluble Proteins Serving Metabolic functions can be Characterized as compactly folded Globular Molecules, such as Myoglobin. The folding Pattern puts Hydrophilic Amino acid side chains on the outside and buries Hydrophobic side Chains in the Interior, making the Protein highly water soluble. (c) Membrane Proteins fold so that Hydrophobic Amino Acid Side Chains are Exposed in their Membrane-associated regions.

Membrane proteins are found in association with the various membrane systems of cells. For interaction with the nonpolar phase within membranes, membrane proteins have hydrophobic amino acid side chains oriented outward. As such, membrane proteins are insoluble in aqueous solutions but can be solubilized in solutions of detergents. Membrane proteins characteristically have fewer hydrophilic amino acids than cytosolic proteins.

THE LEVELS OF PROTEIN STRUCTURE

The architecture of protein molecules is quite complex. Nevertheless, this complexity can be resolved by defining various levels of structural organization.

Primary Structure

The amino acid sequence is the primary (1°) structure of a protein, such as that shown in Figure, for example.

Secondary Structure

Through hydrogen bonding interactions between adjacent amino acid residues, the polypeptide chain can arrange itself into characteristic helical or pleated segments. These segments constitute structural conformities, so-called regular structures,that extend along one dimension, like the coils of a spring. Such architectural features of a protein are designated secondary (2°) structures (Figure). Secondary structures are just one of the higher levels of structure that represent the three-dimensional arrangement of the polypeptide in space.

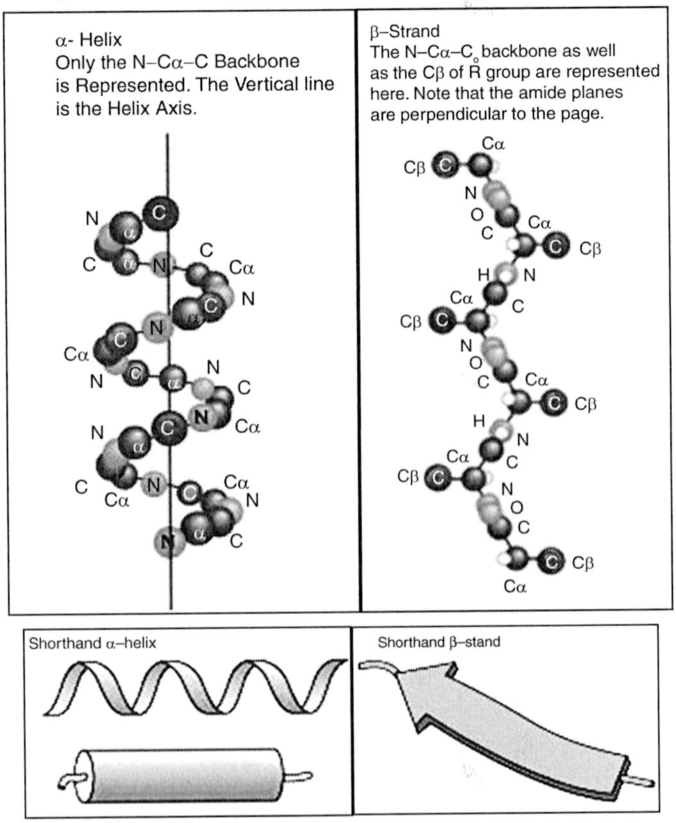

Fig. Two Structural Motifs that Arrange the Primary Structure of Proteins into a higher level of Organization Predominate in proteins: the *a*-helix and the *b*-pleated strand. Atomic Representations of these Secondary Structures are shown here, Along with the Symbols used by Structural Chemists to Represent them: the Flat, Helical ribbon for the *a*-helix and the flat, wide arrow for *b*-structures. Both of these Structures owe their Stability to the formation of Hydrogen bonds between N–H and O=C functions along the Polypeptide Backbone.

Quaternary Structure

Many proteins consist of two or more interacting polypeptide chains of characteristic tertiary structure, each of which is commonly referred to as a subunit of the protein. Subunit organization constitutes another level in the hierarchy of protein structure, defined as the protein's quaternary (4°) structure(Figure). Questions of quaternary structure address the various kinds of subunits within a protein molecule, the number of each, and the ways in which they interact with one another.

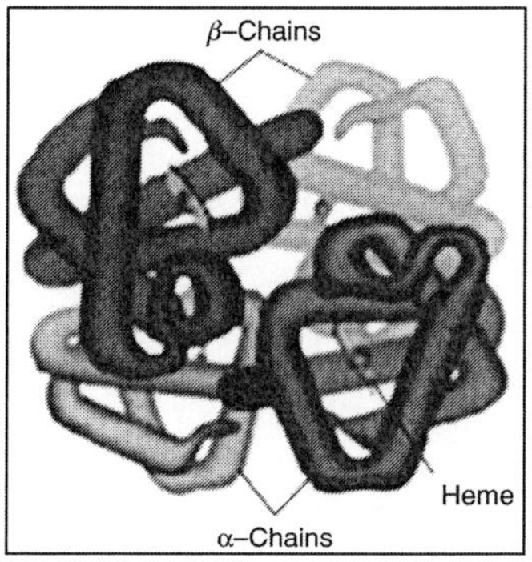

Fig. Hemoglobin, Which Consists of two α and two β Polypeptide Chains, is an example of the Quaternary level of Protein Structure. In this Drawing, the β-chains are the two Uppermost Polypeptides and the two α-chains are the lower half of the Molecule. The two closest chains (darkest coloured) are the β_2-chain (*upper left*) and the α_1-chain (*lower right*). The heme groups of the four Globin Chains are Represented by rectangles with Spheres (the heme iron Atom). Note the Symmetry of this Macromolecular Arrangement.*(Irving Geis)*

Whereas the primary structure of a protein is determined by the covalently linked amino acid residues in the polypeptide backbone, secondary and higher orders of structure are determined principally by noncovalent forces such as hydrogen bonds and ionic, van der Waals, and hydrophobic interactions.

It is important to emphasize that *all the information necessary for a protein molecule to achieve its intricate architecture is contained within its 1° structure,* that is, within the amino acid sequence of its polypeptide chain(s).

PROTEIN CONFORMATION

The overall three-dimensional architecture of a protein is generally referred to as its conformation.This term is not to be confused with configuration, which denotes the geometric possibilities for a particular set of atoms. In going from one configuration to another, covalent bonds must be broken and rearranged.

In contrast, the *conformational possibilities* of a molecule are achieved without breaking any covalent bonds. In proteins, rotations about each of the single bonds along the peptide backbone have the potential to alter the course of the polypeptide chain in three-dimensional space. These rotational possibilities create many possible orientations for the protein chain, referred to as its conformational possibilities. Of the great number of theoretical conformations a given protein might adopt, only a very few are favoured energetically under physiological conditions. At this time, the rules that direct the folding of protein chains into energetically favourable conformations are still not entirely clear; accordingly, they are the subject of intensive contemporary research.

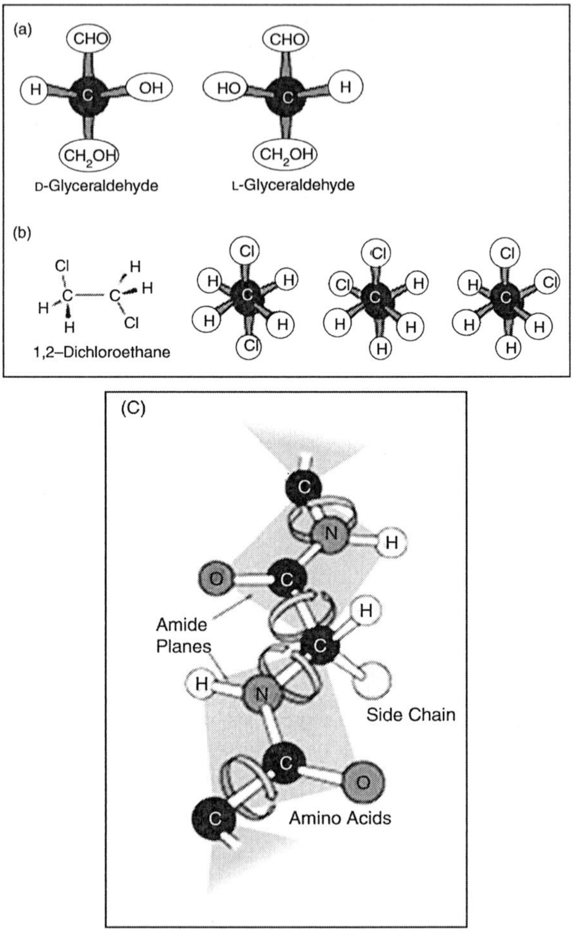

Fig. Configuration and Conformation are *not* Synonymous. (a) Rearrange-ments between Configurational Alternatives of a Molecule can be achieved only by breaking and Remaking Bonds, as in the Transformation between the D- and L-configurations of glyceraldehyde. No possible rotational reorientation of bonds linking the atoms of D-glyceraldehyde yields geometric identity with L-glyceraldehyde, even though they are mirror images of each other. (b) The intrinsic free rotation around single covalent bonds creates a great variety of three-dimensional conformations, even for relatively simple molecules. Consider 1,2-dichloroethane. Viewed end-on in a Newman projection, three principal rotational orientations or conformations predominate. Steric repulsion between eclipsed and partially eclipsed conformations keeps the possibilities at a reasonable number. (c) Imagine the conformational possibilities for a protein in which two of every three bonds along its backbone are freely rotating single bonds.

Later we return to an analysis of the 1° structure of proteins and the methodology used in determining the amino acid sequence of polypeptide chains, but let's first consider the extraordinary variety and functional diversity of these most interesting macromolecules.

PROTEINS ARE THE AGENTS OF BIOLOGICAL FUNCTION

Virtually every cellular activity is dependent on one or more particular proteins. Thus, a convenient way to classify the enormous number of proteins is by the biological roles they fill. Table summarizes the classification of proteins by function and gives examples of representative members of each class.

Table. Biological Functions of Proteins and Some Representative Examples

Functional Class	Examples
Enzymes	Ribonuclease
	Trypsin
	Phosphofructokinase
	Alcohol dehydrogenase
	Catalase
	"Malic" enzyme
Regulatory proteins	Insulin
	Somatotropin
	Thyrotropin
	lac repressor
	NF1 (nuclear factor 1)
	Catabolite activator protein
(CAP)	
	AP1
Transport proteins	Hemoglobin
	Serum albumin
	Glucose transporter
Storage proteins	Ovalbumin
	Casein
	Zein
	Phaseolin
	Ferritin
Contractile and motile proteins	Actin
	Myosin
	Tubulin
	Dynein
	Kinesin
Structural proteins	α-Keratin
	Collagen
	Elastin
	Fibroin
	Proteoglycans
Scaffold proteins	Grb 2

	crk shc stat IRS-1
Protective and exploitive proteins	Immunoglobulins Thrombin Fibrinogen Antifreeze proteins Snake and bee venom proteins Diphtheria toxin Ricin
Exotic proteins Monellin	Resilin Glue proteins

ENZYMES

By far the largest class of proteins is enzymes. More than 3000 different enzymes are listed in *Enzyme Nomenclature,* the standard reference volume on enzyme classification. Enzymes are catalysts that accelerate the rates of biological reactions. Each enzyme is very specific in its function and acts only in a particular metabolic reaction. Virtually every step in metabolism is catalyzed by an enzyme. The catalytic power of enzymes far exceeds that of synthetic catalysts. Enzymes can enhance reaction rates in cells as much as 10^{16} times the uncatalyzed rate.

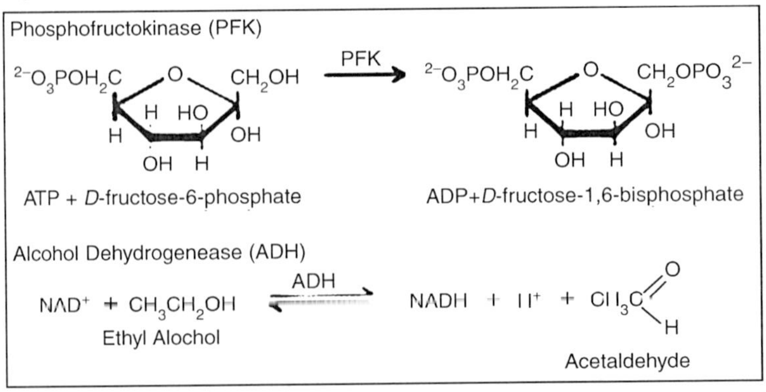

Fig. Enzymes are Classified According to the Specific Biological Reaction that they Catalyze. Cells contain thousands of different Enzymes. Two common Examples drawn from Carbohydrate Metabolism are Phosphofructokinase (PFK), or, more Precisely, ATP:d-fructose-6-Phosphate 1-phosphotransferase, and Alcohol Dehydrogenase (ADH), or Alcohol:NAD^+ oxidoreductase, which Catalyze the Reactions shown here.

Enzymes are systematically classified according to the nature of the reaction that they catalyze, such as the transfer of a phosphate group (*phosphotransferase*) or an oxidation-reduction (*oxidored-uctase*). The formal names of enzymes come from the particular reaction within the class that they catalyze, as in ATP:$_D$-fructose-6-phosphate 1-phosphotransferase and alcohol: NAD^+ oxido-reductase. Often, enzymes have common names in addition to their formal names. ATP:$_D$-fructose-6-phosphate 1-phosphotransferase is more commonly known as *phosphofructokinase* (*kinase* is a common name given to ATP-dependent phosphotransferases). Similarly, alcohol:NAD$^+$ oxidoreductase is casually referred to as *alcohol dehydrogenase*. The reactions catalyzed by these two enzymes are shown in Figure. Other enzymes are known by trivial names that have historical roots, such as *catalase* (systematic name, hydrogen-peroxide:hydrogen-peroxide oxidoreductase), and sometimes these trivial names have descriptive connotations as well, as in *malic enzyme* (systematic name, $_L$-malate:NADP$^+$oxidoreductase).

REGULATORY PROTEINS

A number of proteins do not perform any obvious chemical transformation but nevertheless can regulate the ability of other proteins to carry out their physiological functions. Such proteins are referred to as regulatory proteins. A well-known example is *insulin*, the hormone regulating glucose metabolism in animals. Insulin is a relatively small protein (5.7 kD) and consists of two polypeptide chains held together by disulfide cross-bridges. Other hormones that are also proteins include pituitary*somatotropin* (21 kD) and *thyrotropin* (28 kD), which stimulates the thyroid gland. Another group of regulatory proteins is involved in the regulation of gene expression. These proteins characteristically act by binding to DNA sequences that are adjacent to coding regions of genes, either activating or inhibiting the transcription of genetic information into RNA. Examples include repressors, which, because they block transcription, are considered negative control elements.

A prokaryotic representative is *lac repressor* (37 kD), which controls expression of the enzyme system responsible for the metabolism of lactose (milk sugar); a mammalian example is NF1 (*nuclear factor* 1, 60 kD), which inhibits transcription of the gene encoding the b-globin polypeptide chain of hemoglobin. Positively acting control elements are also known. For example, the E. *coli*catabolite gene activator protein (CAP) (44 kD), under appropriate metabolic conditions, can bind to specific sites along the E. *coli* chromosome and increase the rate of transcription of adjacent genes.

The mammalian AP1 is a heterodimeric transcription factor composed of one polypeptide from the*Jun* family of gene-regulatory proteins and one polypeptide from the *Fos* family of gene-regulatory proteins. AP1 activates expression of the b-globin gene. These various DNA-binding regulatory proteins often possess characteristic structural features, such as helix-turn-helix, leucine zipper, and zinc finger motifs.

TRANSPORT PROTEINS

A third class of proteins is the transport proteins. These proteins function to transport specific substances from one place to another. One type of transport is exemplified by the transport of oxygen from the lungs to the tissues by *hemoglobin* or by the transport of fatty acids from adipose tissue to various organs by the blood protein *serum albumin*.

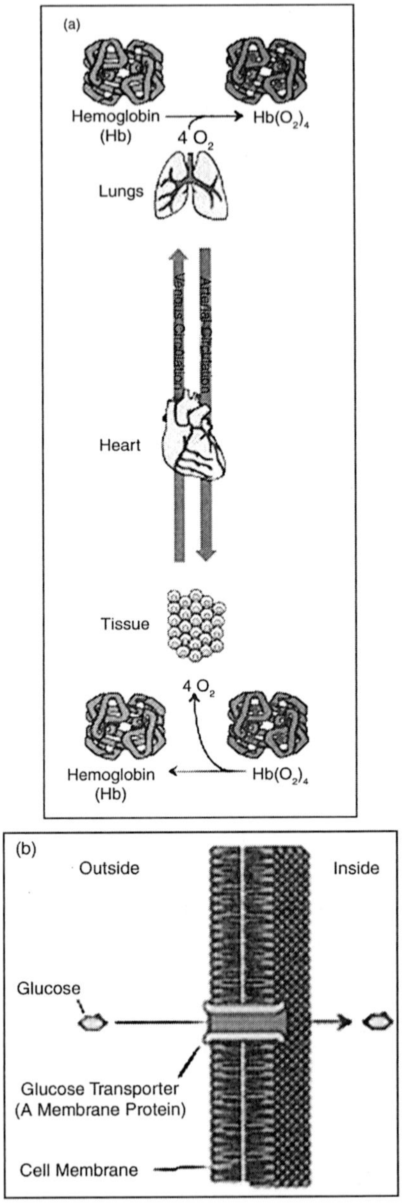

Fig. Two basic Types of Biological Transport are (a) Transport within or between Different cells or Tissues and (b) Transport into or out of cells. Proteins Function in both of these Phenomena. For example, the protein Gemoglobin Transports Oxygen from the Lungs to Actively Respiring Tissues. Transport Proteins of the other type are Localized in Cellular Membranes, where they Function in the uptake of Specific Nutrients, such as Glucose and Amino Acids, or the Export of Metabolites and waste Products.

A very different type is the transport of metabolites across permeability barriers such as cell membranes, as mediated by specific membrane proteins. These *membrane transport proteins* take up metabolite molecules on one side of a membrane, transport them across the membrane, and release them on the other side. Examples include the transport proteins responsible for the uptake of essential nutrients into the cell, such as glucose or amino acids. All naturally occurring membrane transport proteins studied thus far form channels in the membrane through which the transported substances are passed.

STORAGE PROTEINS

Proteins whose biological function is to provide a reservoir of an essential nutrient are called storage proteins. Because proteins are amino acid polymers and because nitrogen is commonly a limiting nutrient for growth, organisms have exploited proteins as a means to provide sufficient nitrogen in times of need. For example, *ovalbumin*, the protein of egg white, provides the developing bird embryo with a source of nitrogen during its isolation within the egg. *Casein* is the most abundant protein of milk and thus the major nitrogen source for mammalian infants.

The seeds of higher plants often contain as much as 60 per cent storage protein to make the germinating seed nitrogen-sufficient during this crucial period of plant development. In corn (*Zea mays* or *maize*), a family of low molecular weight proteins in the kernel called *zeins* serve this purpose; peas (the seeds of *Phaseolus vulgaris*) contain a storage protein called*phaseolin*. The use of proteins as a reservoir of nitrogen is more efficient than storing an equivalent amount of amino acids. Not only is the osmotic pressure minimized, but the solvent capacity of the cell is taxed less in solvating one molecule of a polypeptide than in dissolving, for example, 100 molecules of free amino acids. Proteins can also serve to store nutrients other than the more obvious elements composing amino acids (N, C, H, O, and S). As an example, ferritin is a protein found in animal tissues that binds iron, retaining this essential metal so that it is available for the synthesis of important iron-containing proteins such as hemoglobin. One molecule of *ferritin* (460 kD) binds as many as 4500 atoms of iron (35 per cent by weight).

CONTRACTILE AND MOTILE PROTEINS

Certain proteins endow cells with unique capabilities for movement. Cell division, muscle contraction, and cell motility represent some of the ways in which cells execute motion. The contractile andmotile proteins underlying these motions share a common property: they are filamentous or polymerize to form filaments. Examples include *actin* and *myosin*, the filamentous proteins forming the contractile systems of cells, and *tubulin*, the major component of microtubules (the filaments involved in the mitotic spindle of cell division as well as in flagella and cilia). Another class of proteins involved in movement includes *dynein* and *kinesin*, so-called motor proteins that drive the movement of vesicles, granules, and organelles along microtubules serving as established cytoskeletal "tracks."

STRUCTURAL PROTEINS

An apparently passive but very important role of proteins is their function in creating and maintaining biological structures. Structural proteins provide strength and protection to cells and tissues. Monomeric units of structural proteins typically polymerize to generate long fibres (as in hair) or protective sheets of fibrous arrays, as in cowhide (leather). a-*Keratins* are insoluble fibrous proteins making up hair, horns, and fingernails. *Collagen*, another insoluble fibrous protein, is found in bone, connective tissue, tendons, cartilage, and hide, where it forms inelastic fibrils of great strength. One-third of the total protein in a vertebrate animal is collagen. A structural protein having elastic properties is, appropriately, *elastin*, an important component of ligaments. Because of the way elastin monomers are cross-linked in forming polymers, elastin can stretch in two dimensions.

Certain insects make a structurally useful protein, *fibroin* (a b-keratin), the major constituent of cocoons (silk) and spider webs. An important protective barrier for animal cells is the extracellular matrix containing*collagen*and *proteoglycans*, covalent protein-polysaccharide complexes that cushion and lubricate.

SCAFFOLD PROTEINS (ADAPTER PROTEINS)

Some proteins play a recently discovered role in the complex pathways of cellular response to hormones and growth factors. These proteins, the scaffold or adapter proteins, have a modular organization in which specific parts (modules) of the protein's structure recognize and bind certain structural elements in other proteins through protein-protein interactions.

For example, SH2 modules bind to proteins in which a tyrosine residue has become phosphorylated on its phenolic -OH, and SH3 modules bind to proteins having a characteristic grouping of proline residues. Others includePH modules, which bind to membranes, and PDZ-containing proteins, which bind specifically to the C-terminal amino acid of certain proteins. Because scaffold proteins typically possess several of these different kinds of modules, they can act as a scaffold onto which a set of different proteins is assembled into a multiprotein complex. Such assemblages are typically involved in coordinating and communicating the many intracellular responses to hormones or other signalling molecules

PROTECTIVE AND EXPLOITIVE PROTEINS

In contrast to the passive protective nature of some structural proteins, another group can be more aptly classified as protective or exploitive proteins because of their biologically active role in cell defence, protection, or exploitation. Prominent among the protective proteins are the *immunoglobulins*or *antibodies* produced by the lymphocytes of vertebrates.

Antibodies have the remarkable ability to "ignore" molecules that are an intrinsic part of the host organism, yet they can specifically recognize and neutralize "foreign" molecules resulting from the invasion of the organism by bacteria, viruses, or other infectious agents. Another group of protective proteins is the blood-clotting proteins, *thrombin*and *fibrinogen*, which prevent the loss of blood when the circulatory system is damaged. Arctic and Antarctic fishes have *antifreeze proteins* to protect their blood against freezing in the below-zero temperatures of high-latitude seas. In addition, various proteins serve defensive or exploitive roles for organisms, including the lytic and neurotoxic proteins of snake and bee venoms and toxic plant proteins, such as *ricin*, whose apparent purpose is to thwart predation by herbivores. Another class of exploitive proteins includes the toxins produced by bacteria, such as diphtheria toxin and cholera toxin.

EXOTIC PROTEINS

Some proteins display rather exotic functions that do not quite fit the previous classifications.*Monellin*, a protein found in an African plant, has a very sweet taste and is being considered as an artificial sweetener for human consumption. *Resilin*, a protein having exceptional elastic properties, is found in the hinges of insect wings.

Certain marine organisms such as mussels secrete *glue proteins*, allowing them to attach firmly to hard surfaces. It is worth repeating that the great diversity of function in proteins, as reflected in this survey, is attained using just 20 amino acids.

CHEMICAL GROUPS

Many proteins consist of only amino acids and contain no other chemical groups. The enzymeribonuclease and the contractile protein actin are two such examples. Such proteins are called simple proteins. However, many other proteins contain various chemical constituents as an integral part of their structure. These proteins are termed conjugated proteins.

If the non-protein part is crucial to the protein's function, it is referred to as a prosthetic group. If the non-protein moiety is not covalently linked to the protein, it can usually be removed by denaturing the protein structure. However, if the conjugate is covalently joined to the protein, it may be necessary to carry out acid hydrolysis of the protein into its component amino acids in order to release it. Conjugated proteins are typically classified according to the chemical nature of their non-amino acid component; a representative selection of them is given here and in Table.

Table. Representative Conjugated Proteins

Class	Prosthetic Group (approx.)	Per cent by Weight
Glycoproteins contain carbohydrate		
Fibronectin		
γ-Globulin		
Proteoglycan		
Lipoproteins contain lipid		
Blood plasma lipoproteins:		
High density lipoprotein (HDL) (α-lipoprotein)	Triacylglycerols, phospholipids, cholesterol	75
Low density lipoprotein (LDL) (β-lipoprotein)	Triacylglycerols, phospholipids, cholesterol	67
Nucleoprotein complexes contain nucleic acid		
Ribosomes	RNA	
Tobacco mosaic virus	RNA	60
Adenovirus	DNA	5
HIV-1 (AIDS virus)	RNA	
Phosphoproteins contain phosphate		
Casein	Phosphate groups	
Glycogen phosphorylase α	Phosphate groups	
Metalloproteins contain metal atoms		
Ferritin	Iron	
Alcohol dehydrogenase	Zinc	35
Cytochrome oxidase	Copper and iron	
Nitrogenase	Molybdenum and iron	
Pyruvate carboxylase	Manganese	
Hemoproteins contain heme		
Hemoglobin		
Cytochrome *c*		
Catalase		
Nitrate reductase		
Ammonium oxidase		
Flavoproteins contain flavin		
Succinate dehydrogenase	FAD	
NADH dehydrogenase	FMN	
Dihydroorotate dehydrogenase	FAD and FMN	
Sulfite reductase	FAD and FMN	

GLYCOPROTEINS

Glycoproteins are proteins that contain carbohydrate. Proteins destined for an extracellular location are characteristically glycoproteins. For example, fibronectin and proteoglycans are important components of the extracellular matrix that surrounds the cells of most tissues in animals.Immunoglobulin G molecules are the principal antibody species found circulating free in the blood plasma. Many membrane proteins are glycosylated on their extracellular segments.

LIPOPROTEINS

Blood plasma lipoproteins are prominent examples of the class of proteins conjugated with lipid. The plasma lipoproteins function primarily in the transport of lipids to sites of active membrane synthesis. Serum levels of *low density lipoproteins* (LDLs) are often used as a clinical index of susceptibility to vascular disease.

NUCLEOPROTEINS

Nucleoprotein conjugates have many roles in the storage and transmission of genetic information. Ribosomes are the sites of protein synthesis. Virus particles and evenchr-omosomes are protein-nucleic acid complexes.

PHOSPHOPROTEINS

These proteins have phosphate groups esterified to the hydroxyls of serine,threonine, or tyrosine residues. Casein, the major protein of milk, contains many phosphates and serves to bring essential phosphorus to the growing infant. Many key steps in metabolism are regulated between states of activity or inactivity, depending on the presence or absence of phosphate groups on proteins. Glycogen phosphorylase *a* is one well-studied example.

METALLOPROTEINS

Metalloproteins are either metal storage forms, as in the case of ferritin, or enzymes in which the metal atom participates in a catalytically important manner. We encounter many examples throughout this book of the vital metabolic functions served by metalloenzymes.

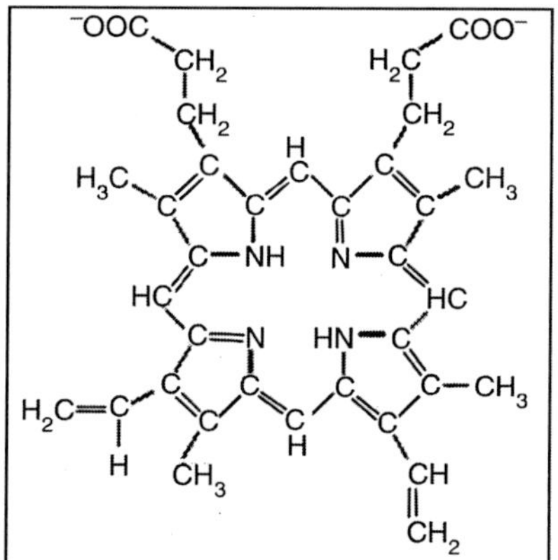

Protoporphyrin IX

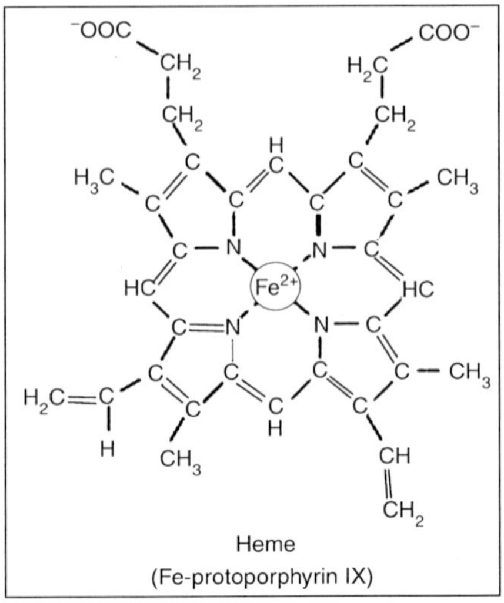

Fig. Heme Consists of Protoporphyrin IX and an Iron Atom. Protoporphyrin, a highly Conjugated System of double bonds, is Composed of four 5-membered Heterocyclic rings (Pyrroles) fused together to form a Tetrapyrrole Macrocycle. The Specific Isomeric Arrangement of methyl, vinyl, and Propionate side chains shown is Protoporphyrin IX. Coordination of an atom of Ferrous iron (Fe^{2+}) by the four Pyrrole Nitrogen Atoms yields Heme.

HEMOPROTEINS

These proteins are actually a subclass of metalloproteins because their prosthetic group is heme, the name given to iron protoporphyrin IX (Figure). Because heme-containing proteins enjoy so many prominent biological functions, they are considered a class by themselves.

FLAVOPROTEINS

Flavin is an essential substance for the activity of a number of important oxidoreductases.

REACTIONS OF PEPTIDES AND PROTEINS

The chemical properties of peptides and proteins are most easily considered in terms of the chemistry of their component functional groups. That is, they possess reactive amino and carboxyl termini and they display reactions characteristic of the chemistry of the R groups of their component amino acids.

PURIFICATION OF PROTEIN MIXTURES

Cells contain thousands of different proteins. A major problem for protein chemists is to purify a chosen protein so that they can study its specific properties in the absence of other proteins. Proteins have been separated and purified on the basis of their two prominent physical properties: size and electrical charge. A more direct approach is to employ affinity purification strategies that take advantage of the biological function or similar specific recognition properties of a protein.

SEPARATION METHODS

Separation methods based on size include size exclusion chromatography, ultrafiltration, andultrace-ntrifugation. The ionic properties of peptides and proteins are determined principally by their complement of amino acid side chains. Furthermore, the ionization of these groups is pH-dependent. A variety of procedures exploit electrical charge as a means of discriminating between proteins, including ion exchange chromatography, electrophoresis, and solubility. Proteins tend to be least soluble at their isoelectric point, the pH value at which the sum of their positive and negative electrical charges is zero. At this pH, electrostatic repulsion between protein molecules is minimal and they are more likely to coalesce and precipitate out of solution. Ionic strength also profoundly influences protein solubility. Most globular proteins tend to become increasingly soluble as the ionic strength is raised. This phenomenon, the salting-in of proteins, is attributed to the diminishment by the salt ions of electrostatic attractions between the protein molecules. Such electrostatic interactions would otherwise lead to precipitation. However, as the salt concentration reaches high levels (greater than 1 *M*), the effect may reverse so that the protein is salted out of solution. In such cases, the numerous salt ions begin to compete with the protein for waters of solvation, and, as they win out, the protein becomes insoluble. The solubility properties of a typical protein are shown in Figure.

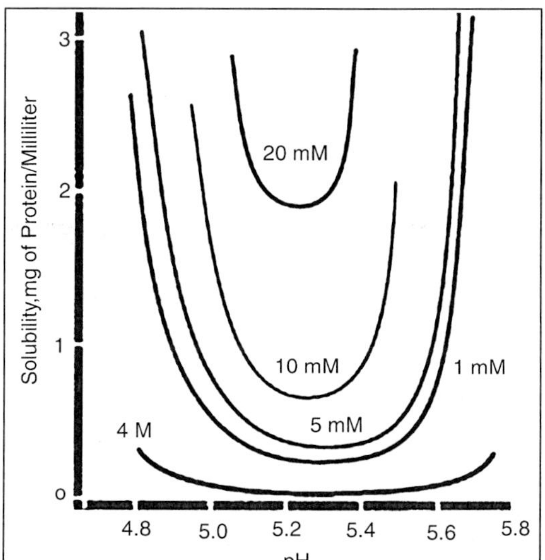

Fig. The Solubility of Most Globular Proteins is markedly Influenced by pH and ionic Strength. This figure shows the Solubility of a Typical Protein as a Function of pH and Various salt Concentrations.

Although the side chains of most non-polar amino acids in soluble proteins are usually buried in the interior of the protein away from contact with the aqueous solvent, a portion of them is exposed at the protein's surface, giving it a partially hydrophobic character. *Hydrophobic interaction chroma-tography* is a protein purification technique that exploits this hydrophobicity.

A TYPICAL PROTEIN PURIFICATION SCHEME

Most purification procedures for a particular protein are developed in an empirical manner, the overriding principle being purification of the protein to a homogeneous state with acceptable yield. Table presents a summary of a purification scheme for a selected protein. Note that the specific activity of the protein (the enzyme xanthine dehydrogenase) in the immuno-affinity purified fraction (fraction 5) has been increased 152/0.108, or 1407 times the specific activity in the crude extract (fraction 1). Thus, xanthine dehydrogenase in fraction 5 versus fraction 1 is enriched more than 1400-fold by the purification procedure.

Table. Example of a Protein Purification Scheme: Purification of the Enzyme Xanthine Dehydrogenase from a Fungus

Fraction	Volume (mL)	Total Protein (mg)	Total Activity*	Specific Activity†	Percent Recovery†
1. Crude extract	3,800	22,800	2,460	0.108	100
2. Salt precipitate	165	2,800	1,190	0.425	48
3. Ion exchange	65	100	720	7.2	29
chromatography	40	14.5	555	38.3	23
4. Molecular sieve	6	1.8	275	152	11
chromatography					
5. Immunoaffinity					
chromatography§					

Note:

- The relative enzymatic activity of each fraction in catalyzing the xanthine dehydrogenase reaction is cited as arbitrarily defined units.
- The specific activity is the total activity of the fraction divided by the total protein in the fraction. This value gives an indication of the increase in purity attained during the course of the purification as the samples become enriched for xanthine dehydrogenase protein.
- The per cent recovery of total activity is a measure of the yield of the desired product, xanthine dehydrogenase.
- The last step in the procedure is an affinity method in which antibodies specific for xanthine dehydrogenase are covalently coupled to a chromatography matrix and packed into a glass tube to make a chromatographic column through which fraction 4 is passed. The enzyme is bound by this immunoaffinity matrix while other proteins

pass freely out. The enzyme is then recovered by passing a strong salt solution through the column, which dissociates the enzyme–antibody complex.

4

Thermodynamic Aspects of Bioenergetics

THERMODYNAMIC MEASUREMENTS

ITC is a quantitative technique that can directly measure the binding affinity (K_a), enthalpy changes (ΔH), and binding stoichiometry (n) of the interaction between two or more molecules in solution. From these initial measurements Gibbs energy changes (ΔG), and entropy changes (ΔS), can be determined using the relationship:

$$\Delta G = -RT\ln K = \Delta H - T\Delta S$$

(where R is the gas constant and T is the absolute temperature).

The Instrument

An isothermal titration calorimeter is composed of two identical cells made of a highly efficient thermal conducting material such as Hastelloy alloy or gold, surrounded by an adiabatic jacket. Sensitive thermopile/thermocouple circuits are used to detect temperature differences between the reference cell (filled with buffer or water) and the sample cell containing the macromolecule. Prior to addition of ligand, a constant power (<1 mW) is applied to the reference cell. This directs a feedback circuit, activating a heater located on the sample cell. During the experiment, ligand is titrated into the sample cell in precisely known aliquots, causing heat to be either taken up or evolved (depending on the nature of the reaction). Measurements consist of the time-dependent input of power required to maintain equal temperatures between the sample and reference cells.

In an exothermic reaction, the temperature in the sample cell increases upon addition of ligand. This causes the feedback power to the sample cell to be decreased (remember: a reference power is applied to the reference cell) in order to maintain an equal temperature between the two cells. In an endothermic reaction, the opposite occurs; the feedback circuit increases the power in order to maintain a constant temperature (isothermic/isothermal operation).

Observations are plotted as the power in μcal/sec needed to maintain the reference and the sample cell at an identical temperature. This power is given as a function of time in seconds. As a result, the raw data for an experiment consists of a series of spikes of heat flow (power), with every spike corresponding to a ligand injection. These heat flow spikes/pulses are integrated with respect to time, giving the total heat effect per injection. The pattern of these heat effects as a function of the molar ratio [ligand]/[macromolecule] can then be analysed to give the thermodynamic parameters of the interaction under study. It should be noted that degassing samples is necessary in order to minimize data interference due to the presence of air bubbles within the cells. The presence of such bubbles will lead to abnormal data plots in the recorded results. The entire experiment takes place under computer control.

GENERAL ASPECTS OF ITC

Calorimetric methods have been an invaluable tool for understanding the forces that stabilize the folded conformations of proteins. Recently, the advent of several highly sensitive titration calorimeters has generated much interest in this technique. A thorough description of the theoretical concepts of ITC has been provided in the above and other reports to which the reader is referred. An ITC instrument consists of two identical cells composed of a highly efficient thermal conducting material (Hasteloy or gold) surrounded by an adiabatic jacket. The jacket is usually cooled by a circulating water bath. Sensitive thermopile/thermocouple circuits detect temperature differences between the two cells and between the cells and the jacket. Heaters located on both cells and the jacket are activated when necessary to maintain identical temperatures between all components.

In an ITC experiment, the macromolecule solution is placed in the sample cell. The reference cell contains buffer or water minus the macromolecule. Prior to the injection of the titrant, a constant power (,1 mW) is applied to the reference cell. This signal directs the feedback circuit to activate the heater located on the sample cell. This represents the baseline signal. The direct observable measured in an ITC experiment is the time-dependent input of power required to maintain equal temperatures in the sample and reference cell.

During the injection of the titrant into the sample cell, heat is taken up or evolved depending on whether the macromolecular association reaction is endothermic or exothermic. For an exothermic reaction, the temperature in the sample cell will increase, and the feedback power will be deactivated to maintain equal temperatures between the two cells. For endothermic reactions, the reverse will occur, meaning the feedback circuit will increase power to the sample cell to maintain the temperature. ITC instruments should be routinely calibrated by applying specified electrical pulses of approximately 5–10 mcal/s. The total heat as determined by the area under the pulse should be within 2 per cent of the expected value. The performance of the calorimeter can also be tested by measuring the heat of a standard chemical reaction, such as the protonation of tris(hydroxymethyl)aminomethane (THAM) by HCl. To ensure optimal performance of the calorimeter both calibration methods are recommended.

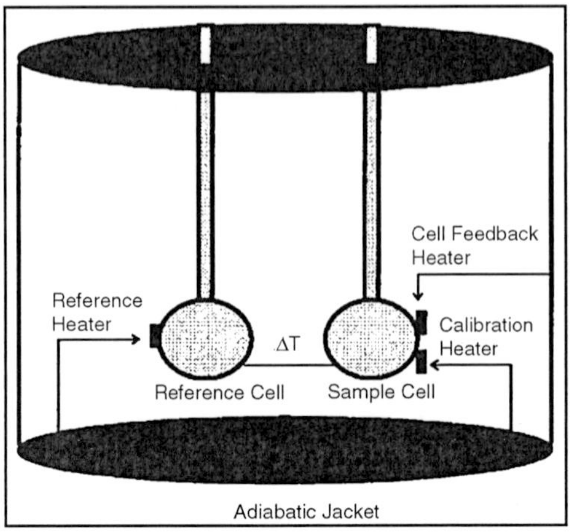

Fig. Schematic of an ITC Instrument

The heat absorbed or evolved during a calorimetric titration is proportional to the fraction of bound ligand. Thus, it is of extreme importance to determine accurately the initial concentrations of both the macromolecule and the ligand. For the initial injections, all or most of the added ligand is bound to the macromolecule, resulting in large endothermic or exothermic signals depending on the nature of the association. As the ligand concentration increases, the macromolecule becomes saturated and subsequently less heat is evolved or absorbed on further addition of titrant. The amount of heat evolved on addition of ligand can be represented by the equation $Q=V_0\Delta H_b[M]_t K_a[L]/(1+K_a[L])$ where V_0 is the volume of the cell, D*H*b is the enthalpy of binding per mole of ligand, [M]t is the total macromolecule concentration including bound and free fractions, *K*a is the binding constant, and [L] is the free ligand concentration. For a more general model of binding, the multiple independent sites model, the macromolecule contains multiple ligand binding sites that are non-interacting. The cumulative heat of binding can be described by

$$Q=V_0[M]_t \sum(n_i\Delta H_i K_{ai}[L]/(1+K_{ai}[L])).$$

The method of analysis will be specific to the system under investigation and the reader is advised to attempt fitting according to a general model before the use of more specific models.

RUNNING A TYPICAL ITC EXPERIMENT

ITC instruments can be of either the single- or dualinjection variety. In a single-injection instrument titrant is added to the sample cell containing the macromolecule. In a separate (control) experiment titrant is added to the sample cell in the absence of the macromolecule. In the dual-injection instrument, the titrant is simultaneously added to a reference cell containing buffer and to the sample cell containing macromolecule. The advantage of this type of instrument is that the control reaction to measure the heat of dilution of the ligand is eliminated. For this review, we describe the use of the single-injection Omega titration calorimeter manufactured by Microcal. Extreme care must be taken in all aspects of an ITC experiment, from sample preparation to data analysis.

Initial Considerations and Sample Preparation

The concentrations of macromolecule and ligand are critical, especially when one or both partners of a complex are difficult to obtain in large quantities or are of minimal solubility. Additionally, for measurement of the association constant, the initial concentrations of both the ligand and the macromolecule should be determined with a high degree of accuracy. The experimental binding isotherm can be characterized by the unitless value *c*, which is the product of the association constant, *K*a, the concentration of macromolecule [M], and the stoichiometry of the reaction, n: $C=K_a[M]n$.

For an accurate determination of the binding constant, a *c* value between 1 and 1000 is recommended. Large *c* values prohibit the determination of *K*a since the transition is very sharp and too few points are collected near equivalence (saturation may be achieved in a single injection of ligand). Binding isotherms with low *c* values lose the characteristic sigmoidal shape and are very broad transitions that approach linearity and the equivalence point cannot be identified. For tight binding complexes such as antibody–antigen interactions with *K*a values in the range 109–1010 M it is difficult to obtain accurate association constants since lowering the antibody concentration to obtain an appropriate *c* value extends beyond the sensitivity of the calorimeter.

The choice of buffer is also critical when planning an ITC experiment. If on complex formation protons are taken up or released, the equivalent number of protons will be taken up or released by the buffer. If the selected buffer has a large enthalpy of ionization, the measured enthalpy will reflect both buffer ionization and complex formation. This can be quite informative if one wishes to determine whether protons are taken up or released on complex formation and is accomplished simply by performing equivalent experiments in a buffer with negligible enthalpy of ionization (e.g., sodium phosphate) and one with a large enthalpy of ionization (e.g., Tris–HCl).

For initial experiments we recommend the use of a buffer such as sodium phosphate with a low ionization enthalpy. Both titrant and macromolecule should be exhaustively dialyzed in buffer (preferably in the same flask) to minimize artifacts arising from mismatched buffer components. The final dialysis buffer should be saved and used for any necessary concentration adjustments of the macromolecule or titrant solutions. Macromolecule and ligand are filtered or centrifuged to remove any precipitated material. Immediately prior to loading the sample cell and injection syringe, the ligand and macromolecule solution are degassed to remove residual air bubbles. Even the smallest air bubbles remaining in the cell or injection syringe can interfere with the feedback circuit. Air bubbles can additionally lead to poor baselines.

Loading the Sample and Reference Cells

The reference cell usually contains water or buffer with 0.01 per cent sodium azide and need not be changed after every experiment. We recommend changing the solution on a weekly basis if the instrument is routinely used. The macromolecule (usually but not necessarily the larger component of the interaction) is added to the sample cell of the calorimeter using a long needle glass syringe. Typically, 1.5 to 2 ml of the solution is prepared to fill a cell with a volume of 1.3–1.5 ml. The utmost care is required to fill the sample cell without introducing air bubbles.

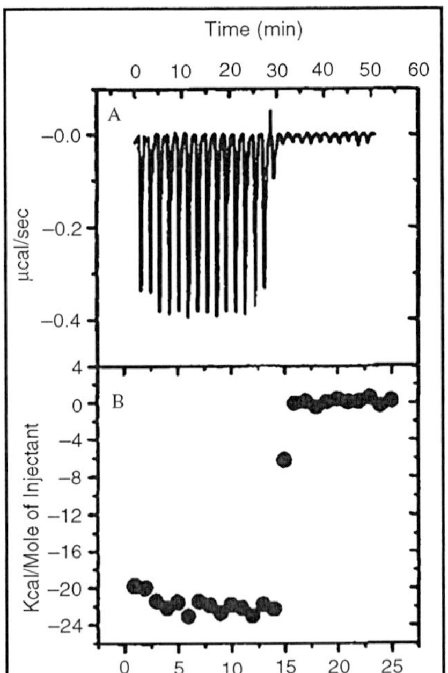

Fig. Calorimetric Titration of MAb 5F8 with Cytochrome c in 0.1 M sodium Phosphate, pH 7.0.

Once the cell is completely filled, several rapid additions of solution will dislodge any residual air bubbles that cling to the side of the cell. Any excess solution remaining in the reservoir is removed.

Filling and Attachment of the Injection Syringe

Filling the injection syringe with the titrant or ligand solution also requires great care. The concentration of ligand solution should be such that the molar ratio of ligand to macromolecule, following the last injection, is approximately 2. Typically, a complete titration will involve approximately fifteen to twenty 5- to 10-ml injections of ligand.

Handling of the injection syringe is extremely critical. Great care must be taken to avoid bending of the needle while the injection syringe is loaded into place. Bending of the injection syringe needle can result in some of the titrant solution being expelled into the macromolecule solution, causing the first injection to be unusable. Any minor bending in the syringe can also result in poor baselines when the injection apparatus is stirring.

Experimental Parameters

The parameters of the titration are input into the software programme controlling data acquisition. To determine accurately the enthalpy of binding, it is critical that the first several shots define a baseline region where all added ligand is bound to the macromolecule. The equivalence region should also be well defined by the concentration range spanned by the injections, to determine an accurate value of the association constant. It is necessary that concentrations be chosen so that measurable amounts of free and bound ligand are in equilibrium within the titration zone defined by the titrant injections.

For characteristically tight binding affinities such as those exhibited by antibody–antigen complexes this can be an impossible task.

In the tight binding limit, even when a binding constant cannot be determined, it is still possible to determine an accurate value for the binding enthalpy. Several injections should be performed after complete saturation of the macromolecule by ligand.

The heat evolved or absorbed following saturation represents the heat of dilution of the titrant. The length of time of injection should be such that proper mixing is achieved. Typically, we recommend 7- to 10-s injections of ligand. To ensure proper mixing the injection syringe is fitted with a Teflon paddle and attached to a stirring motor by a rubber belt. Stirring at approximately 400 rpm should ensure good mixing.

Control Experiments to Determine Heats of Dilution

The observed binding isotherm is usually normalized as kilocalories per mole of ligand injected and plotted versus the molar ratio of ligand to macromolecule. The observed heats of binding include contributions from the dilution of the titrant (ligand) and dilution of the macromolecule.

A small contribution arising from stirring is also included in the observed binding enthalpy. Several control experiments must be performed to correct for the heats of dilution. The heat of dilution of the ligand is usually the most significant. This is generally true since the initial concentration and the dilution factor for the ligand are typically 10–20 times greater than those of the macromolecule. The heat of dilution of the ligand can be determined by performing an identical titration experiment in which ligand is injected into a sample cell containing buffer only (no macromolecule).

The heat of dilution of the macromolecule (which is typically less significant than the heat of dilution of the ligand) is determined by titrating buffer solution into the sample cell containing the macromolecule. The two heats of dilution are used to correct the concentration-normalized binding isotherm. Since injection of ligand following saturation of the molecule is essentially a measurement of the heat of dilution of the ligand, the measured enthalpies of the last several injections can be averaged and subtracted to correct for heats of dilution. We recommend performing the control titrations for the most precise determination of the heats of dilution.

Data Analysis

The method of data analysis depends on the system of interest. For this article we briefly describe the procedure for fitting data to the multiple independent binding site model using the analysis software origin provided with the Omega ITC. Prior to peak integration, the heats of binding are normalized as a function of ligand concentration. Additionally, a volume correction is also performed due to dilution of the macromolecule during each injection. The areas under the peaks are integrated in either a manual peak-by-peak fashion or automatically by routines provided in the software package. Baseline selection is an important factor in ITC data analysis and user input in the automated integration routine is limited. Therefore, we recommend manual peak-by-peak integration in which the operator defines the baseline regions used in the integration step. To determine *n*, *K*a, and D*H*b Eq.is represented in terms of the binding constant and total ligand concentration [L]T to obtain the quadratic:

$$Q=(n[M]_t\Delta HV_0)/2\{1+[L]_t/(n[M]_t)+1/(nK_a[M]_t)-$$
$$[(1+[L]_t/(n[M])_t)+1/(nK_a[M]_t))^2$$
$$-4[L]_t/(n[M]_t)]^{1/2}\}$$

The *n*, *K*a and D*H*b parameters are then optimized using the standard Marquardt method with routines provided in the ORIGIN software.

Troubleshooting

Routine problems are expected during ITC experiments, most of which are easily corrected through practice. One common problem frequently observed is that the enthalpy of binding measured for the initial injection is less than that of subsequent injections. This is due to ligand solution slowly leaking from the injection syringe or due to the syringe plunger not being exactly flush with the driving piston. To avoid slow leakage of ligand from the injection syringe simply reduce the length of time that the injection syringe is in contact with the macromolecule prior to the first injection. Some equilibration following attachment of the injection apparatus is required. It is recommended that a baseline be initiated following the insertion of the injection syringe into the sample cell. The signal will level off on thermal equilibration of sample, reference cell, and jacket. The experiment should be started after the baseline has leveled and remained steady for several minutes. Low binding enthalpies measured after the initial injection can be observed if the injection syringe and drive piston are not exactly aligned. Aligning the drive piston exactly flush with the injection syringe can be a difficult task.

The use of a magnifying glass will greatly aid in the proper alignment of the injection syringe and drive piston. Fortunately, more recent versions of the OMEGA instrument have solved this problem by the development of an automatic alignment device in which an infrared beam guides the alignment of the injection syringe and drive piston. These problems can also be corrected by injecting a small volume of the ligand for the first addition and discarding the observed data. If a preinjection is used, the concentration of added ligand must be taken into consideration during analysis. Baseline stability is also a common problem occurring during ITC experiments and can arise for several reasons. As mentioned previously, a bent injection syringe can lead to poor baselines. Air bubbles can also result in reduced quality of baselines and can be corrected by degassing solutions longer (5–10 min should be sufficient) as well as by taking additional care in loading the sample cell. Condensation around the adiabatic jacket may also lead to poor baselines. For experiments below room temperature, the jacket must be purged with dry nitrogen prior to low-temperature equilibration.

THERMODYNAMIC PARAMETERS OF Β-LACTOGLOBULIN

Isothermal titration calorimetry (ITC) was used to determine the binding constant, stoichiometry, enthalpy, and entropy of β-lactoglobulin (β-lg)/low- or high-methoxyl pectin (LM- or HM-pectin) complexes at 22°C and at pH 4. The binding isotherms revealed the formation of soluble intrapolymer complexes further followed by their aggregation in interpolymer complexes. The interaction between β-lg and LM- or HM-pectin in intra and interpolymer complexes occurred spontaneously with a Gibbs free energy around -10 kcal·mol-1. The intrapolymer complexes were enthalpically driven, whereas enthalpic and entropic factors were involved in the formation of interpolymer complexes. Because ITC did not allow dissociating different enthalpic contributions, the values measured as pectin and β-lg interacted could partially be attributed to conformational changes.

The β-lactoglobulin (β-lg) is a small globular protein of 18,350 Da. The β-lg has a well-known structure containing mainly β-sheets, some β-turns, and one á-helix. Its quaternary structure is influenced by pH, mineral content, and temperature. This protein is monomeric at pH below 3 or above 8, dimeric at neutral pH, and found in both states between pH 4 and 6. The β-lg may form a few octamers around pH 4.5 but some authors are doubtful about the existence of such an assembly. Pectin is a cell-wall polysaccharide, generally extracted from apple pomace or citrus fruits. This anionic polysaccharide is essentially made up of D-galA residus. High-methoxyl pectin (HM-pectin) occurs when more than 50 per cent of its carboxylic groups are esterified, whereas low-methoxyl pectin (LM-pectin) is found when less than 50 per cent of these groups are esterified.

The interactions between proteins and polysaccharides can lead to the formation of complexes. These complexes, or coacervates, have many applications, including fat substitution protein separation, microencapsulation and enzyme immobilization. The nature of protein/polysaccharide complexes is influenced by entropic factors, such as the structure, and the molecular weight of biopolymers. The complexes are also influenced by enthalpic forces, which are regulated by the protein/polysaccharide ratio and the nature and density of charges on the biopolymers. Many studies on protein/polysaccharide interactions have been carried out. Ultrafiltration used with destabilizing agents showed that interactions between β-lg and pectin occurred through ionic and hydrogen bonds, whereas the hydrophobic interactions were negligible. The measurement of thermodynamic parameters, such as binding constant, enthalpy, entropy, and binding stoichiometry of interactions, is essential to accurately evaluate binding conditions. Binding parameters from biopolymer systems are sometimes hard to quantify because of the heterogeneity of the natural material. Consequently, few quantitative studies have been carried out on the interactions between β-lg and polysaccharides. Thermodynamic quantities involved in biopolymer interactions are often indirectly determined through non-calorimetric experimental methods. These methods involve the calculation of thermodynamic parameters from theoretical relationships.

Enthalpy variations (i.e., gain and loss) are the easiest thermodynamic parameters to measure during complexation. Isothermal titration calorimetry (ITC) is currently the only technique used to directly measure the enthalpy for many ternary mixtures. ITC was used to study the interaction between β-lg and LM- or HM-pectin at pH 4 in 5 mM of sodium phosphate buffer. To our knowledge, this technique has not yet been used to carry out a binding study involving two biopolymers having heterogeneous molecular weights.

MATERIALS AND METHODS

Materials

Bovine β-lg A and B (3X crystallized) was purchased from Sigma-Aldrich as 95 per cent pure protein. Low-methoxyl pectin (DE 28.3 per cent, apparent molecular weight 94.3 kDa) and high-methoxyl pectin (DE 73.4 per cent, apparent molecular weight 118 kDa) were donated by Hercules Copenhagen A/S. All reagents were of analytical grade and used as received.

Preparation of Solutions

To eliminate excess salts, β-lg and LM- and HM-pectin were dialyzed against deionized water for 24 hours at 4°C. Afterwards, β-lg and LM- or HM-pectin were dialyzed extensively in the same beaker against a 5-mM phosphate buffer (pH 4) to equilibrate the pH and ionic strength of the solutions. The weak concentration of the buffer was used to prevent dissociation of β-lg and pectin complexes, which are very sensitive to ionic strength as demonstrated earlier.

The dialysis buffer was preserved for ITC experiments. The dialysis was critical to avoid inaccuracies inherent to buffer mismatching. Ionic strength or pH differences could have lead to modifications in the dimer/monomer equilibrium of the β-lg. The β-lg, LM- and HM-pectin solutions and buffer were filtrated through a 0.22-μm membrane to eliminate insoluble particles. The concentration of dialyzed β-lg was measured with the bicinchoninic acid protein assay. The LM- and HM-pectin concentrations were assayed with the phenol-sulfuric acid method.

A dialysis buffer was used as solvent to adjust the LM- or HM-pectin and β-lg solutions around 10 and 1500 μM (monomeric equivalent) respectively. All solutions were degassed 10 minutes at 20°C prior to ITC measurements by means of a device provided with the ITC apparatus.

Isothermal Titration Calorimetry (ITC)

An isothermal titration calorimeter was used to measure enthalpy values. After setting the temperature of solutions at 20°C in the degassing device, the apparatus was equilibrated at 22°C for at least half an hour before the measurement. The LM- or HM-pectin solution in the 1.445-mL reaction cell was titrated with 50 successive 5-μL injections of β-lg solution. Each addition lasted 12 s with an interval of 800 s between consecutive injections. This long resting time between injections was necessary to reach thermodynamic equilibrium between injections. This equilibrium was attained when no more energy was released or absorbed in the reaction cell. The solution in the reaction cell was stirred at 310 rpm throughout the experiments. The heats of dilution from the blank titration of the β-lg into the buffer solution and the buffer into pectin solution were subtracted from the raw data. Measurements were carried out in duplicates.

Thermodynamic parameters, including binding constant (K), enthalpy (ΔH), entropy (ΔS), and binding stoichiometry (N) (molar ratio monomeric β-lg/pectin), were calculated by iterative curve fitting of the binding isotherms. The

"2-binding-site" model from the software Microcal ORIGIN provided with the ITC apparatus was used. The Gibbs free energy was calculated from the equation $\Delta G = \Delta H - T\Delta S$.

Overlapping Binding Site Model

An earlier binding study, involving the β-lg and pectin under similar experimental conditions, was carried out using a capillary electrophoresis (CE) method. The free β-lg was separated from the complexed β-lg by means of an electric field. Raw data obtained with CE were subsequently treated with the mathematical overlapping-binding-site model. According to this model, complexation between ligand and macromolecule including interacting ligand molecules can be expressed with the equation.

The binding density of the overlapping binding site model has to be lower or equal to 1. Since more than one protein molecule is expected to complex on a pectin molecule, the D-galA residue was used as the interacting unit instead of the whole pectin molecule used with the ORIGIN software.

ITC calorimetric data were transformed to obtain the binding density "ν" and free b-lg concentration "L" values. For each injection, the volume in the cell "vol" (L), the energy released "Q" (J), the enthalpy "H ($J \cdot mol^{-1}$), the b-lg amount "Xc" (mol), the b-lg concentration ""Xt" (M), and the pectin concentration "Mt" (M) were calculated. Consequently, the maximum energy value "Qmax" (J), the complexed ligand concentration "Xb" (M), and the complexed b-lg concentration were calculated with eq.

$$Qmax = \Delta Xt_1 \times \Delta H \times Vol$$

Where "Xt1 is the β-lg concentration after the first injection

$$XB = (\Delta Xt \times Q) / Qmax$$

$$Xt = Xc / vol$$

Given the apparent molecular weights and the fact that pectin contains 80 per cent of D-galA residues, the average content of D-galA residues was estimated at 420 and 507 per LM- and HM-pectin molecule, respectively. The binding parameters L (M) and ν (β-lg molecules complexed/D-galA residue) were calculated with equations 26 and 27.

$$L = Xt - Xb$$

$$\nu = (Xb \times N_A \times vol) / Mt \times vol \times N_A \times D - galA$$

Where NA is the Avogadro number and D - galA is the average content of D-galA in a pectin molecule.

The non-linear curve fitting of ν/L) plotted against ν was used to obtain K, n and ω values. The JMP software was used for the curve fittings and statistics. The ITC binding data according to the overlapping binding site model were then compared to CE data. Raw results from ITC measurements of the β-lg/ LM-pectin complexes are presented on the thermogram at the top of. The peak areas correspond to the energy released by the cell containing the pectin at each β-lg injection. Because all the injections throughout the analysis released energy, the complexation between β-lg and LM-pectin is considered as exothermic.

Saturation of the LM-pectin with the β-lg was reached at a β-lg/LM-pectin molar ratio of about 26:1, as the energy released in the cell became null. The injections liberated about 2-3 μcal·s-1 in the first third of the analysis. The energy released was relatively low considering that ITC analysis should have had at least 5-10 μcal·s-1 peaks in the first two-thirds of the thermogram. However, the 0.1-μcal sensitivity of the VP-ITC model used provided accurate results with only 3-5 μcal·s-1 liberated. It was hard to improve signal intensity through higher component concentrations because of the high viscosity brought on by the pectin.

Moreover, the β-lg/pectin complexes increased the overall viscosity of a solution as it was demonstrated earlier. This higher viscosity required a faster stirring leading to unacceptable baseline distortion.

The higher viscosity also reduced the solution of β-lg in the cell, which, in turn, drastically slowed down the thermodynamic equilibrium between injections. Decreasing the number of injections to increase their volume did not improve results as complex binding isotherms were better described with many small injections rather than a few large ones.

The binding isotherm of β-lg/LM-pectin complexes is the integrated areas of heat peaks plotted against the molar ratios of components. Although the software integrates the peak areas automatically, it was recommended to adjust the baseline and to integrate the peaks manually for more accurate results. This binding isotherm had two inflexion points, which are generally indicative of two types of independent binding sites on the macromolecule. Two types of binding sites have been observed in an earlier binding study involving the protein p85, a subunit of phosphoinositide 3-kinase, and the tyrosyl phosphopeptides pY740 and pY751.

Because of the non-specific nature of interactions and because the carboxyl group was the only interacting group on the pectin, the two inflexion points are more likely indicative of two different steps involved in the β-lg/pectin complexation rather than two different binding sites. For that reason, the expression "2-site model" was replaced here with the more appropriate "2-step model". In the first step, the amount of β-lg complexed on pectin decreased with each injection until pectin was saturated corresponding to the first plate. Then, the second step begins with a plate where the same amount of β-lg molecules interact with each injection. Then, the pectin became gradually saturated with β-lg until no more energy was released. The x coordinates of inflexion points correspond to the sum of stoichiometry values of β-lg/LM-pectin complexes at each step. ITC measurements showed that the titration of β-lg with salt solutions was exothermic due to the shift in dimer/monomer equilibrium towards the dimer state. Blank titrations carried out were negligible showing that titration did not modify the equilibrium of dimeric β-lg at pH 4 or the pectin aggregation. Thus, the prior dialysis was effective to equilibrate the ionic strength and pH of phosphate buffer, β-lg and pectin solutions.

The thermogram at the top of raw data obtained for β-lg/HM-pectin complexes. First injections of β-lg in HM-pectin released between 2-3 μcal·s-1, as with LM-pectin. The complexation between β-lg and HM-pectin was also exothermic. The binding isotherm of β-lg/HM-pectin had two inflexion points, which indicates that two steps are also involved in the complexation process, as it was the case for LM-pectin. The saturation of the HM-pectin with the β-lg was reached at a molar ratio of about 24:1.

There were no significant differences between the binding isotherms of β-lg interacting with LM- or HM-pectin. About 45 per cent of the total interaction time was needed to achieve the first step, while 55 per cent was required for the second step. The difference between the analysis lengths is explained by the initial HM-pectin concentration (9.43 μM), which was higher than that of LM-pectin (7.33 μM). Therefore, more injections were needed for the β-lg/HM-pectin mixture to reach equivalent molar ratios to β-lg/LM-pectin.

The curve fitting of the binding isotherms of β-lg/LM- or HM-pectin with the Microcal ORIGIN software was also used to calculate the binding stoichiometry of complexes, the binding constant, the enthalpy, the entropy, and the Gibbs free energy. The enthalpy is the only parameter that is model independent and corresponds to the energy released per mole of interacting β-lg. At high "c" value the enthalpy on the binding isotherm can be approximated by the energy released on the first injections where all the β-lg molecules injected are complexed on the pectin. In the subsequent injections, pectin molecules become increasingly saturated with β-lg and only a part of the protein molecules injected is complexed, which explains the lowering of the energy released. Enthalpy values correspond roughly to -14 and -10 kcal·mol-1 of β-lg injected for the LM- and HM-pectin respectively.

Because of the two inflexion points, both binding isotherms obtained were characteristic of two different events in the reaction cell. Thus, the model of the ORIGIN software appropriate to fit binding isotherms with two inflexion points was used in order to calculate the thermodynamic parameters of the two steps. There were about 8 β-lg molecules complexed on the LM-pectin at the first step and 16 at the second step. 6 β-lg were complexed on the HM-pectin at the first step, which was significantly lower than what was obtained at the first step on the LM-pectin. The amount of β-lg complexed at the second step was not significantly different from the value obtained for the LM-pectin.

The binding constants with LM- and HM-pectin at the first step were not statistically different. However, the second step had lower binding constants, which were about 10 times lower than the ones obtained during the first step. Enthalpy values were negative on the first and second steps for LM- and HM-pectin. The enthalpies, as a result of the interaction between β-lg and LM- or HM-pectin respectively were higher at the first steps (–12.6 ± 1.94 and –10.0 ± 0.6 kcal mol–1) than at the second steps (–5.5 ± 0.7 and –3.6 ± 0.5). The entropy values of complexes containing respectively LM- and HM-pectin multiplied by the temperature were slightly negative at the first step (–2.5 ± 2.8 and –0.2 ± 0.04 kcal mol–1) and positive at the second step (3.5 ± 0.5 and 4.1 ± 0.4 kcal mol–1).

The Gibbs free energy values were all negative and no significant differences were observed between interactions at the first or at second step or between LM- or HM-pectin. Given its lower molecular weight, the LM-pectin would have reacted more with the β-lg than the HM-pectin did. However, there was no statistical evidence for that tendency.

When the binding constant and stoichiometry are known prior to an ITC experiment, the c value is calculated to determine the initial macromolecular concentration in the cell. The c value can be calculated using the binding constants of the β-lg/LM- or HM-pectin complexes to verify whether the initial concentrations of pectin in the cell were adequate. The c value is unitless and is calculated as follows:

$$c = K \times M_{tot} \times N$$

Where K is the binding constant, Mtot is the pectin concentration in the analysis cell at the start of the experiment and N is the binding stoichiometry. The ideal c value should be between 1 and 1000 and preferably between 10 and 100. At high c values, the binding isotherm has a square shape, which limits the use of fitting models. As a result, the slope of the curve can only be used to calculate the enthalpy of the reaction. Results obtained with ITC were used to verify if the pectin concentrations was appropriate. The c value was calculated for each step. Calculated c values for the β-lg/LM-pectin complexes were 1437 and 498 for the first and second steps respectively and 905 and 93 for the HM-pectin for the first and second steps respectively. All the c values were in the suitable range, except for the first step of the β-lg/LM-pectin, which exceeded 1000.

However, the high LM-pectin concentration in the cell was justified by the need to reach an acceptable signal on the ITC apparatus. The c values around 1000 could explain the relatively high error margin observed in the binding constants *K* calculated with the non-linear fitting model. Moreover, these results show that higher concentrations used would have increased the error margin on the binding parameters calculated. The binding isotherms generated by the ORIGIN software were also treated with the overlapping binding site model.

Using this model, the binding stoichiometry, binding site size, binding constant, and cooperativity values obtained from the ITC measurements for the second step of β-lg/LM- or HM-pectin complexation were calculated. Binding parameters for the first step were not calculated because not enough points were available for the non-linear fitting. Therefore, the non linear least square fitting was achieved for values of ν between 0.015 and 0.025 and between 0.010 and 0.030 for complexes with LM- and HM-pectin respectively.

Results obtained with the overlapping binding site model followed the same tendencies observed with the ORIGIN software. The overlapping model made it possible to calculate the binding site size covered by 1 β-lg molecule, which was 14 and 22 D-galA residues for LM- and HM-pectin respectively. The interpolymer complexes had a binding constant of 2128 M-1 with LM-pectin which was higher than the value 1233 M-1 obtained with the HM-pectin. One can see that the binding constants calculated with the overlapping binding model were much lower than those obtained with the ORIGIN software. In fact, the overlapping binding site model uses the main repetitive unit of the pectin (D-galA residues) in the calculation as the interacting unit, whereas the ORIGIN software considers the whole pectin molecule. Consequently, the amount of interacting units is much higher in the overlapping binding model, which leads to lower binding constants. The β-lg complexed with LM- or HM-pectin did not show any significant cooperative behaviour.

The negative Gibbs free energy values show the spontaneous character of the interaction between β-lg and LM- or HM-pectin. However, the relatively long period needed for the system to reach equilibrium between injections showed that the β-lg/pectin complexes have a slow binding kinetics.

The existence of two steps involved in the β-lg/LM- or HM-pectin complexation process is in agreement with the Tainaka theory which states that the complexes are formed in two steps passing from the molecular to the aggregated state. The Tainaka theory is adapted from the Veis-Aranyi theory and can be applied to systems regardless of their charges density.

The first step would correspond to the formation of soluble intrapolymer complexes (C1) between β-lg and pectin molecules. The second step would imply the aggregation of these intrapolymer complexes forming insoluble interpolymer complexes (C2). The isoenthalpic plateau observed in the binding isotherms ends when the β-lg/pectin ratio required to induce intrapolymer complex aggregation is reached. A light scattering study carried out on poly(vinyl alcohol) sulfate/proteins salt free systems at different ratios revealed that intrapolymer complexes did not aggregate as long as proteins were in excess.

The lower stoichiometries and higher binding constants of C1 compared to C2 are the result of lower levels β-lg in the cell that can interact with pectin and the higher proportion of complexed β-lg on the pectin, respectively. The first step was obviously enthalpically driven principally because of the direct interaction between the β-lg and pectin ionized carboxylic groups, which were shown to be located mainly in the smooth region of the pectin. The unfavourable entropy observed at the first complexation step is in agreement with the Tainaka theory, which states that C1 have poor conformational entropy. It was previously shown that the formation of protein/polyelectrolyte complexes was driven by electrostatic interactions that could be partially encountered by a loss in polyelectrolyte conformational freedom.

According to the Tainaka theory, the process called complexation begins at the second step where C1 aggregate to form C2. Results show that enthalpic and entropic contributions support this second step. The favourable enthalpic contribution would result from the interaction of β-lg with the residual net negative charge on the C1. The overlapping of intrapolymer complexes leads to an electrostatic energy gain resulting from the increased ion density in the overlapped domain.

The positive entropic contribution at the second step essentially the product of the rearrangement of C2 in the solution and of the dilution of the concentrated phase containing the C2. The presence of C1 and C2 was also observed with the BSA/PDMDAAC (bovine serum albumin/ poly(dimethyldiallylammonium chloride) mixture. The enthalpy values measured with ITC are the sum of contributions coming from the interaction between β-lg and pectin, conformational changes, and aggregation of molecules. Hence, the enthalpy value measured could have been either overestimated or underestimated depending on the possible conformational changes in the molecules.

Monte Carlo simulations showed that at low ionic strengths, a polyelectrolyte chain would wrap around an oppo-sitely charged spherical particle.The entropically favo-rable release of counterions and water molecules at the time of complexation may have been compensated by the negative entropy arising from conformational changes in the pectin. The exothermic reaction observed between maltodextrin and SDS can be explained by conformational changes in the maltodextrin.

Measures with circular dichroism revealed the partial disappearance of the β-lg á-helix as it interacts with the acacia gum at pH 4.2. Conformational changes are also suspected on the ribulose diphosphate carboxylase and on the 11s broad bean globulin once complexed with the LM-pectin. A contribution to enthalpy could also come from the possible aggregation of proteins around pH 4. To verify the accuracy of the binding parameters obtained with ITC, the calorimetric data treated with the overlapping binding site model were compared.

The total stoichiometry values for ITC and CE were respectively 25 and 23 monomers of β-lg complexed per LM-pectin molecule. However, the stoichiometry values calculated for the β-lg/HM-pectin complexes were higher with ITC than with CE. This could be explained by a partial dissociation of β-lg/HM-pectin complexes during the CE analysis due to the electric field. The lower binding constant obtained at the second step is indicative of weaker interactions in β-lg/HM-pectin interpolymer complexes compared to those obtained with LM-pectin.

A previous study using ultrafiltration and destabilizing agents demonstrated that β-lg/HM-pectin interpolymer complexes were more likely dissociated with higher pH values or ionic strengths than β-lg/LM-pectin complexes. The binding constant at the second step of the β-lg/LM- or HM-pectin complexes obtained with ITC were respectively 2128 and 1233 M-1 compared with 1431 and 857 M-1 calculated with capillary electrophoresis for the same mixture. The binding constant values obtained by ITC, were higher than those found with capillary electrophoresis.

This binding parameter calculated from enthalpy values may have been overestimated, because of conformational changes involved in pectin and β-lg molecules as they are complexing. The binding site sizes calculated with ITC (14 and 22 D-galA) were similar to those obtained with capillary electrophoresis (12 and 20 D-galA) for the LM- and HM-pectin respectively. Contrary to what was observed with capillary electrophoresis, the β-lg complexed with LM- or HM-pectin did not show any significant cooperative behaviour. The complexes analysed with capillary electrophoresis were "T" as the β-lg and pectin solutions were mixed before the acidification. The ITC complexes were "M" since β-lg and pectin solutions were at pH 4 before mixing. The "T" and "M" complexes were shown to have cooperative and non-cooperative behaviours respectively.

Isothermal titration calorimetry is a useful tool for carrying out binding studies with biopolymers. The low cost of the food biopolymers makes them ideal molecules for studies in ITC, which requires a substantial amount of sample. ITC was used to characterize the formation of β-lg/LM- or HM-pectin complexes. The complexation is believed to occur in two steps as predicted by the Tainaka theory. Enthalpic contribution from conformational changes in pectin and β-lg molecules may have lead to an overestimation of binding parameters. Static light scattering at low angles was used to determine the diameter and the phase separation kinetics as β-lg/pectin complexes are formed in order to validate the β-lg/pectin intrapolymer and interpolymer complexes formation.

PROTEIN SOLUBILITY

At the surfaces of proteins are amino acid residues that interact with water. The amino acids are referred to as hydrophilic amino acids and include arginine, lysine, aspartic acid, and glutamic acid. At pH 7 the side chains of these amino acids carry charges—positive for arginine and lysine, negative for aspartic acid and glutamic acid. As the pH increases, lysine and arginine begin to lose their positive charge, and at pHs greater than about 12 they are mainly neutral. In contrast, as pH decreases, aspartic acid and glutamic acid begin to lose their negative charges, and at pHs less than 4 they are mainly neutral.

The surface of a protein has a net charge that depends on the number and identities of the charged amino acids, and on pH. At a specific pH the positive and negative charges will balance and the net charge will be zero. This pH is called the isoelectric point, and for most proteins it occurs in the pH range of 5.5 to 8. A protein has its lowest solubility at its isoelectric point. If there is a charge at the protein surface, the protein prefers to interact with water, rather than with other protein molecules. This charge makes it more soluble. Without a net charge, protein-protein interactions and precipitation are more likely.

The solubility of proteins in blood requires a pH in the range of 7.35 to 7.45. The bicarbonate–carbonic acid buffer system of blood ($HCO_3^- + H^+ \leftrightarrow H_2CO_3$), in which the bicarbonate is in excess of the carbonic acid, helps to maintain the correct pH. Exhalation of carbon dioxide from the lungs causes some of the bicarbonate ions in blood to combine with protons, and this would raise the pH. However, because there is an excess of bicarbonate ions and protons, the loss of a small number of protons does not influence the pH significantly.

The proteins of protein mixtures can be separated using a technique known as isoelectric focusing. A mixture is placed in a polyacrylamide gel that has a pH gradient. An anode (positive electrode) and a cathode (negative electrode) are positioned at the low and high ends of the pH gradient, respectively. If a protein is located in the high pH region, it will be negatively charged and will move towards the anode. As the protein moves to a lower pH region, its surface charge will become less negative, and a pH region will be reached at which the protein net charge is zero (the isoelectric point). The protein will stop moving and, because different proteins have different isoelectric points, separation can be achieved.

THERMODYNAMICS AND INTERACTIONS

DIFFERENTIAL SCANNING CALORIMETRY

Differential scanning calorimetry is a technique we use to study what happens to polymers when they're heated. We use it to study what we call the *thermal transitions* of a polymer. And what are thermal transitions? They're the changes that take place in a polymer when you heat it. The melting of a crystalline polymer is one example. The glass transition is also a thermal transition.So how do we study what happens to a polymer when we heat it? The first step would be to heat it, obviously. And that's what we do in differential scanning calorimetry, or DSC for short. We heat our polymer in a device that looks something like this:

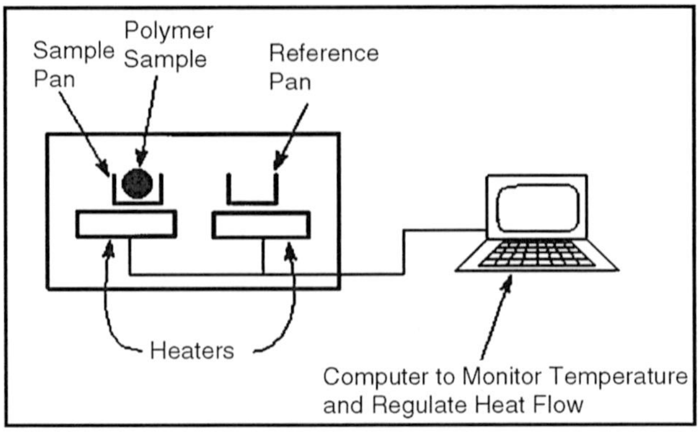

Fig. Scanning of Calorimetry

It's pretty simple, really. There are two pans. In one pan, the sample pan, you put your polymer sample. The other one is the reference pan. You leave it empty. Each pan sits on top of a heater.

Then you tell the nifty computer to turn on the heaters. So the computer turns on the heaters, and tells it to heat the two pans at a specific rate, usually something like 10 °C per minute. The computer makes absolutely sure that the heating the rate stays exactly the same throughout the experiment. But more importantly, it makes sure that the two separate pans, with their two separate heaters, heat at the same rate as each other.

Why wouldn't they heat at the same rate? The simple reason is that the two pans are different. One has polymer in it, and one doesn't. The polymer sample means there is extra material in the sample pan. Having extra material means that it will take more heat to keep the temperature of the sample pan increasing at the same rate as the reference pan. So the heater underneath the sample pan has to work harder than the heater underneath the reference pan. It has to put out more heat. By measuring just *how much* more heat it has to put out is what we measure in a DSC experiment. Specifically what we do is this: We make a plot as the temperature increases. On the x-axis we plot the temperature. On the y-axis we plot difference in heat output of the two heaters at a given temperature.

HEAT CAPACITY

We can learn a lot from this plot. Let's imagine we're heating a polymer. When we start heating our two pans, the computer will plot the difference in heat output of the two heaters against temperature. That is to say, we're plotting the heat absorbed by the polymer against temperature. The plot will look something like this at first.

The heat flow at a given temperature can tell us something. The heat flow is going to be shown in units of heat, q supplied per unit time, t. The heating rate is temperature increase T per unit time, t. Got it?

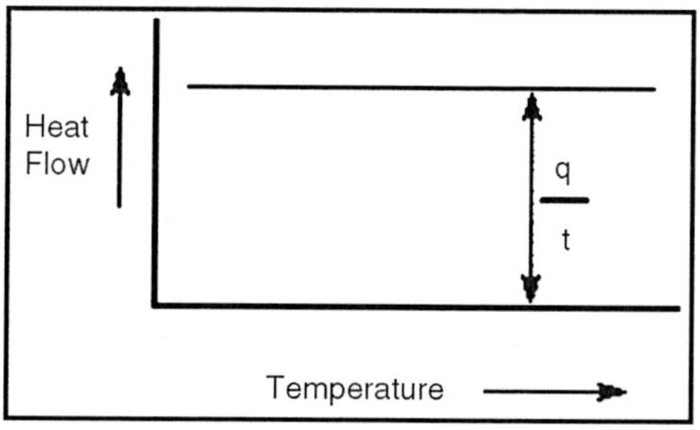

$$\frac{\text{heat}}{\text{time}} = \frac{q}{t} = \text{heat flow}$$

$$\frac{\text{temperature increase}}{\text{time}} = \frac{\Delta T}{t} = \text{heating tate}$$

Let's say now that we divide the heat flow *q*/*t* by the heating rate *T*/*t*. We end up with heat supplied, divided by the temperature increase.

$$\frac{\frac{q}{t}}{\frac{\Delta T}{t}} = \frac{\Delta T}{t} = C_p = \text{heat capacity}$$

Remember from the glass transition page that when you put a certain amount of heat into something, its temperature will go up by a certain amount, and the amount of heat it takes to get a certain temperature increase is called the *heat capacity*, or C_p.We get the heat capacity by dividing the heat supplied by the resulting temperature increase.And that's just what we've done in that equation up there. We've figured up the heat capacity from the DSC plot.

THE GLASS TRANSITION TEMPERATURE

Of course, we can learn a lot more than just a polymer's heat capacity with DSC. Let's see what happens when we heat the polymer a little more. After a certain temperature, our plot will shift upward suddenly, like this:

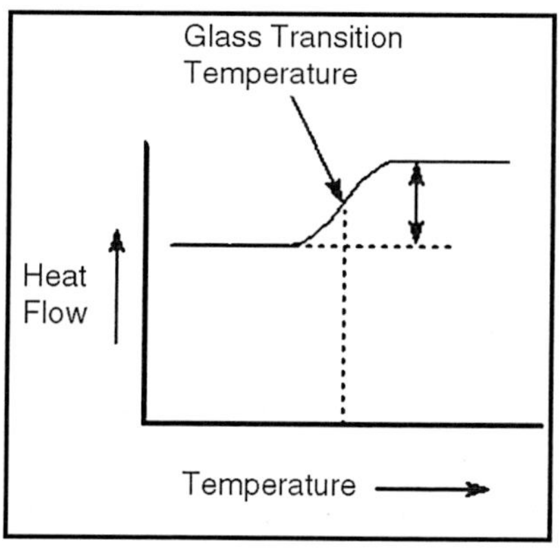

This means we're now getting more heat flow. This means we've also got an increase in the heat capacity of our polymer. This happens because the polymer has just gone through the glass transition. And as you learned on the glass transition page, polymers have a higher heat capacity above the glass transition temperature than they do below it. Because of this change in heat capacity that occurs at the glass transition, we can use DSC to measure a polymer's glass transition temperature. You may notice that the change doesn't occur suddenly, but takes place over a temperature range. This makes picking one discreet *T*g kind of tricky, but we usually just take the middle of the incline to be the *T*g.

CRYSTALLIZATION

But wait there is more, so much more. Above the glass transition, the polymers have a lot of mobility. They wiggle and squirm, and never stay in one position for very long. They're kind of like passengers trying to get comfortable in airline seats, and never quite succeeding, because they can move around more. When they reach the right temperature, they will have gained enough energy to move into very ordered arrangements, which we call crystals, of course.

When polymers fall into these crystalline arrangements, they give off heat. When this heat is dumped out, it makes the little computer-controlled heater under the sample pan really happy. It's happy because it doesn't have to put out much heat to keep the temperature of the sample pan rising.

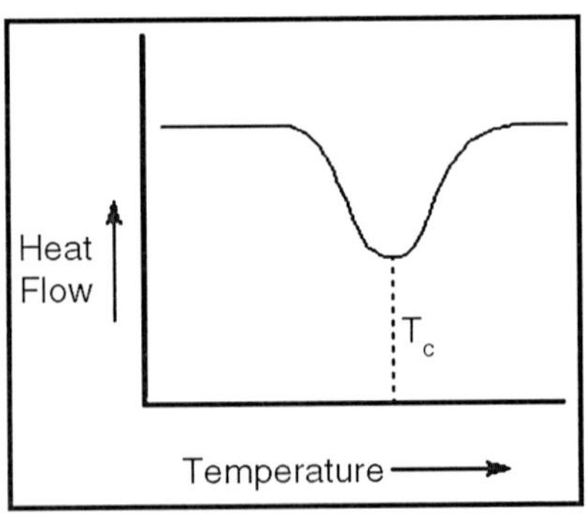

This dip tells us a lot of things. The temperature at the lowest point of the dip is usually considered to be the polymer's crystallization temperature, or T_c. Also, we can measure the area of the dip, and that will tell us the latent energy of crystallization for the polymer. But most importantly, this dip tells us that the polymer can in fact crystallize.

If you analysed a 100 per cent amorphous polymer, like atactic polystyrene, you wouldn't get one of these dips, because such materials don't crystallize. Also, because the polymer gives off heat when it crystallizes, we call crystallization an *exothermic* transition.

MELTING

Heat may allow crystals to form in a polymer, but too much of it can be their undoing. If we keep heating our polymer past its T_c, eventually we'll reach another thermal transition, one called melting. When we reach the polymer's melting temperature, or T_m, those polymer crystals begin to fall apart, that is they melt. The chains come out of their ordered arrangements, and begin to move around freely. And in case you were wondering, we can spot this happening on a DSC plot.

Remember that heat that the polymer gave off when it crystallized? Well when we reach the T_m, it's payback time.

There is a latent heat of melting as well as a latent heat of crystallization. When the polymer crystals melt, they must absorb heat in order to do so. Remember melting is a first order transition.

This means that when you reach the melting temperature, the polymer's temperature won't rise until all the crystals have melted. This means that the little heater under the sample pan is going to have to put a lot of heat into the polymer in order to both melt the crystals *and* keep the temperature rising at the same rate as that of the reference pan. This extra heat flow during melting shows up as a big peak on our DSC plot, like this:

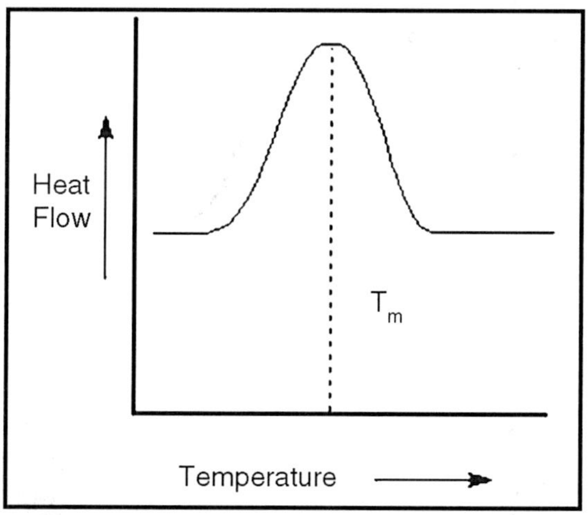

We can measure the latent heat of melting by measuring the area of this peak. And of course, we usually take the temperature at the top of the peak to be the polymer's melting temperature, T_m. Because we have to add energy to the polymer to make it melt, we call melting an *endothermic* transition.

PUTTING IT ALL TOGETHER

So let's review now: we saw a step in the plot when the polymer was heated past its glass transition temperature. Then we saw a big dip when the polymer reached its crystallization temperature. Then finally we saw a big peak when the polymer reached its melting temperature. To put them all together, a whole plot will often look something like this:

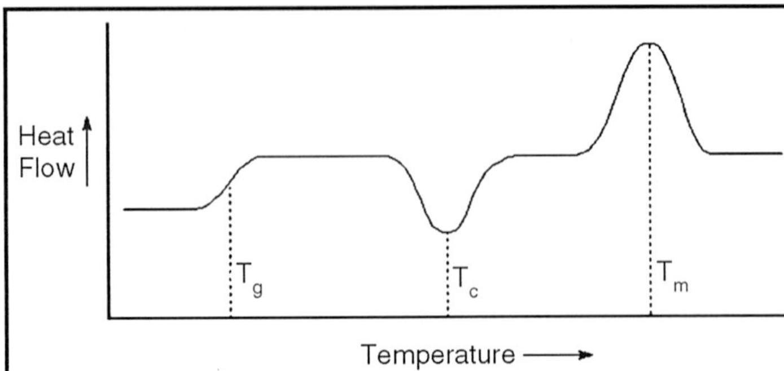

The entire DSC plot,right before your very eyes!

The crystallization dip and the melting peak will only show up for polymers that can form crystals. Completely amorphous polymers won't show any crystallization, or any melting either. But polymers with both crystalline and amorphous domains.

If you look at the DSC plot you can see a big difference between the glass transition and the other two thermal transitions, crystallization and melting. For the glass transition, there is no dip, and there's no peak, either. This is because there is no latent heat given off, or absorbed, by the polymer during the glass transition.

Both melting and crystallization involve giving off or absorbing heat. The only thing we do see at the glass transition temperature is a change in the heat capacity of the polymer.

Because there is a change in heat capacity, but there is no latent heat involved with the glass transition, we call the glass transition a *second order transition*. Transitions like melting and crystallization, which do have latent heats, are called *first order transitions*.

LEVEL OF CRYSTALLINITY

DSC can also tell us how much of a polymer is crystalline and how much is amorphous. If you read the page dealing with polymer crystallinity, you know that many polymers contain both amorphous and crystalline material. But how much of each? DSC can tell us. If we know the latent heat of melting, ΔH_m, we can figure out the answer.

The first thing we have to do is measure the area of that big peak we have for the melting of the polymer. Now our plot is a plot of heat flow per gram of material, versus temperature. Heat flow is heat given off per second, so the area of the peak is given is units of heat × temperature × $time^{-1}$ x $mass^{-1}$. We usually would put this in units such as joules × kelvins × $(seconds)^{-1}$ × $(grams)^{-1}$:

$$\text{area} = \frac{\text{heat} \times \text{temperature}}{\text{time} \times \text{mass}} = \frac{\text{JK}}{\text{sg}}$$

Got that? Don't worry. It gets simpler. We usually divide the area by the heating rate of our dsc experiment. The heating rate is in units of K/s. So the expression becomes simpler:

$$\frac{\text{area}}{\text{heating}} + \frac{\frac{\text{JK}}{\text{sg}}}{\frac{\text{K}}{\text{s}}} = \frac{\text{J}}{\text{g}}$$

Now we have a number of joules per gram. But because we know the mass of the sample, we can make it simpler. We just multiply this by the mass of the sample:

$$\left(\frac{\text{J}}{\text{g}}\right) \times \text{g} = \text{J}$$

Now we just calculated the total heat given off when the polymer melted. Neat, huh? Now if we do the same calculation for our dip that we got on the DSC plot for the crystallization of the polymer, we can get the total heat absorbed during the crystallization. We'll call the heat *total* heat given off during melting $H_{m,\ total}$, and we'll call the heat of the crystallization $H_{c,\ total}$.

Now we're going to subtract the two: $\text{H}_{\text{m,total}} - \text{H}_{\text{c,total}} = \text{H}'$

Why did we just do that? And what does that number *H*' mean? *H*' is the heat given off by that part of the polymer sample which was already in the crystalline state *before* we heated the polymer above the T_c. We want to know how much of the polymer was crystalline before we induced more of it to become crystalline.

That's why we subtract the heat given off at crystallization. Is everyone following me? Now with our magic number *H*' we can figure up the per cent crystallinity. We're going to divide it by the specific heat of melting, H_c^*. The specific heat of melting? That's the amount of heat given off by a certain amount, usually one gram, of a polymer. *H*' is in joules, and the specific heat of melting is usually given in joules per gram, so we're going to get an answer in grams, which we'll call m_c.

$$\frac{\text{H}'}{\text{H}^*_{\text{m}}} = \text{m}_{\text{c}} \frac{\text{J}}{\frac{\text{J}}{\text{g}}} = \text{g}$$

This is the total amount of grams of polymer that were crystalline below the T_c. Now if we divide this number by the weight of our sample, m_{total}, we get the fraction of the sample that was crystalline, and then of course, the per cent crystallinity:

$$\frac{m_c}{m_{total}} = \text{crystalline fraction}$$

$$\text{crystalline fraction} \times 100 = \%\,\text{crystallinity}$$

And that's how we use DSC to get per cent crystallinity.

Copolymer Sequence Distribution

It has been demonstrated previously1,2,3,4,5 that the melting temperature and degree of crystallinity of a series of copolymers are, as a first approximation, independent of the nature of the comonomer (except for methyl branches), while being largely determined by the crystallizable sequence length distributions. It is important to address this aspect of the copolymers before proceeding to the details of the calorimetric experiments.

This observation is further verified by the plots of weight per cent crystallinities and peak melting temperatures as functions of comonomer content respectively the crystallinities of all three types of copolymers can be represented by a single curve, irrespective of the comonomer type present.

Due to differences in the size of the three types of comonomers compared, the per cent of a comonomer required to yield a copolymer of given density is significantly different for the three comonomer types. As a result, the crystallinities derived for copolymers with a given mole per cent comonomer, are identical to each other only when they are normalized with respect to the weight per cent of crystallizable ethylene sequences available in the copolymer.

Furthermore, the crystallinities of hydrogenated polybutadienes (HPBs) obtained from the literature6,7 are also found to fall on the same curve. The HPBs are considered as model ethylene/1-butene random copolymers, whose compositional homogeneity is assured by the synthetic methods adopted,8,9,10 and is known to be accurately determined. It follows that the copolymers examined conform to the random statistics expected of materials synthesized using the single-site constrained geometry catalyst technology. In addition, the fact that all three types of copolymers display an identical crystallinity behaviour, in spite of their significantly different branch types - namely ethyl, propyl and butyl - implies that these branches are predominantly excluded from the crystal lattice.

This result is in accordance with the assumption of comonomer exclusion made in Flory's crystallization theory for copolymers. Exclusion of the branch types mentioned above has also been established in a number of previous investigations.15,16,17 The random statistics of the copolymers are further exemplified where the melting temperatures of the copolymers as well as those of the HPBs are shown to be represented by a single curve. Effect of Cooling Rate on the Crystallization Process The crystallization behaviours of the ethylene/1-butene, ethylene/1-pentene and ethylene/1-hexene copolymers on cooling from the melt

(180°C) to -20°C (using a dry ice/isopropanol mixture) at different cooling rates were examined.

The rates studied ranged from 5 to 20°C/min. The corrected heat capacities estimated for the sample, EB- 912, at two different cooling rates (5 and 20°C/min.). It has been previously demonstrated4,18,19 that the crystallization process in such copolymers occurs over a broad temperature range, with significant increases in crystallinity even at temperatures as low as -20°C. In agreement with those observations the presence of two distinct exotherms in the crystallization curves of the copolymer - a sharp exotherm at high temperatures and a much broader exotherm extending to very low temperatures. The temperature corresponding to the peak of the sharp crystallization exotherm is seen to be strongly rate-dependent, shifting to lower temperatures at higher cooling rates.

A major consideration in the estimation of heat capacities and degrees of crystallinity from the areas of crystallization exotherms is the accuracy of the baseline used in the calculation of peak areas. Incorrectly determined baselines can lead to significant errors in the crystallinity values. This concern is particularly significant in the case of copolymers, in view of the broad crystallization ranges involved.

In order to verify the accuracy of the heat capacity calculations, the experimental values were compared to the Cp(T) data for linear polyethylene. It is reassuring to note that the experimental data in the melt display excellent agreement with the Cp values of linear PE in the melt. This result is also consistent with previous reports in the literature,4,20 indicating that the Cpmelt of the ethylene/a-olefin copolymers with branch concentrations less than ~15 mole per cent, are coincident with those for LPE.

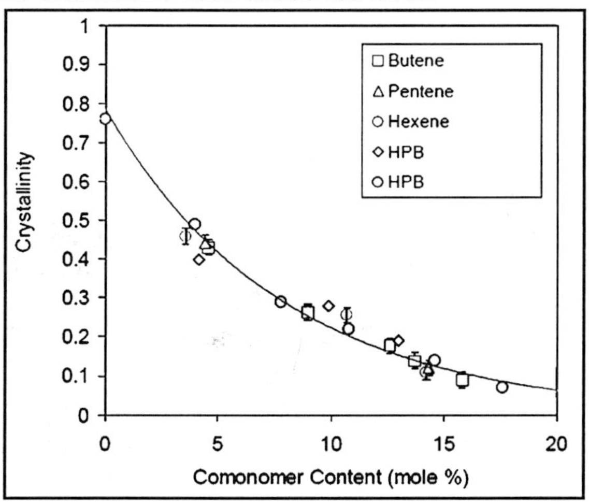

Fig. Degrees of Crystallinity at -20°C for Ethylene/1-Butene, Ethylene/1-Pentene and Ethylene/1-Hexene Copolymers (after Cooling from the Melt at 5°C/min) as a Function of Comonomer Content. The Crystallinity data for HPBs are also Included.

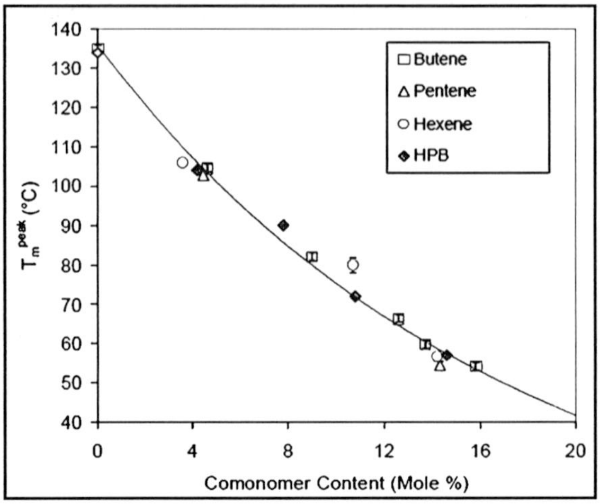

Fig. Peak Melting Temperatures for Ethylene/1-Butene, Ethylene/1-Pentene and Ethylene/1-Hexene Copolymers as a Function of Comonomer Content.

The Peak Melting Temperatures for HPBs are also Included. Partial areas were derived from these heat capacity curves and subsequently converted into degrees of crystallinity (χc) through the use of appropriate scaling factors, depending on the mass of the sample used. It is also noted there was no effect of sample mass on the results of the cooling experiments, within the range of sample masses (3-5 mg) utilized in the present study.

Thus, the crystallinity values as a function of temperature, χc(T), were obtained for all the samples, at different cooling rates. The variation of χc with temperature for the various EB, EP and EH copolymers. The effect of cooling rate on the crystallization behaviour of linear polyethylene (LPE) was reported as part of an earlier investigation carried out in this laboratory.

In the case of LPE, the crystallinities display a strong cooling rate dependence at all temperatures.22 Furthermore, the χc values increase significantly at high temperatures (during cooling), and appear to reach a nearly constant value extending over a broad temperature range (at the lower temperatures).

This is in sharp contrast to the trend exhibited by the copolymers. The crystallinity curves for these materials clearly depict the presence of two distinct regions of crystallization, demarcated by a cross-over temperature, T*. On the low temperature side of T, the crystallinity values increase over a wide temperature range, while being cooling rate-independent.

On the high temperature side, however, the crystallinities increase appreciably over a relatively narrow range and exhibit a strong dependence on cooling rate.

The contribution of the increase in χc in the high temperature region to the total crystallinity is significantly larger for the lower branched materials.

As the comonomer content is increased, this region becomes progressively less significant and becomes indiscernible beyond a certain branch content.

The comono-mer content also influences the cross-over temperature, T*, which systematically shifts to lower temperatures with increasing branch concentrations.

Another significant difference between LPE and the copolymers is the fact that the copolymer crystallinities continue to increase noticeably even at temperatures as low as –20°C, as opposed to the nearly constant values for LPE at such low temperatures. This observation is consistent with previous studies4,23 whose results indicate that the level of crystallinity in such copolymers continues to increase appreciably until the glass transition temperature is reached.

The degrees of crystallinity for the three types of copolymers at –20°C obtained from the crystallinity curves are plotted as a function of comonomer the crystallinity data for HPBs from the literature6,7 are also included to verify the random statistics of the copolymers studied. The cc's for the copolymers are found to systematically decrease with increasing branch content, even at –20°C. This observation is directly related to the partitioning of the polymer backbone into shorter crystallizable segments by the branches.

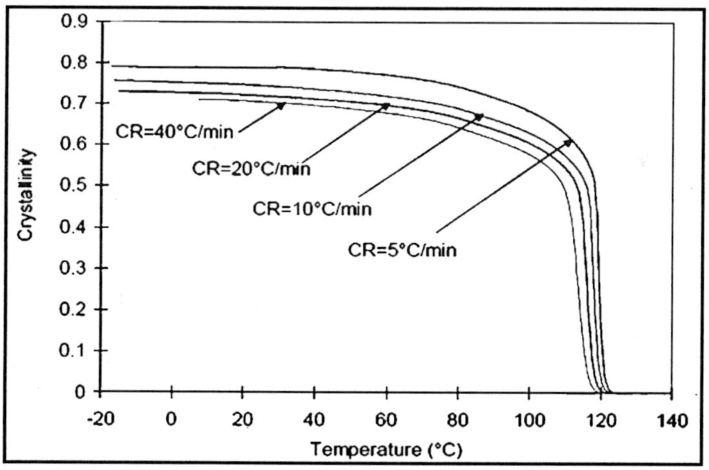

Fig. Degrees of Crystallinity as a Function of Temperature Obtained at Different Cooling Rates from the Melt for Linear Polyethylene.

Melting Behaviour Subsequent to Crystallization during Cooling at Constant Rate

The reversibility of the crystallization and melting processes in the copolymers was investigated by obtaining the melting traces of the samples immediately after quenching from the melt to –20°C. The peak melting temperatures (Tm's) from the melting traces of the samples are plotted as a function of comonomer. It is observed that Tm decreases progressively with increasing branch concentrations for all the copolymers studied. The degrees of crystallinity calculated from the heating traces for EB-894 (with ~ 9 mole per cent branches) are plotted as a function of temperature. The $\chi c(T)$ values estimated during the prior quenching of this sample from the melt to –20°C. Comparison of the two curves clearly indicate that the crystallinities obtained from the heating and cooling traces are identical to each other at low temperatures, and depart significantly from each other at higher temperatures. Two distinct regions of temperature dependence of the crystallinities are also found to exist.

There is a high temperature region that exhibits a significant degree of hysteresis, and a lower temperature region with a remarkably reversible behaviour. Similar observations were made on the other copolymers used in the present study, as well as on ethylene/1- octene copolymers that were studied previously in our laboratory21.

It has been observed that the crystallinity change in the hysteretic region decreases progressively with increasing branch concentrations. This is parallel to the increase in the crystallinity change in the reversible region, with increasing comonomer content.

Furthermore, the temperature delineating the two regions (hysteretic and reversible) is found to be identical to the temperature, T*, that marks the cross-over from a rate-dependence to a rateindependence of the degree of crystallinity from the cooling experiments. This behaviour of the copolymers in sharp contrast to that of LPE. It has been demonstrated21 that LPE predominantly exhibits hysteresis between the crystallinity curves derived from the crystallization and melting processes.

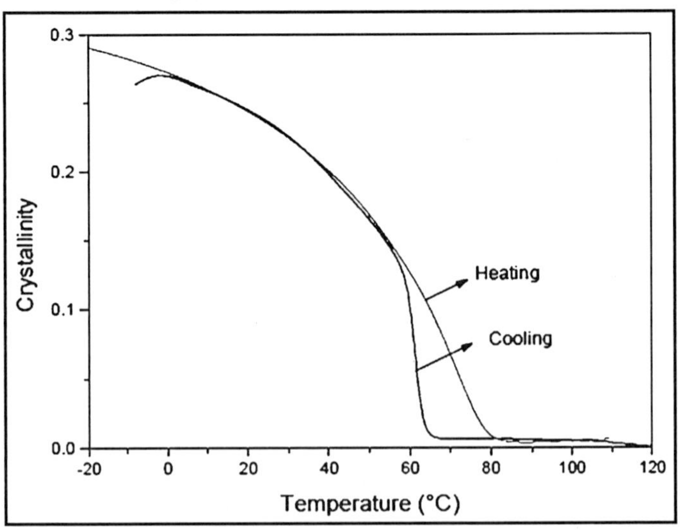

Fig. Degrees of Crystallinity as a Function of Temperature for EB-894 During Cooling at 20°C/min from the Melt and During Subsequent Heating at 20°C/min.

AVRAMI EQUATION

In this study, the temporal evolution of the enthalpies of fusion of the low endotherms was modeled using a double logarithmic plot. Such a representation is based on the Avrami equation in the context of the free growth approximation. It is useful to briefly review the underlying theory before proceeding with the details of the enthalpy analysis.

Theory

The theoretical as well as experimental aspects related to the crystallization of homo- and copolymers of ethylene. It is clear from that the equilibrium relationships proposed are satisfied only to a very limited extent in polymer crystallization, the main reason being the restrictions imposed by the kinetic factors involved with the process. A better understanding of these factors can be gained by studying the rates and mechanism of crystallization in the samples.

In the present study, this was achieved by adopting an Avrami-type analysis of the heats of fusion. The important aspects of this type of analysis are presented below. The Avrami equation29,30,31 was originally derived based on the theory that the emergence of a crystalline phase in a semi crystalline polymer occurs via two steps:

- Initial formation of a nucleus or a growth centre as a result of random fluctuations in the bulk liquid (homogeneous nucleation) or at the interface of foreign bodies (heterogeneous nucleation).
- Subsequent growth of the above nuclei determined by factors including temperature, chain mobility etc. A number of different crystallization mechanisms were considered and an equation of the form shown below, common to all of them, was obtained.

$$\theta_r(t) = 1 - \exp(-Kt'')$$

where $\theta r(t)$ = the relative crystallinity at time t
n = the Avrami exponent, and
K = the Avrami rate constant.

The relative crystallinity, θr, at any time, t is given by the ratio of the crystallinity at that time to that at the end of crystallization, i.e., $t = \infty$.

$$\theta_r(t) = \frac{\theta_c(t)}{\theta_c(\infty)}$$

The rate constant, K, is a cluster of numerical and growth constants, and provides information regarding the growth rate, G, and the nucleation density, N. For instance, in the case of a predetermined nucleation of growth centers followed by a three-dimensional growth of crystallites, the rate constant, K, is given by the expression,

$$K = \frac{4}{3}\pi G^3 N$$

The exponent, *n*, expresses the power dependence on time of the crystallization process, and can assume integral as well as non-integral values ranging from 0.5 to 432,33. The value of this exponent is directly related to the processes governing the nucleation and growth mechanisms,34 some of which are listed in Table.

It is evident from this table that there exists a multitude of possibilities for the crystallization mechanism in a selected polymer, depending on the crystallization conditions. Although the values of the exponent derived from a few different mechanisms are identical to each other, the method of using the Avrami exponent as a means of characterizing the crystallization process is still valuable.

This is particularly true in instances where the processes of crystallization in two or more samples or in a given sample under different conditions, are compared to each other. In the present study, the behaviours during the crystal formation of the copolymers with three different branch types were compared to each other.

Furthermore, for a given branch type the variations with differing branch concentrations and crystallization temperatures as well as times were considered. Therefore, the use of the Avrami method in such an analysis can provide useful information regarding the crystallization behaviours of the copolymers.

DETECTION OF PHASE TRANSITIONS

The basic principle underlying this technique is that, when the sample undergoes a physical transformation such as phase transitions, more or less heat will need to flow to it than the reference to maintain both at the same temperature. Whether more or less heat must flow to the sample depends on whether the process is exothermic or endothermic. For example, as a solid sample melts to a liquid it will require more heat flowing to the sample to increase its temperature at the same rate as the reference. This is due to the absorption of heat by the sample as it undergoes the endothermic phase transition from solid to liquid.

Likewise, as the sample undergoes exothermic processes (such as crystallization) less heat is required to raise the sample temperature. By observing the difference in heat flow between the sample and reference, differential scanning calorimeters are able to measure the amount of heat absorbed or released during such transitions. DSC may also be used to observe more subtle phase changes, such as glass transitions. DSC is widely used in industrial settings as a quality control instrument due to its applicability in evaluating sample purity and for studying polymer curing.

DTA

An alternative technique, which shares much in common with DSC, is differential thermal analysis (DTA). In this technique it is the heat flow to the sample and reference that remains the same rather than the temperature. When the sample and reference are heated identically phase changes and other thermal processes cause a difference in temperature between the sample and reference. Both DSC and DTA provide similar information; DSC is the more widely used of the two techniques.

DSC Curves

The result of a DSC experiment is a curve of heat flux versus temperature or versus time. There are two different conventions: exothermic reactions in the sample shown with a positive or negative peak; it depends by the different kind of technology used by the instrumentation to make the experiment. This curve can be used to calculate enthalpies of transitions. This is done by integrating the peak corresponding to a given transition. It can be shown that the enthalpy of transition can be expressed using the following equation:

$$\Delta H = KA$$

where ΔH is the enthalpy of transition, K is the calorimetric constant, and A is the area under the curve. The calorimetric constant will vary from instrument to instrument, and can be determined by analyzing a well-characterized sample with known enthalpies of transition.

Applications

Differential scanning calorimetry can be used to measure a number of characteristic properties of a sample. Using this technique it is possible to observe fusion and crystallization events as well as glass transition temperatures (T_g). DSC can also be used to study oxidation, as well as other chemical reactions. Glass transitions may occur as the temperature of an amorphous solid is increased. These transitions appear as a step in the baseline of the recorded DSC signal. This is due to the sample undergoing a change in heat capacity; no formal phase change occurs.

As the temperature increases, an amorphous solid will become less viscous. At some point the molecules may obtain enough freedom of motion to spontaneously arrange themselves into a crystalline form. This is known as the crystallization temperature (T_c). This transition from amorphous solid to crystalline solid is an exothermic process, and results in a peak in the DSC signal. As the temperature increases the sample eventually reaches its melting temperature (T_m). The melting process results in an endothermic peak in the DSC curve. The ability to determine transition temperatures and enthalpies makes DSC an invaluable tool in producing phase diagrams for various chemical systems.

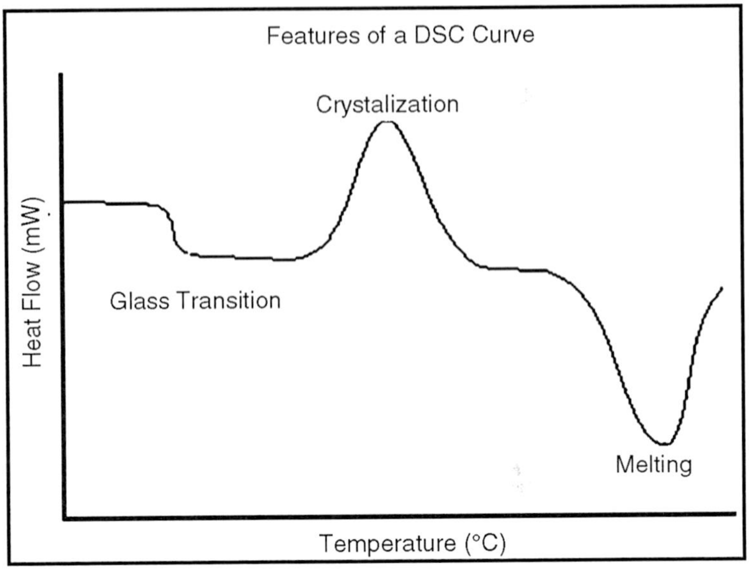

Fig. A Schematic DSC Curve Demonstrating the Appearance of Several Common Features

EXAMPLES

The technique is widely used across a range of applications, both as a routine quality test and as a research tool. The equipment is easy to calibrate, using low melting indium for example, and is a rapid and reliable method of thermal analysis.

Liquid Crystals

DSC may also be used in the study of liquid crystals. As matter transitions between solid and liquid it often goes through a third state, which displays properties of both phases. This anisotropic liquid is known as a liquid crystalline or mesomorphous state. Using DSC, it is possible to observe the small energy changes that occur as matter transitions from a solid to a liquid crystal and from a liquid crystal to an isotropic liquid.

Oxidative Stability

Using differential scanning calorimetry to study the oxidative stability of samples generally requires an airtight sample chamber. Usually, such tests are done isothermally (at constant temperature) by changing the atmosphere of the sample. First, the sample is brought to the desired test temperature under an inert atmosphere, usually nitrogen. Then, oxygen is added to the system. Any oxidation that occurs is observed as a deviation in the baseline. Such analysis can be used to determine the stability and optimum storage conditions for a compound.

Drug Analysis

DSC is widely used in the pharmaceutical and polymer industries. For the polymer chemist, DSC is a handy tool for studying curing processes, which allows the fine tuning of polymer properties.

The cross-linking of polymer molecules that occurs in the curing process is exothermic, resulting in a positive peak in the DSC curve that usually appears soon after the glass transition.

In the pharmaceutical industry it is necessary to have well-characterized drug compounds in order to define processing parameters. For instance, if it is necessary to deliver a drug in the amorphous form, it is desirable to process the drug at temperatures below those at which crystallization can occur.

Food Science

In food science research, DSC is used in conjunction with other thermal analytical techniques to determine water dynamics. Changes in water distribution may be correlated with changes in texture. Similar to material science studies, the effects of curing on confectionery products can also be analysed.

Polymers

DSC curves may also be used to evaluate drug and polymer purities. This is possible because the temperature range over which a mixture of compounds melts is dependent on their relative amounts. This effect is due to a phenomenon known as freezing point depression, which occurs when a foreign solute is added to a solution. (Freezing point depression is what allows salt to de-ice sidewalks and antifreeze to keep your car running in the winter.) Consequently, less pure compounds will exhibit a broadened melting peak that begins at lower temperature than a pure compound.

DSC is used widely for examining polymers to check their purity and composition. Melting points and glass transition temperatures for most polymers are available from standard compilations, and the method can show up possible polymer degradation by the lowering of the expected melting point, Tm, for example. Tm depends on the molecular weight of the polymer, so lower grades will have lower melting points than expected.

Metals

In last few years this technology has been involved in metallic material study. The characterization of this kind of material with DSC is not easy yet because of the low quantity of literature about it. It is known that it is possible to use DSC to find solidus and liquidus temperature of a metal alloy, but the widest application is, by now, the study of precipitations, Guiner Preston zones, phase transitions, dislocations movement, grain growth etc.

Isothermal Titration Calorimetry

Isothermal titration calorimetry (ITC) is a biophysical technique used to determine the thermodynamic parameters of (biochemical) interactions. It is most often used to study the binding of small molecules (such as medicinal compounds) to larger macromolecules (proteins, DNA etc.).

FUNDAMENTALS OF ENZYME KINETICS

Catalysts increase the rate of otherwise slow or imperceptible reactions without undergoing any net change in their structure. The early development of the concept of catalysis in the 19th century went hand in hand with the discovery of powerful catalysts from biological sources. These were called enzymes and were later found to be proteins. They mediate all synthetic and degradative reactions carried out by living organisms.

They are very efficient catalysts, often far superior to conventional chemical catalysts, for which reason they are being employed increasingly in today's high-technological society, as a highly significant part of biotechnological expansion. Their utilization has created a billion dollar business including a wide diversity of industrial processes, consumer products, and the burgeoning field of biosensors. Further applications are being discovered constantly.

Enzymes have a number of distinct advantages over conventional chemical catalysts. Foremost amongst these are their specificity and selectivity not only for particular reactions but also in their discrimination between similar parts of molecules(regiospecificity) or optical isomers (stereospecificity). They catalyse only the reactions of very narrow ranges of reactants (substrates), which may consist of a small number of closely related classes of compounds (e.g. trypsin catalyses the hydrolysis of some peptides and esters in addition to most proteins), a single class of compounds (e.g. hexokinase catalyses the transfer of a phosphate group from ATP to several hexoses), or a single compound (e.g. glucose oxidase oxidises only glucose amongst the naturally occurring sugars). This means that the chosen reaction can be catalysed to the exclusion of side-reactions, eliminating undesirable by-products.

Thus, higher productivities may be achieved, reducing material costs. As a bonus, the product is generated in an uncontaminated state so reducing purification costs and the downstream environmental burden.

Often a smaller number of steps may be required to produce the desired end-product. In addition, certain stereospecific reactions (e.g. the conversion of glucose into fructose) cannot be achieved by classical chemical methods without a large expenditure of time and effort.

Enzymes work under generally mild processing conditions of temperature, pressure and pH. This decreases the energy requirements, reduces the capital costs due to corrosion-resistant process equipment and further reduces unwanted side-reactions.

The high reaction velocities and straightforward catalytic regulation achieved in enzyme-catalysed reactions allow an increase in productivity with reduced manufacturing costs due to wages and overheads.

There are some disadvantages in the use of enzymes which cannot be ignored but which are currently being addressed and overcome. In particular, the high cost of enzyme isolation and purification still discourages their use, especially in areas which currently have an established alternative procedure. The generally unstable nature of enzymes, when removed from their natural environment, is also a major drawback to their more extensive use.

5

The Cell

THE SPECIALIZED CELLS

We have been dealing with cells in a rather general way, using a variety of cell types to illustrate points concerning cell physiology. It is proposed now in the last stages of this book to consider two or three specific types of cells. To begin this chapter we will consider the glandular cell and, since there are a wide variety of both exocrine and endocrine glands in the body, comments will be restricted to the type of secretion process which is demonstrated particularly well either in the salivary glands or the pancreas and particularly we will refer to the pancreas, although some details of the salivary glands will be included as well.

The mechanism of secretion in externally secreting glands has been a subject of study for a good many years and very detailed work was published on this subject by Nassonow in 1923; Bowen in a series of publications between 1924 and 1929 fully confirmed and extended Nassonow's work. The general trend of studies by Nassonow and Bowen was the relationship of the Golgi material in particular to the production of secretion droplets. They both demonstrated that the first droplets of secretion which are visible under the optical microscope appeared in the interstices of the Golgi network, that as the number and size of these droplets increased, the Golgi network became hypertrophied and that when the cell carried its full load of secretion the apparatus appeared to break up and small pieces of it appeared to become attached to some of the secretory granules.

Studies have been made on the pancreatic cells of external secretion (as distinct from the islets of Langerhans) from the point of view of fine structure. The pancreatic cell structure can thus be taken as a general model for the cells of external secreting glands.

Two groups have made a special study of the fine structure of the acinar cells of the pancreas. These two are Ekholm and Edlund who investigated the human pancreas and Palade and Siekewitz who studied the guinea pig pancreas. Generally speaking there was little difference between the two types of pancreas. The same cytoplasmic elements--the endoplasmic reticulum with associated ribonucleoprotein granules, the mitochondria and the zymogen granules were present in both.

In guinea pigs which had been starved, Siekewitz and Palade demonstrated that there was a well-formed endoplasmic reticulum (ergastoplasm) and that the cisternae (spaces between the pairs of membranes) were very small and had special orientation within the cell. There were very few granules in the cisternal spaces. The apical parts of the cell contained a number of zymogen granules and the lumina of the acini were very small and appeared to be occupied by an amorphous material which had a density similar to that of the zymogen granules. This finding agrees with the statement that secretion of zymogen does occur from the pancreas in small amounts even in long starvation. In guinea pigs that had been killed one hr. after feeding, there were many granules within the cisternae of the endoplasmic reticulum which were greatly distended. The number of zymogen granules found in the apex of the cell was generally less than in the starved animals. The lumina of the acini were distended and contained irregular and illdefined masses of electron dense material which was probably discharged zymogen.

Siekewitz and Palade subsequently made a very interesting study of the functional variations in the microsomal fractions in the pancreas of the guinea pig. The microsomal fraction is mentioned before represents the endoplasmic reticulum plus its attached ribonucleic acid particles. Their experiments started from the fact that they observed that, when the microsomal fraction from starved guinea pigs was analyzed for enzyme activity, it was found that the ribonuclease and TAPase (trypsin activatable protease) activities varied a great deal in different preparations and in some cases reached as high as 30 per cent of the activities found in the zymogen fraction.

The zymogen fraction is that part of the homogenate of the pancreatic cells which is formed from the granules in the cell that are described as the zymogen granules and which are primarily comprised of digestive enzymes and precursors of enzymes-hence the name "zymogen." Since the zymogen fraction is composed almost exclusively of zymogen granules and since these have a higher total RNAase and TAPase activity than any of the other cell fractions, the authors wondered whether this similarity indicated some functional connection and they initiated a series of experiments in order to test this out.

In other investigations they had found that in some cases dense granules had been found within the cisternae of the endoplasmic reticulum in electron microscope preparations of the exocrine cells of the pancreas of the guinea pig. These intracisternal granules could be compared, both from the point of view of fine texture which they demonstrated and their density, with the zymogen granules. However, there was a considerable difference in size and the zymogen granules appeared to be outside the endoplasmic reticulum whereas the smaller particles were actually in the cisternae between the double membranes. Further studies have demonstrated that these intracisternal granules are present in large numbers in this position about 1 and 2 hrs. after starved guinea pigs have been fed but that by 3 to 4 hrs. they have more or less disappeared.

The authors consider that, since it is the endoplasmic reticulum which forms the microsomes in homogenization, perhaps some of the high enzyme activity demonstrated in the microsomal fraction might have been due to the enclosure of some of these granules which were present within the microsomal cavity. They investigated this problem by taking various fractions from starved and fed animals, collecting the glands about 1 hr. after feeling, the time when they expected to obtain the highest concentration of granules within the cisternae of the endoplasmic reticulum. They were able to obtain subfractions an hour after feeding which consisted largely of cisternal granules together with some detritus derived from the microsomal fraction. They found that the TAPase and the RNAase activities present in this granular fraction were actually higher than those of the microsomes from which they were presumably derived and were sometimes as high as the values for the zymogen fraction as such. A similar increase in activity was also described for the enzyme amylase.

The granules present in the cisternae were already in a finished state and were probably destined therefore to become zymogen granules, but their precise position in the secretory circle is not clearly seen at the moment. It is known that in the late stages of the secretory cycle of the pancreas the large vacuoles in the Golgi zone become filled up with a material which has a high electron density and it is possible that these intiracisternal granules feed into the Golgi material and there condense into large granules which are passed to the apical part of the cell. Hirsch has said that in many cases it appears that the granules which are shed into the lumen of the pancreas appear to be enclosed in a structure resembling a membrane of cytoplasmic material. This might be explained in the following way.

Assuming that there is a continuous passage from the endoplasmic reticulum lumen through the cisternae in the Golgi apparatus and so to the exterior, it is possible that Siekewitz and Palade's intracisternal granules pass to the Golgi apparatus and there coalesce to form large granules which then pass away from the Golgi material towards the apical part of the cell along those parts of the endoplasmic reticulum which lead to the exterior of the cell. Since these granules are very big when the leave the Golgi apparatus, they no doubt cause great dilation of the cisternae and it is even possible that as they are pushed out of the cell they tear part of the double membranes from the ergastoplasm and carry this as an investment as they pass out into the exterior. Even if there is no continuous passage it is quite possible that the droplets might be excreted from the cell covered with a submicroscopic skin derived from the endoplasmic reticulum or the Golgi membranes or simply composed of protein secreted on to them by one of these organelles.

Siekewitz and Palade point out that in their preparations, although they found a very high proteolytic RNAase activity in the isolated intracisternal granules, quite considerable proteolytic activity and RNAase activity were also found in the less dense fragments of the microsomes which presumably are composed simply of endoplasmic reticulum and the small RNA particles attached to them. It appears very likely that this activity is localized in the RNA protein particles. It is possible that these individual particles sypthesize enzymes, become greatly enlarged and pass through the membrane of the endoplasmic reticulum into the cisternae.

This represents a scheme for the production of secretion granules by the cell and the joint role of the endoplasmic reticulum, the mitochondria and the Golgi apparatus in this process. According to this scheme, in the pancreas amino acids become associated with the RNA granules of the endoplasmic reticulum membrane (left of diagram) and are passed as preprotein (polypeptides) into the cisternae where they grow in size to become the intracisternal granules of Palade and Siekewitz this stage is headed "1 hr. after stimulation" in the diagram).

In the text of this book the previous observations of Hirsch, in which he has seen "secretion" droplets in the pancreatic acinar cell moving through the cytoplasm to the Golgi zone have been mentioned, these droplets are approximately the same size as the intracisternal granules described by Palade and Siekewitz. Hirsch now believes that his moving droplets represent the intracisternal granules passing along the endoplasmic reticulum. The large secretion droplets of gland cells appear in the Golgi apparatus and if the intracisternal granules represent presecretory material their transference from the endoplasmic reticulum to the Golgi material requires some explanation. There is no evidence that the cisternae of the endoplasmic reticulum are continuous with the Golgi material but this possibility must be considered.

Hirsch believes that the mitochondria provide the energy required to dissolve the cisternal granules, to pass them through the membranes of the endoplasmic reticulum and through the Golgi membranes inside which they become concentrated into droplets of secretion, growing larger and larger until they are converted into zymogen granules, surrounded by a membrane and pushed to apex of the cell ready for discharge. This is the morphological story to date as told by the electron and light microscopes.

The metabolic story has been studied, using salivary gland material by Junquiera and his colleagues with rats as their experimental animals. Normal cats and rats with ligated salivary ducts were used. The significance of ligating the excretory duct is that it leads to cessation of secretion, disappearance of granules of secretion, decrease in size of the cells and the gland as a whole and there is a corresponding decrease in activities of the enzymes. The gland does not undergo any degenerative process. Junquiera and his colleagues used these types of glands to study not only the histochemical and biochemical changes in the cells but they tried also to study the morphological changes as demonstrated by the light microscope in an attempt to equate the various activities and morphological elements in the cell with the processes of secretion.

These results are given in summary as follows. In the control gland it was found that there were many mitochondria, they were mainly in the form of short rods through some round specimens were present. As a result of ligation they were very greatly reduced in numbers. Millon's histochemical test for proteins demonstrated a strong reaction in the normally secreting gland and only a very weak reaction in the gland which had been ligated. Studies in basophilia (to demonstrate degree of accumulation of RNA) showed a very strong basophilia in the control gland, but in the ligated gland which was apparently not secreting at all the basophilia was very weak. Biochemical investigation of the ratio of ribonucleop- rotein over deoxyribonucleoprotein (RNAP/DNAP) showed this ratio to be 2.18 in cells of the control gland but only 0.84 in the ligated gland, so there was obviously a very profound decrease of RNA by comparison with DNA in the nonsecreting gland.

Biochemical studies of protease activity demonstrated that in the control gland it was 560.0g. phenol/100 mg. tissue and in the ligated gland it was 266.0 ?g. phenol/100 mg. tissue. In the mouse, the cathepsin activity was 16.9g. tyrosine/1g. DNA protein and in the ligated gland it was not reduced quite as much as some of the other enzymes but was down to 10.3 ?g. tyrosine/1g. DNAP. Alkaline phosphatase varied a little between rats and mice. In rats the figure was 2.8 mg. phenol/100 mg. tissue and in the ligated gland it was 1.7 mg. phenol/100 mg. tissue so this was a reduction of something like a third. In mice curiously enough, in the control gland there was 2.0 mg. phenol/100 mg. tissue of alkaline phosphatase and in the ligated gland it actually went up to 2.4 mg. phenol/100 mg. tissue. So it seems that if the alkaline phosphatase is playing a significant part in the secretory processes of rats, it is a specific process and it does not play the same part in mice.

The acid phosphatase results were interesting because in rats the control gland showed 1.8 mg. phenol/100 mg. tissue activity and in the ligated glands there was no change at all. In mice, however, in the control gland there was 2.0 mg. phenol/100 mg. tissue activity and in the ligated gland 0.42 mg. phenol/100 mg. tissue, so that, perhaps, if phosphatase is playing a part in the secretory cycle, then in the rat it is alkaline phosphatase that is involved and acid phosphatase in the mouse.

Succinic dehydrogenase was studied by Thunberg's methylene blue method and there was found to be high activity (50 per cent reduction in 12 min. of methylene blue) in the control gland and in the ligated gland there was a 12 per cent reduction in colour, which was decreased in mice to less than a quarter. There was not quite as much reduction in rats; using the Warburg apparatus there was a 63 per cent increase of the QO2 after the addition of succinate and in the ligated glands only 28 per cent increase. So there was a significant reduction in oxidative activities which was confirmed by the cytochemical indophenol oxidase test for cytochrome oxidase, the control gland cells giving a dark blue and ligated glands only a light blue colour with this reagent. The oxygen consumption in the rat cells gave a QO2 of 4.2 in the control gland and only 1.8 in the ligated gland. The figure for glycolysis and the glycolytic quotient in relation to nitrogen was 3.2 in the control gland and 3.3 in the ligated gland, so there was significant different in the glycolytic rate. Pyruvate utilization similarly dropped by about two-thirds and thus it appears from these figures that not only is there a drop in respiratory activity of the cells when they are not secreting but there is also a drop, as one might expect, in oxidative phosphorylation.

It is of interest, however, that glycolysis is maintained at a constant rate in both ligated and non-ligated cells and, as Junquiera and his colleagues point out, it appears that the energy derived from glycolysis itself is used by the cell largely for its basic needs and not for its specialized function as a secreting cell. It appears that the ATP, ADP and phosph-ocreatine are really the compounds that are the sources of energy for the secretory process in the cell since there is such a spectacular decrease in these following ligation. It is of interest that the activity of cathepsin in the secretion of the salivary glands seems to be controlled in vivo by the male sex hormones. There is a parallel between cathepsin activity and protein synthesis and the drop in activity of cathepsin in the cells of ligated glands may be significant from this point of view.

Junquiera and Hirsch have also described the light-microscope changes that take place in cells of ligated glands. The Golgi apparatus is undoubtedly related to the processes of secretion if the early work which we have mentioned before can be taken at its face value. Junquiera and Hirsch summarized the evidence for the participation of Golgi material in the process of secretion as follows. The Golgi apparatus changes in size and structure very considerably during the secretory cycle. Droplets appear in the substance of the Golgi material and gradually appear to be transformed into secretory granules. This fact has been observed not only in fixed and stained preparations but also in the living pancreas, for instance, by Hirsch in 1932.

In 1939 he suggested that the Golgi bodies are the sites in the cell where cytoplasmic products congregate and become formed into zymogen granules. There is no biochemical data, as has been pointed out both by the present author and by Hirsch, that the Golgi bodies are actually the site of protein synthesis since they contain no RNA and do not appear to contain oxidative enzymes or other enzymes concerned with energy production, although in some cells such as the absorptive cells of the gut they do contain phosphatase, ATPase (heat stable), alkaline phosphatase, acid phosphatase and so on. In 1954, Sjöstrand and Hanson demonstrated by ultrastructural studies that there was an intimate relationship between the Golgi vacuoles, the zymogen granules, the ground substance of the Golgi apparatus and the alpha cytome-mbranes (endoplasmic reticulum): "there are all transition stages observed between, on the one hand, bodies with a pronounced elongated form and with the most direct topographic relations to the Golgi membranes and Golgi granules embedded in the ground substance on the other.

The impression when observing these pictures is that they show snapshots of the conversion of membranes into granules and vice versa. The small granules seem to coalesce to bigger granules which gradually gain the size, form and opacity of the zymogen granules."

This work appears to confirm the conception of the Golgi material as a condensing or aggregating region of the cell and reminds one of the statement of Kirkman and Severinghaus in 1938. "A great deal of work strongly suggests that the Golgi apparatus neither synthesizes secretory substances nor is transformed directly into them but acts as a condensation membrane for the concentration, into droplets or granules, of products elaborated elsewhere and diffused into the cytoplasm. These elaborated products may be lipoids, yolk, bile constituents, enzymes, hormones or almost any other form of substance." We do not know, of course, to what extent these synthetic processes are controlled by the mitochondria. There are papers in the literature suggesting that the mitochondria actually produce the zymogen granules in the pancreas and it seems very likely that in parts of the endoplasmic reticulum the mitochondria come in very intimate contact with the nucleoprotein granules on the outside of the membrane and that they supply the energy for the synthetic processes taking place in these particles. At any rate the particles of enzyme protein that are produced pass into the spaces and presumably undergo some benefit from contact with the Golgi material and possibly even from contact with the nucleus.

One could conceive of them as all passing into the cisternal space which surrounds the nucleus, but to what extent the metabolic activity of the nucleus might affect any enzyme protein passing through this space one cannot say for certain. Perhaps the passage of the granules to the Golgi region serves not only as a mechanism whereby the granules are aggregated to a large zymogen granule but possibly something is done to prepare the enzymes for activity upon secretion. Also one should consider the possibility that vitamin C, despite the known defects of localization in the technique, may occur, as the present author has claimed, in the Golgi apparatus at times of great synthetic activity and there is a possibility that in passing through into an area which is saturated with a reducing substance, oxidation of the contents of these granules prior to their excretion may be prevented.

So we can see a division of labour very clearly in the process of secretion in the gland cell but the interesting thing, as we have been emphasizing right through this book, is that we are not dealing with an isolated division of labour but an interdigitating division so that each job fits in very well with the job done by another part of the cell.

THE STRIATED MUSCLE FIBRE

The muscle fibre is composed of four fundamental constituents:

- The sarcolemma which is the equivalent of the cell membrane;
- The fibrils which represent the structural elements responsible for contraction;
- The sarcosomes which are really the mitochondria and which contribute the supply of energy for muscular contraction;
- The sarcoplasm which is the ground cytoplasmic substance in which the other structures of the muscle fibre are embedded.

In addition, there is a specialized region of the fibre called the motor end-plate at the point where the motor nerve fibre makes junction with the sarcolemmal membrane and which is responsible for initiating the contraction of the fibre. For the moment we will discuss very briefly the structure of each of these elements and we will include the description of the structure of the caveolae introcellulares with an account of the structure of the sarcolemma.

Muscle fibres themselves have three main shapes. They may be cylindrical with ends which are conical, they may be spindle-shaped with extremities tapering off very finely, or they may be conical with one end long and attenuated and a broad base at the other. This depends, of course, on their position in the muscle itself. The first of these types of fibres usually runs the whole length of the muscle, the second type is usually situated within the main part or belly of the muscle, the third type is usually attached to a tendon at one end and the other end terminating somewhere in the interior of the muscle itself. The shape of a transverse section of muscle is oval or spherical when it is fresh but it shrinks considerably following fixation and one usually finds rather angular cross sections of the fibres. Voluntary muscle fibres vary from about 10-100? in diameter. Presumably this difference in size is associated in some way with the amount of work which the muscle has to perform. There are in any single muscle, fibres of many different diameters. Fibres vary very much in length and may in some cases extend only a few millimeters and in others, fibres in excess of 34 cm. in length have been seen.

Schwann in 1839 and Bowman in 1840 described a thin membrane which surrounded the voluntary muscle fibre. It was described by Schwann as the cell membrane of the fibre and this was later agreed to by Bowman. It can be demonstrated by placing a muscle fibre in fresh water or by causing a sudden coagulation of the contents which then contract. The membrane or sarcolemma, as demonstrated in this way, seems to be a pale, colourless and apparently, under the optical microscope, structureless membrane and it is semipermeable since if placed in water it swells by imbibition and if subsequently placed in concentrated sugar solution this water is removed. It was thought to be about 0.1 across and to have a slight infolding at each Z band.

Others have shown that the sarcolemma is more complicated, i.e., under the electron microscope there are two dark osmiophilic lines separated by a light osmiophilic line which together are approximately 300 A across, each of the lines in this triple structure being about 100 A thick; also there is a thick layer of dense material about 500 A thick which extends inward from the inner of the dark lines of the membrane.

Under the light microscope the early workers had described the sarcolemmal membrane as being about 0.1? thick which is in fact 1000 A. We will see that, if we add the 500-A thick complex to the three lines which themselves total about 300 A, we get a figure of about 800 A which is pretty close to the 0.1? size of the sarcolemma described by the optical microscopists. As Robertson points out we are faced, of course, with the decision as to which of these many structures can be regarded as the true membrane. What we are in fact dealing with here is really a membrane complex rather than a single membrane and this, strictly speaking, applies to the membranes of all other cells as well. Every now and again the sarcolemma is penetrated by holes which are the ends of the caveolae intracellulares.

Some of these little openings or caveolae expand into vesicles and in some cases both the vesicles and the caveolae themselves may extend in a complicatd fashion, branching and ramifying over and in between the myofibrils. It has been suggested by Bennett and also by Palade that such caveolae are not a standardized feature of the structure of the sarcolemma of the muscle fibre, in general, but that they represent a mechanism whereby ions or macromolecules such as carbohydrates or proteins can obtain access into the interior of the fibres without violating the osmotic properties of the membrane.

According to the nature of the binding sites there, the membrane would then infold, form a caveola and this would seal off to form a vesicle and the vesicle would then pass into the interior and might then be destroyed so that the substances become free in the cytoplasm. This is a pinocytotic type of interpretation of these caveolae and the possibility exists, of course, that this type of activity may occur in many types of cells. Compounds could be passed out of the muscle fibre in a similar way, that is, by becoming attached to the internal surface of the plasma membrane which would then surround it and open up to form a caveola and the contents would become liberated to the exterior. Thus large macromolecules could get through the membrane without actually piercing it. It is of interest that in connection with this theory Whatman and Mostyn noted droplets of fat which had been stained with fat dyes in the process of passing through the sarcolemma of cardiac muscle fibres. Quite a number of cells and perhaps all cells appear to have attached on the outside some sort of polysaccharide coating to the plasma membrane. Such a coating occurs on the surface of mammalian erythrocytes where it is extremely thin and it has been pointed out that this may represent the surface coating which causes the ABO agglutination reaction.

Plant cells are, of course, expert in putting polysaccharides around the outside of their cell membranes and bacteria consistently do this too. Outside the sarcolemma there also appears to be a cloud of material, approximately 500 A thick which is possibly polysaccharide material. It is of interest that there is a positive periodic acid-Schiff (PAS) reaction in the sarcolemma and this may be the cause of the reaction. This significance of the polysaccharide complex may be its ability to bind various molecules which are to be taken into the interior of the muscle membrane.

Fig. Cholinesterase in Human Neuromuscular Junctions

In the region of the motor end-plates there is considerable modification of the sarcolemma and Robertson in 1956 and Couteaux in a series of papers have published detailed accounts of this. The sarcolemma in the region of the motor end-plate is very much thicker, denser and more folded particularly into a series of troughs which penetrate deeply into the sarcoplasm of the muscle fibre. The external polysaccharide-like material just described is present everywhere and separates the actual endings of the neural components of the motor end-plate from the sarcolemmal membrane itself. The sarcoplasm of the muscle fibre was originally described by Spidel in 1939 as being in a gelled condition, this can change to a sol if there is any damage to the muscle.

The sarcoplasm not only surrounds the nucleus and is obvious in that part of the fibre but it also penetrates and surrounds and fills in the spaces between the myofibrils themselves and between the myofibrils and sarcolemma. The relative amounts of sarcoplasm in the fibres varies very considerably and it is of interest that Golarz and Bourne have demonstrated great activity for the enzyme acetyl phosphatase in the sarcoplasm of human muscle. The sarcoplasm also contains most of the soluble proteins of the muscle fibre. Under the electron microscope it is less dense than the other components of the fibre. There are also a number of small granules scattered about the sarcoplasm which show a great range in size, something between ten and a couple of hundred angstroms across. They are distributed without any particular pattern and in some regions of the sarcoplasm there do not appear to be any at all. They appear to be more abundant in certain areas of the sarcoplasm in heart muscle fibres and, according to Fawcett and Selby they may, in fact, be composed of glycogen. Such areas give a strong PAS reaction and, if the section is treated with amylase prior to application of the PAS reaction, the reaction disappears. This is strong evidence that these granules are, in fact, glycogen.

Beckett and Bourne have demonstrated that PAS positive material was present which was dispersed irregularly in human muscle fibres, sometimes as irregularly distributed aggregates and sometimes as fine granules. To some extent the difference in size of these granules depends on the changes which take place following death. A biopsy specimen, for instance, shows only small granules and aggregations but by 12 hrs. the aggregates have reached their optimum size and after 48 hrs. autolytic procedures have removed them altogether. Beckett and Bourne have demonstrated that the amount of stainable material with the PAS technique in normal muscles is extremely variable and that there is no correlation between the amount present and the anatomical site of the muscle concerned. Also in muscular and neuromuscular disorders the amount of stainable material is variable and inconstant and shows no particular relationship with the neuromuscular diseases which were studied.

It is of interest that Dempsey and his colleagues and Beckett and Bourne were unable to remove all this PAS positive material with either diastase or saliva. This suggests that we are dealing either with glycogen which is linked to and protected by protein in some way or with a mucopolysaccharide type of substance or possibly even aldehyde groups are being produced by the PAS reaction by lipids or fat. Beckett and Bourne also applied the McManus PAS technique to formalin fixed, frozen sections in conjunction with the Sudan black test for fats. These were made both on rat and human muscle. By this frozen section method it was found that the muscle fibre was stained in a regular diffuse positive PAS background staining and there were also droplets of the material giving a strikingly intense positive PAS reaction. These droplets are often considerably increased in pathological muscles and appear to accumulate at points of mechanical damage in muscle fibres. These strongly positive droplets do not appear to be glycogen. Paraffin embedding of the muscle, however, removes these droplets. It is possible because of these facts that they may be a lipid of some sort. They are not, however, sudanophilic (thus are probably not fat) and do not give a Schultz reaction for cholesterol.

Many muscles fibres contain droplets of fat in the sarcoplasm. These vary very considerably in individual muscle fibres, in individual muscles and in individual muscles as a whole. Some muscles have more fat than other muscles. This was first described as long ago as 1841 by Henle and subsequently investigated by Kölliker and others. Fat was recorded in human muscle as long ago is 1889. Beckett and Bourne using the Sudan black technique in human muscle found that fat could occur in droplets of very variable size, distributed at random, but that a few were situated at the poles of the nuclei. Also there was some staining in the regions of the cross striations with Sudan black.

Some muscle fibres have a red colour and some have a white colour. Red and white muscles are well known amongst mammals and in animals such as the rabbit and guinea pig and also in the turkey both red and white muscles occur separately, but generally speaking, particularly in humans, muscles are a mixture of both red and white fibres. The red colour of the fibre depends upon the fact that it contains myoglobin or muscle hemoglobin. The red fibres contract more slowly and remain contracted for a longer period than the white muscles. There are more of them in those muscles which are concerned with posture. Generally speaking too, the red muscle fibres contain more sarcoplasm than the white and they also appear to contain more fat which presumably functions as a reserve supply of energy for the muscle fibre.

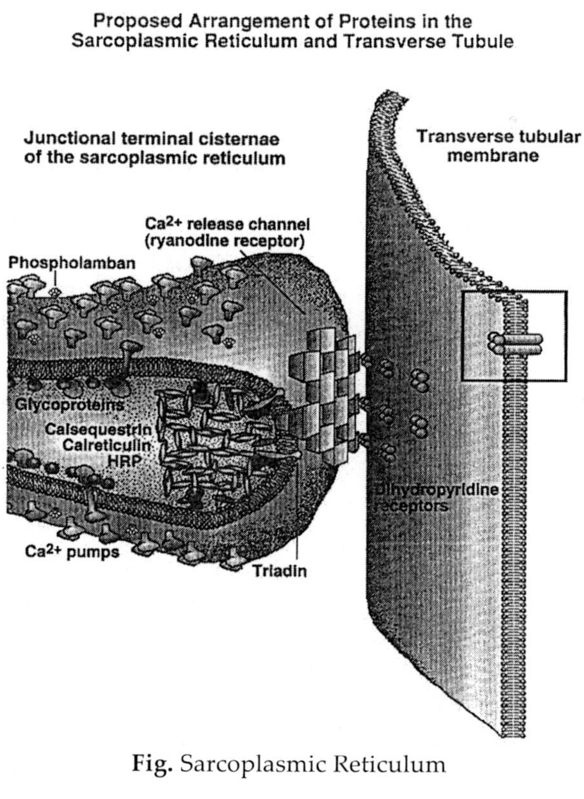

Fig. Sarcoplasmic Reticulum

Ribonucleic acid granules have not been identified in the sarcoplasm of adult muscle. An interesting new development in the structure of muscle fibres is the delineation of the sarcoplasmic reticulum described by Bennett in his chapter on "Structure and Function of Muscle." As Bennett points out, for more than 100 years papers have reported that there appears to be some sort of a network or series of tubules associated with muscle fibres. Retzius in 1881, had described a sarcoplasmic network and Kölliker had referred to a network in 1888, however, the most detailed treatment was given by Veratty in 1902. He described a "reticular apparatus" in muscle, this is a confusing term since the Golgi apparatus had often been referred to in these terms. Bennett and Porter in 1953, found this component in electron-microscope pictures of muscle and described it as a sarcoplasmic reticulum.

Sjöstrand and his colleagues subsequently referred to them as sarcotubules. It is of interest that the work on this reticulum which was described and understood very well by these early workers appeared to be completely ignored in later books of histology where one finds scarcely any reference at all to what now appears to be a very important structure in the muscle fibres.

The work of Bennett and Porter and a number of other investigators has demonstrated that this reticulum really resembles in its fine structure the endoplasmic reticulum found in other cells. Its membranes have the triple structure comparable with what Robertson described as the "unit membrane." These tubules appear to extend inward from the sarcolemma. The work by A. F. Huxley and his colleagues has indicated that there is probably a structure in the muscle fibre which is capable of transmitting depolarization of an ion aligned membrane into the interior of the muscle. These authors, in fact, found that you could get a localized contraction of a muscle fibre by stimulation in certain spots and, if the sarcoplasmic reticulum is, in fact, the component in the fibre which has this function, then presumably these are the spots on the sarcolemma where the sarcoplasmic reticulum or at least one unit of it begins. Porter and Palade made a very detailed study of the sarcoplasmic reticulum with the electron microscope.

The reticulum appears to be arranged in the form of alternating sets of tubes which anastomose with each other and surround the fibre like a bracelet. The anastomosing of the tubules gives the general impression of a lacelike type of structure. One set of these tubules has a plane of symmetry at the Z band and extends on teristic both of it to about the junction between the A and I bands, alternating between these is a series of tubules which is in the plane of the M bands. It should be emphasized that these structures which are being described are characteristic both of skeletal muscle and cardiac muscle.

Characteristic Golgi apparatus is also found in the sarcoplasm of muscle close to the nucleus as one might expect. Although we have referred to a striated muscle fibre as a cell, probably it is really a number of cells since any one striated muscle fibre may contain a number of nuclei and in a very long fibre, which measures several centimeters in length, some hundreds of nuclei may be present. They lie at the surface of the fibres (except in cardiac muscle where they are in the centre) just under the sarcolemma surrounded by sarcoplasm. The nuclei are generally oval in shape, with their long axes parallel with the long axis of the fibre, they are approximately 8-10? long and may be extend as much as 17?. They are hard to see in unstained fresh muscle fibres but come out very well in fixed and stained preparations. In muscle diseases and in degenerating muscle, they appear to migrate into the centre of the fibre and one can see long strings of muscle nuclei aligned end-to-end along the centre of the fibre. It is of interest that in embryonic muscle fibres the nuclei occupy a middle rather than a hypolemmal (close to the sarcolemma) position.

The fine structure of muscle nuclei does not differ from those of other cells. The nuclei are surrounded by two unit membranes analogous to the membranes of the endoplasmic reticulum of other cells. In fact, Bennett thought the outer membrane of these pairs may be continuous with the cytomembranes which are found in the perinuclear cytoplasm or even continuous with those of the sarcoplasmic structure is fundamentally the same in muscle fibres is in other cells. The membrane of the muscle nucleus also contains a number of pores.

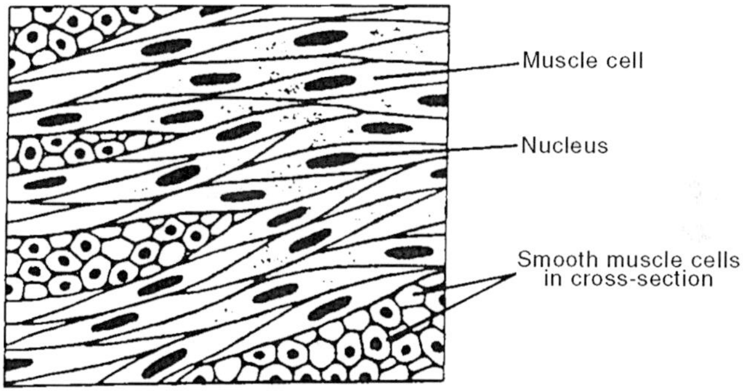

Fig. Cross Striations in Skeletal Muscle Fibres

The fibrils of the muscle fibres are the contractile elements. They appear to extend for the entire length of the fibre and they are approximately 1-3? in diameter. Muscle fibres under the light microscope show a well-determined cross striation and this, of course, is visible in the fibrils and it is of interest that the individual fibrils are aligned so that the appropriate segments of their bands coincide with each other. Between the fibrils are the sarcosomes or muscle mitochondria, to which we will return in a little while.

It is interesting to note that the fibrils consist simply of orientated protein molecules and do not have a surface membrane, therefore the substances in solution in the cytoplasm can have a direct influence on the protein of the fibrils. They have unrestricted access to each other.

The transverse striations of the fibrils of the muscle fibre are due to a difference in density. These cross striations form a pattern and each pattern repeats along the whole of the length of the fibre and each repeating part of the pattern is known as a "sarcomere." In vertebrate muscle each sarcomere is about 2-3 in length. The fibril itself is constructed of filaments which are made of protein and which lie parallel to the long axis of the fibre and overlap each other. There are two kinds of these filaments, one being about twice the diameter of the other and they are also different in length. These two types alternate along the length of the fibril and they overlap at parts and interdigitate with each other. The regions where the thick filaments are located are known as the "A" bands, these are extremely dense and are anisotropic under the polarizing microscope. Where the thin filaments are present there is less density and there is less birefringence, these are known as the I or isotropic bands. There is a band which appears across the middle of these which is called a Z membrane or band.

At the centre of the A or anisotropic band, thick filaments are present alone but at both ends of the A band there is interdigitation of the thin and the thick filaments and this is the densest part of the fibril. The part of the A band where the thick filaments occur alone is called the H band and is naturally not as dense as the rest of the A band. The Z membrane is due to the presence of a band of amorphous material which occupies the spaces between the filaments at the mid-point of the area where the thin filaments are located (the I band). In the case of the A bands, thickening of the thick filaments themselves about the middle of the band causes them apparently to be cut across by what is described as the M line or as the M strip by some authors.

The filaments which compose the fibrils of the muscle fibre are made up of protein and there appear to be three proteins concerned: myosin, actin and tropomyosin. There is more myosin than any other protein; it constitutes approximately 54 per cent of the total protein of the fibrils. There is only about 11 per cent of tropomysin but about 20-25 per cent of actin. This leaves a deficiency about 10 per cent in total protein and it is possible, according to Huxley and Hanson in their chapter in the "Structure and Function of Muscle," that the figures for the various proteins are a little too low particularly for tropomyosin and probably for actin too.

It appears that myosin is the protein of the thick filaments which constitute the A bands of the fibres. Actin, on the other hand, appears to be the protein which makes up the thin filaments of the I bands. Tropomyosin also appears to be present where the actin is situated. In vitro actin and myosin form a complex called actomyosin and this can be persuaded to contract under the influence of ATP just as the muscle fibre contracts under appropriate stimulation in vivo. Each filament of myosin appears to contain 425 molecules. If myosin is treated with trypsin, it splits into two types of meromyosins--one which is called light and the other heavy meromyosin. These meromyosins apparently occur as distinct units in the myosin molecule. Huxley and Hanson suggest that the backbone of a filament of myosin is made up of light meromyosin units which are longitudinally arranged and staggered and the heavy meromyosin units, which are attached to the light meromyosins and project from the filaments, these projections can be seen under the electron microscope.

The number of molecules in each filament of actin can also be calculated and they amount to approximately 600. There is some evidence that the actin filaments contain tropomyosin and, according to Huxley and Hanson, there are about 1.7 molecules of actin to 1 of tropomyosin. The way these two proteins are linked in the filaments is not known.

Huxley and Hanson believe that the actin molecules are exposed at the surface of the filament so that they may form actomyosin links with the neighbouring myosin filaments where they interdigitate. The difficult thing to explain is the fact that a muscle can contract very strongly without significant change in the length of its filaments. There are a number of experiments which have been carried out to elucidate this point. However, it appears that what probably happens is that the shortening of the muscle fibre takes place by a sliding of the actin filaments in between the myosin filaments of the A band; they already interdigitate at their ends and the actin filaments simply slide further in. In the non-contracting muscle the little bumps which have been mentioned before (projections on the heavy meromyosin filaments) come in contact with the actin filaments at the regions of interdigitation and these lock to form an actomyosin complex. Now this happens when no ATP is present, but when ATP is present these locks break apart and the actin filaments can slide up into the myosin filaments.

When ATP is dephosphorylated then the locks will form again and it is possible that in contraction the actin filaments slip into the myosin filaments point-by-point, there being a locking and releasing corresponding to the production and dephosphorylation of ATP.

Then when all the ATP formed is dephosphorylated the bands lock again. As soon as ATP is re-formed then the locks break and the actin filaments can slide out from the A bands again and the muscle becomes extended. Thus ATP is necessary both for contraction and relaxation of muscle. It is a very interesting fact that the enzyme ATPase which is responsible for the dephosphorylation of ATP is actually the myosin molecule itself.This is a very interesting example of a structural molecule which is also performing a metabolic function as an enzyme.

One should mention here just briefly one or two of the important chemical reactions which take place in muscle and which play a part in the contraction cycle. We have seen that the sliding of the actin filaments into the myosin filaments is dependent on the periodic formation and dephosphorylation of ATP.

The re-formation of ATP is a result of the famous Lohnman reaction: creatine phosphate + ADP = creatine + ATP. The ATP, as has been mentioned before, is a high-energy phosphate and its dephosphorylation results in the liberation of energy which can be used for the mechanical movements necessary by the molecules in the contraction cycle. We do not propose to give full details here of the chemical reactions that take place in muscles in contraction.

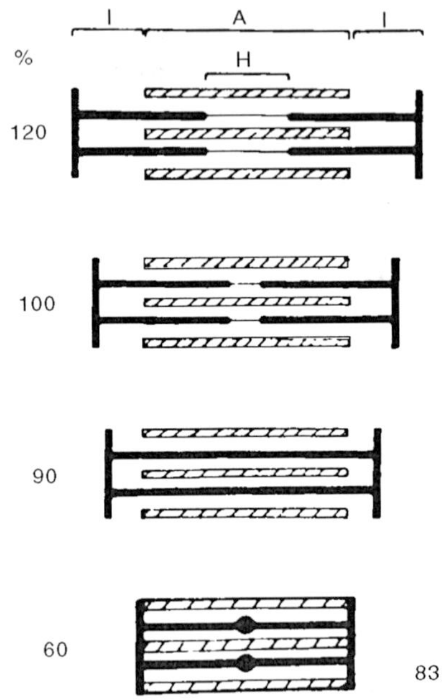

Fig. Molecular Basis of Contraction of Muscle

We might note that the rephosphorylation of ATP requires energy to reconstitute a high-energy bond and this energy can be provided by oxidative phosphorylation produced as a result of coupling respiration with such a process. There are elements in the muscle which are capable of doing this and these elements are the muscle mitochondria, the sarcosomes, which are strategically placed to perform this function. We will say a word now about their structure and position.

Electron-microscope studies of the sarcosomes demonstrate that they are identical in their fine structure with that of mitochondria. For example, they have a double membrane and also internal folds of the inner membrane to form double membrane cristae. They are located in the muscle fibres, at least in the case of heart muscle, exactly opposite the A bands. Slater in his chapter on sarcosomes in "Structure and Function of Muscle" points out that Holmgren had originally classified muscles as falling into two types which depended on the type of distribution of the granules (sarcosomes) and he believed that the two classes corresponded to the two physiological types of muscles. Muscles that had granules at the level of the I bands of the muscle fibrils were those which acted only intermittently, such as the skeletal muscles of most vertebrates and those which had granules at the level of the A bands were required for continuous activity and in this group would come the flight muscles of birds and insects and, of course, the heart muscles of vertebrates.

It is of interest that, for instance, in the flight muscles of the humming bird and of the hovering insects where there is a phenomenal rate of contraction cycles per second, the mitochondria are relatively enormous in comparison with any other structural elements in the fibre, indicating the high rate of oxygen consumption that must be necessary when the fibres are contracting at such high speeds. Also interesting to note is that sarcosomes, like the sarcoplasmic reticulum, were discovered a good many years ago, in the 19th century in fact. Regaud in 1909 had claimed that they were, indeed, identical with mitochondria because they showed similar staining reactions. Yet, in histology books published since then and up to recent times, sarcosomes have been largely ignored in accounts of the histological structure of muscle. Thus, there are two curious facts about the history of muscle structure--the sarcoplasmic reticulum and the sarcosomes were both discovered more than half a century ago and then dropped out completely from the histological picture and from the textbooks.

Not only do the sarcosomes resemble the mitochondria of other cells in staining reactions and in their fine structure, but they also contain the same series of respiratory enzymes, i.e., those enzymes concerned with the Krebs cycle and cytochrome system found in mitochondria. It is of interest that in red muscle fibres, where, as had been mentioned the red colour is due to myoglobin, there are more mitochondria than in white fibres and that the myoglobin assists in the rapid transfer of oxygen from the blood to the respiratory enzymes found in the mitochondria.The chemical reactions which take place in muscle can be stated as follows:

$$C_6H_{12}O_6 + 2\ DPN+ + 2\ ADP + 2\ P\ (inorganic)$$
$$\rightarrow 2\ CH_3 \cdot CO \cdot COOH + 2\ DPNH + 2\ H + 2\ ATP\ (I)$$

$$2\ CH_3 \cdot CO \cdot COOH + 5\ O_2 + 30\ ADP + 30\ P\ (inorganic)$$
$$\rightarrow 6\ CO_2 + 4\ H_2O + 30\ ATP\ (II)$$

$$2\ DPNH + 2\ H+ + O_2 + 6\ ADP + 6\ P\ (inorganic)$$
$$\rightarrow 2\ DPN+ + 2\ H_2O + 6\ ATP\ (III)$$

Reaction I appears to be restricted to the sarcoplasm, in non-muscle cells it is in the cytoplasm; reactions II and III take place in the sarcosomes as they do in the mitochondria. If we sum up those three reactions, we see that the sum of the activities of these processes in the muscle is the synthesis of 38 molecules of ATP for each time the cycle occurs.

$$C_6H_{12}O_6 + 6\ O_2 + 38\ ADP + 38\ P\ (inorganic)$$
$$\rightarrow 6CO_2 + 6\ H_2O + 38\ ATP$$

The function of the sarcosomes in the muscle fibres appears to be the provision of ATP for the functioning of the contractile elements. The myofibrils have no membrane, the sarcosomes are virtually in contact with the myofibrils and the ATP can diffuse almost instantaneously into the sarcofibrillar filament. If an inadequate supply of oxygen is available or in muscles with an insufficient amount of sarcosomes, the DPNH and the pyruvic acid which are formed as a result of reaction I produce lactic acid. This reaction is

$$CH_3 \cdot CO \cdot COOH + DPNH + H^+$$
$$\rightarrow CH_3CHOH \cdot COOH + DPN^+$$

It is also of interest that sarcosomes can bring about oxidation of fatty acids and in the oxidation of stearic acid as much as 147 molecules of ATP are formed with each reaction. Also sarcosomes, similarly to mitochondria and other cell organelles, may become unstable if ATP is not present and, according to Slater, this suggests that some energy provided by the ATP is necessary to stabilize the structure of these organelles. It is believed that ATP is also used to some extent to maintain a difference in the concentration of ions as between the sarcosomes and the rest of the cytoplasm. Here then, in the muscle fibre we have probably a division of labour as well defined as one could find in any cell in the body.

There is a special part of the cell, the contractile part, which is exclusively used for this purpose, not only that but we have the contractile protein acting as an enzyme which can trigger off the whole process. The ATP which is necessary for the process is provided by the oxidative phosphorylations coupled with the respiration of the mitochondria which are themselves in close physical contact with the fibrils. The changes in ionic polarization of the membranes which are initiated by the motor endplate and which pass along the sarcolemma can be carried into the fibrils of the muscle fibre by means of the sarcoplasmic endoreticulum. Furthermore, the sarcolemma appears to be adapted to undergo pinocytosis and can take macromolecules directly into the sarcoplasm. This is the perfect example of division of labour in cells and yet it should be pointed out again how much these individual labours depend on and are integrated with each other.

THE NERVE FIBRE

In conclusion, we will consider the nerve cell and the nerve fibre. The nerve fibre itself is not a cell but it is a prolongation of a cell. It is a process of a neuron which is adapted for carrying nervous impulses. The functioning of the nerve fibre is a little more mysterious at the moment than is, for instance, the muscle fibre since we have little in the way of moving parts to investigate and a great deal of the information which is available is theoretical. Most of the studies on the physiology of the nerve fibre are devoted to the movements of ions, particularly sodium and potassium ions, in and out of the nerve fibre membrane since it is believed that it is the polarization and depolarization of the membrane produced by these ions which constitutes the passage of a nerve impulse along the fibre. The nerve fibre originates from the neuron, from a conical area known as the cone of origin. The nerve cell itself contains a welldefined nucleus and nucleolus. There are numerous Nissl bodies.

Among the characteristic structures classically described in the nerve cell are:

- Neurofibrils,
- Nissl bodies also called tigroid bodies because of their striped appearance which suggested a resemblance to the coat of a tiger to earlier investigators,
- Golgi apparatus,
- Mitochondria,
- Fats, phospholipids,
- Pigments.

Neurofibrils have been claimed to be present in all nerve cells and in the axon, but there is some doubt as to whether they exist as structural elements in the living cells although up to the beginning of the century they had been regarded as the basis of nerve impulse propagation. There is evidence from microdissection that in the living nerve cells of invertebrates they exist as discrete threadlike structures which are interwoven with each other and appear to be more viscous than the surrounding neuroplasm. In fixed and stained preparations, the neurofibrils are extremely fine and it is not surprising therefore they are hard to see in the living cell.

Fibrous structures have, however, been seen in the cytoplasm of nerve cells in tissue cultures and, if fresh nervous tissue is treated with salts such as $CaCl_2$ or various fixatives, the neurofibrils rapidly come into view. One view is that the neurofibril is a rodlet sol, that it is composed of orientated molecular particles which are not, in fact, sufficiently strongly attached to each other to form a definite fibril, but that a variety of agents may cause these rodlets to precipitate as fibrils.

Electron-microscope preparations also show Nissl bodies in nerve cells. They may be stained by basic aniline dyes and in specially fixed nervous material, histochemical tests have shown that they contain phosphoric acid and iron and they also appear to contain ribonucleic acid. In certain acute pathological conditions the Nissl bodies may disappear. This is a process known as chromatolysis and, if there is injury to the axon, the appropriate nerve cells show central chromatolysis, that is, the dissolution of the Nissl bodies surrounding the nucleus. There is also a loss of Nissl material from the central neurones of birds in long transmigratory flights. Under ultraviolet light Nissl bodies have the appearance of a flocculant precipitate and it has been shown that they can be displaced in the cell prior to fixation by ultracentrifugation.

Classic Golgi preparations show that the Golgi apparatus exists as a network close to and wholly or partly surrounding the nucleus, in fact, it was in nerve cells, as we have stated earlier, that Golgi first discovered the apparatus which is called after him.

Mitochondria are fairly obvious in properly stained nerve cells and in cell bodies they tend to be spherical but in the dendrites they are more elongated while in the axon they may appear as long filaments. Near the nucleus there may be an accumulation of fatty-like droplets which may be Golgi material and in a similar position two types of pigment may also be found, e.g., in the substantia migra of the midbrain a black-to-red melanin pigment occurs in this site in the cell. Many neurons show the presence of yellow droplets of lipochrome pigment.

The cell membrane of the nerve cells is an extremely delicate structure much finer than the nuclear membrane. Electron microscopy demonstrates that it cannot be more than a few hundredths of a micron thick, whereas the nuclear membrane is about 1/10 ? thick, probably due to its nature as a double fold of the endoplasmic reticulum. There is some evidence from the electron microscope that the cytoplasm of nerve cells is of high water content and that the material of the axon is still more dilute. The axon itself arises from an elevated portion

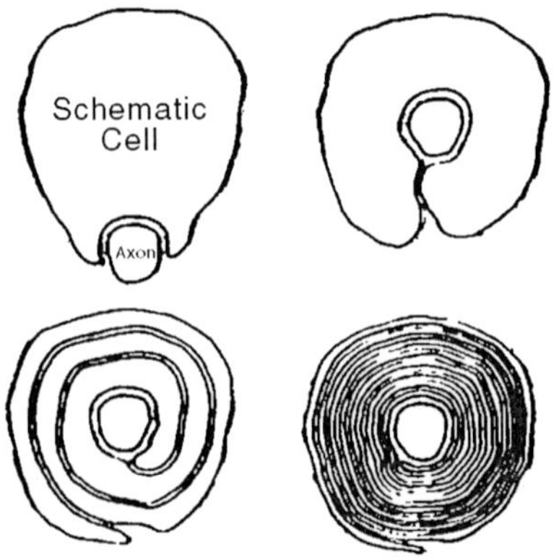

Fig. Schematic Representation of the Membrane Theory of the Origin of Nerve Myelin

of the neuron calles the axon hillock or cone of origin and it may give off branches or collateral but it usually ends in arborizations known as telodendria. A short way from the cell body, the axon develops a number of sheaths and becomes a nerve fibre.

If it has a well-defined myelin sheath, it is known as a myelinated nerve and if not, as an unmyelinated nerve fibre. Studies with polarized light however and with the electron microscope lead us to believe that all nerve fibres possess some type of lipoprotein sheath similar to the myelin sheath. Studies of the myelin sheath using polarized light, x-ray diffraction and later electron microscopy (particularly in the last case the work of FernandezMoran should be mentioned) have shown that lipoid and protein are laid down essentially in a series of alternating concentric layers. These lipoprotein layers appear to be about 130 A thick and the protein sections to be about 30 A thick. The lipid and the protein molecules of which these layers are composed are oriented radially, that is at right angles to the surface of the fibre.

When hematoxylin and eosin preparations are made of nerves fixed by conventional fixatives in the usual way and embedded in wax, a reticulate structure can be seen in the myelin sheaths which is known as the neurokeratin network. We shall have more to say about this later. Every now and again the myelin sheath becomes divided at the node of Ranvier, the neurilemmal sheath is continuous at this point but the myelin sheath itself is interrupted. The reason for this is not known but it is possible that the myelin may be a type of liquid crystal. This would probably be unstable in a tube longer than that represented by the internodes between the two adjacent nodes of Ranvier. It has been mentioned that the axon is also in a semifluid condition and, under the polarizing microscope, it can be seen to be birefringent suggesting a regular molecular orientation which is not incompatible with fluidity since it represents another form of liquid crystal similar to that of the myelin sheath but its chemical composition is different from that of the latter.

It has been found that the concentraton of ions is different inside and outside the cell and the nerve fibre; sodium and chloride, for instance, have a higher concentration extracellularly and potassium intracellularly. An electrical potential exists across the cell membrane because of this difference. Both sodium and potassium diffuse across cell membranes so that presumably they would eventually become equally distributed on either side of the membrane and the membrane potential would then disappear. But this does not happen and a factor which prevents it is a metabolic "pump" which maintains the difference in ion distribution. The level of the membrane potential depends upon the ratio of (K^+) and (CI-) ions and the rate at which the sodium ion (Na^+) is pumped.

The presence of a membrane potential along a nerve membrane is essential for the passage of an impulse. In the resting condition the membrane is known as the K^+ membrane. During the passage of an impulse the potential across the nerve membrane changes, this is known as the action potential and is due to a change in the permeability of the membrane and a sudden influx of sodium ions. The nerve fibre membrane is then called a (Na^+) membrane. The recovery of the membrane and the re-establishment of a (K^+) membrane starts with an increase in permeability to (K^+) ions and a decrease in permeability to (Na^+) ions. These changes finally bring the potential back to the resting value. The studies on the conductance of (Na^+) and (K^+) ions in nerve were carried out by Hodgkin and Katz with the introduction of the membrane voltage clamping technique. In effect what they demonstrated was that the movement of (Na+) into the fibre during the passage of an impulse depolarizes the membrane by carrying electrical charges through it. The re-establishment of a resting potential across the fibres (the K membrane) requires the sodium ions to be pumped out against a gradient of (Na+). This can only be done by the use of energy. That metabolic processes are involved in this action is indicated by the relatively high oxygen uptake of nerve.

By using optical methods, it has also been demonstrated that changes in the electrical activity of the nerve fibre which are associated with the passage of a stimulus are associated with changes in the optical density or in the light scattering capacity of the fibre. The scattering of light by the fibre indicates change of volume. It is of interest that this swelling is maintained for several seconds after the summated action of several series of impulses. The rate of change in optical density is not comparable to the velocity of propagation of the action potential, but there appears to be a connection between the wave of excitation and the optical changes. However, the nature of this connection is obscure at the moment.

The fluid substance of the axon is believed to be under pressure. As evidence for this one might quote the experiments of Paul Weiss in Chicago, who found that if a nerve fibre is ligated, the axoplasm is dammed up to the point where the constriction occurs and when pressure is released it flows forward again. He also showed that, if a nerve fibre is cut, the axonal material will stream out from the cut surface. It has also been shown that, if such a severed nerve is stimulated, the speed of outflow of axoplasm is decreased. This appears to be due to gelatinization of the axoplasm and a consequent increase in its viscosity.

In our laboratory at Emory University Professor Portela and his colleagues have been making studies of the correlation of physicochemical properties of the axoplasm in a state of excitation. Studies are being made with an interference microscope of surface movements of the nerve in response to "stimulation" and these movements are being recorded simultaneously with change in action potential. The surface movements of the nerve are indicated by the reflection of interference fringes in the interference microscope of the light from the nerve surface.

It is interesting to note that the Russian authors Kayushin and Lyudkovskaya believed that the surface movements (as expressed by shifts in the interference bands) result from changes in the fibre volume and since these shifts disappeared when there was no stimulation, it appeared that these volume changes had returned to normal and the movement was probably of an elastic nature. These mechanical waves (surface movements, volume changes) are presumably connected with changes of the molecular micellar structure in the nerve. Thus, it appears that the whole of the structural system of the nerve takes part in the conduction of an impulse. This way of looking at the relationship between molecular structure and the excitationconduction-recovery cycle in the nerve is a new approach to the investigation of nerve physiology.

It is possible that the mechanical waves in the nerve fibre push acetylcholine down towards the myoneural junction and there is some evidence this compound may be formed either in the axon or possibly in the cell body itself. There is evidence of considerable synthesis of protein by nerve cells but the significance of this is not known.

In our laboratory attempts are being made to correlate the lipid and protein structure and organization of the nerve and function of the nerve as a conductor. The approach is similar to that of Tobias. Proteolytic and lipolytic enzymes are used to produce structural changes in the nerve fibre and afterwards records of interference and electrical measurements and fine structural changes are made. The removal of phospholipids from the nerve membrane results in the nerve becoming inexcitable but, following treatment with proteolytic enzymes, nerve conduction is not blocked. It is of interest that pressure (about 11½ lbs.) on the nerve will cause an interruption of the normal function of a motor but not of a sensory nerve and this may last some two weeks before coming completely back to normal. This is particularly curious when such pressure produces no microscopically visible degenerative changes in the nerve fibre.

The cell membrane in nerve fibres is complicated because of the close association of the axon with the Schwann cell. The Schwann cell is associated with the axon in both myelinated and non-myelinated fibres. In the case of the non-myelinated fibres, the axon is embedded to varying degrees into the protoplasm of the Schwann cell, but is still surrounded closely by the cell membrane of the Schwann cell which it has invaginated with it. The actual distance between the outer parts of the two membranes under these circumstances is approximately 150 A. The membrane of the axon and the membrane of the Schwann cell are each approximately 75 A and are made up of three parts. There are two rather dense areas which are about 25 A across and these are separated from each other by a less dense area also about 25 A across. So that we may consider that the membrane of the nerve fibre is enlarged to this extent by the membrane of the Schwann cell itself.

In myelinated fibres the axon instead of just being embedded in the cytoplasm of the Schwann cell is surrounded by many layers of Schwann cell cytoplasm and cell membrane. The mechanism which produces this has been demonstrated by Geren in tissue cultures of developing nerve fibres. She demonstrated that the Schwann cell winds in a spiral fashion around the axon thus building up many layers of Schwann cell membrane around the axon itself. Many degrees of myelination of nerve fibres are known and the degree of myelination depends on the number of times the Schwann cell winds around the axon. Electron-microscope studies of myelinated nerve fibres have demonstrated very complex arrangements of the myelin so produced. The first of the high-resolution studies of such fibres were produced by Fernandez-Moran and Sjöstrand. If a myelinated nerve is first fixed in osmium tetroxide (osmic acid), it could be seen that the myelin was made up of a series of lines which absorbed the osmium very strongly.

These lines were approximately 25 A thick and repeated at a period of up to 120 A. Each of these dark lines was separated by a light line of approximately the same size and running across the centre of each light line was a line which was more dense than the light line but less dense than the main 25-A repeating lines. In interpreting the significance of these lines, Robertson has pointed out that the gap, which in the unmyelinated fibres was between the two cell membranes and measured 75 A across, is largely eliminated in myelinated fibres and the outer part of the membrane of the Schwann cell comes into direct contact with the membrane of the axon.

Thus we can say that, as Robertson points out, each of the repeating units which are obvious in electron micrographs of the myelin sheath is two Schwann cell membranes in contact along their outside surfaces. The Schwann cell membrane probably consists of a single bimolecular leaflet of lipids of which the hydrophilic ends are associated with monolayers of a non-lipid material and which is probably protein, although there is a possibility that either some polysaccharide or glycoprotein of some sort may be associated with it, possibly on the side directed towards the cytoplasm.

It has recently been shown by Tewari and present author that, in addition to the complex structure demonstrated by the electron microscope, the neurokeratin network previously considered to be a fixation artifact of myelin may indicate the presence of some differentiation of the myelin sheath in the living nerve fibre and the presence of various enzymes including a large variety of phosphatases as well as ATPase and creatine phosphatase (acid phosphatase is restricted to the axon). There is also a good deal of phospholipid as demonstrated by the Baker phospholipid technique. This network has been found, in fact, from these preparations to consist of areas resistant to damage by fixatives and other reagents and apparently arranged in the form of hexagonal prisms which radiate from the axon to the surface of the fibre (actually to the neurilemma) and that it is along these areas that the enzymes and the phospholipid we have mentioned appear to be concentrated. The areas appear to contain at least unsaturated fat and lipid and it is of interest that, in nerve which has been osmicated for 5 days and extracted with turpentine, it is the material from the interior of these active areas that is extracted, whereas the similarly osmicated material on the faces of the active areas remains.

It is possible that the reason why it is retained in the faces of the prisms is due to the fact that the osmicated lipids are bound strongly to a protein and cannot be extracted by turpentine. There is also some evidence that it is held in place by polysaccharide. It is of interest, too, that in these same areas techniques which would normally demonstrate mitochondria and which are presumably really demonstrating a variety of lipoprotein show a localization of similar compounds in the neurokeratin network. It is difficult to reconcile this work with that of electron microscopy. Although if we assume that the protein of the myelin lamellae is enzyme protein in the regions of the neurokeratin network but not in between them, then we can reconcile the two suggested types of structure. We also have to ask what happens to the mitochondria of the Schwann cell when they wrap around the axon.

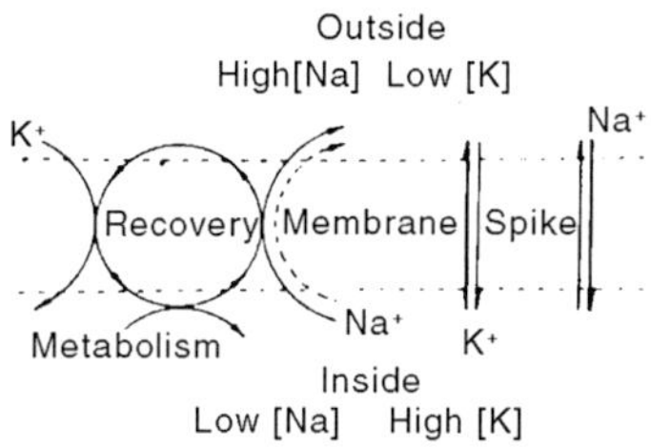

Fig. Diagrammatic Illustration of Ion Movements Through the Nerve

Do they disintegrate or do they become enormously elongated so that their membranes constitute some of the lamellae described by the electron microscopists. We have mentioned that there is a high concentration of potassium ions inside the nerve fibre in the axon and much less concentration of sodium. Now it appears that sodium continuously diffuses into the nerve fibre and it must be excreted from the axon to the exterior again. The difficulty about this is that the sodium has to be excreted against a gradient, that is to say, sodium has to be passed out of the axon into a medium where there are more sodium ions than there were in the axon and vice versa with potassium. This process requires energy and there have been two principal views as to how this could be done. The two theories are the redox pump and the sodium pump theory.

Conway has suggested the following way in which the outward movement of sodium ions in the nerve fibre could be explained. Sodium ions could be continuously passed on from unsaturated phospholipid molecules to the next unsaturated phospholipid molecules through the membrane to the outside and there would be alternate oxidation and reduction of the phospholipid molecules as this happened.

The other view is that sodium is pumped out by energy produced by presence of ATP, creatine phosphate and oxidative enzymes and possibly also with the aid of phospholipid. What is not certain yet is whether Na and K ions need to be passed through the myelin sheath to maintain the physiological functioning of the nerve fibre. This is a problem for which we will no doubt get the answer in due course. It is difficult to draw many conclusions about the division of labour in the nerve fibre. First of all, in the nerve cell we have great production of protein and possibly the production of acetyl choline and the passage of one or both these compounds down the axon towards the myoneural junction; the passage of the nerve impulse is provided for by the axon membrane itself.

In myelinated nerves we may have the metabolic mechanism of the neurokeratin network which might be used for pumping ions away from the area of the axon membrane. However, the precise relationship of the myelin sheath to the propagation of the nerve impulse is not known. In myelinated nerves, the impulse appears to be faster than in non-myelinated or poorly myelinated nerves, anti one does not know quite in what way the myelin sheath is able to do this.

At the termination of the nerve fibre there is a modification in the form of a complex branching in association with the sarcolemma to build up the myoneural junction. There is an accumulation here of acetyl choline which presumably passes through the membrane of the neural part of the mechanism to stimulate the formation of a contraction wave by its influence on the sarcolemmal membrane. The presence of cholinesterase in these junctions, which is necessary for inactivation of acetyl choline after it has done its job, is of interest but its precise origin is unknown. So we can see evidence of division of labour in the nerve cell and its axon but the nature of the labour performed by each part is still partly speculative.

We have studied the structure and to some extent the chemical composition and the biochemical composition of the cell in general and of three specialized types of cells. We have demonstrated that various parts of the cell perform very important but different functions, that, in other words, there is a division of labour. Each part performs the job for which it is chemically and structurally suited; thus the nucleolus synthesizes a considerable amount of protein and ribonucleic acid. We know that the chromosomes of the nucleus contain the genes which produce hereditary effects in the cell. We have seen that the oxidative activities of the cell are carried on in the mitochondria and that, as a result of respiratory activity of these structures, ATP is produced which is available for energy and structural purposes in the cell. In the glandular cell, the ATP provides energy for the synthesis of the products of secretion.

In the muscle fibre, it provides the energy for muscular contraction; and presumably in the nerve fibre the myelin sheath provides the energy for the extrusion of Na ions to the neurilemma (although in this latter case this is an example of division of labour between cells. The cytoplasmic (endoplasmic) reticulum is still a very debatable structure; the presence of ribonucleoprotein granules attached to the outside of these membranes indicates that they are probably regions of protein synthesis. If, as is believed by some authors, the cistemae in the endoplasmic reticulum are continuous with the outside we have, in fact, an enormous increase of area of the membrane of the cell. We have the outside world actually penetrating deep into the cytoplasm and even surrounding the nucleus so that nuclear products could be secreted directly to the exterior of the cell without actually passing through the cytoplasm itself. In this way glucose could pass directly into the nucleus as well as directly into the interior of the cytoplasm.

What is so striking in the cytoplasm is not that there is a division of labour but, what is more fantastic, the degree to which all the various parts of the cell cooperate with each other to build up a metabolic picture which is ordered and controlled. This co-operation rarely gets out of gear and when it does disorders of growth and metabolism occur of which one variety can be cancer. The mechanism of cell function is under control of the endocrine system and we have mentioned earlier in this book how the fine structure of the prostate cell is under control of the male sex hormone and how the localization of acid phosphatase in the same cells is also under control of the same hormone. This is a field which is still scarcely touched and is likely to produce in the future an enormous amount of interesting and valuable information about the function of the cell. The relationship of vitamins and nutritive factors in general to cell structure is also only beginning to be investigated. Studies on starvation have demonstrated a great increase in the number of mitochondria and our own studies on scurvy have demonstrated the more intimate association of the endoplasmic reticulum with the mitochondria with a simultaneous increase in the number of mitochondria. Here is a vast field in nutrition where electron microscopy would probably produce immense contributions to a study not only of the physiology of the cell but of the function of vitamins and other nutrients in cell metabolism.

The precarious nature of the cell structure is demonstrated by the fact that the structural integrity of mitochondria is affected in the absence of ATP and it seems possible that this whole complicated system of energy production is essential not only to provide for cell activity but also energy to maintain the structural integrity of the cell and its organelles; if this is so, then any serious interference with the ATP production cycle will lead to a breakdown of the complex cellular structure as well as interference with the metabolism.

New techniques of cytological investigation have in the last few years made fundamental alterations in our outlook on cell structure and function and we can expect that the next few years will shed much more light on the intriguing subject of the division of labour in cells.

THE STRUCTURE OF THE CELL MEMBRANE

Let us trace briefly the origin of our present knowledge of the structure of cell membranes. Overton in 1895, was originally responsible for the idea that lipids were important in cell membrane structure. He made this suggestion because of the relative ease with which lipid soluble substances penetrated the cells. At that time there was no clear idea as to what types of lipid molecules were present or as to how they were arranged.

In 1917, Langmuir and other workers made deductions about the arrangement of molecules in molecular films spread on water surfaces and these have been described as being of fundamental importance to the modern concepts of the cell membrane structure. Langmuir suggested (as mentioned in the introduction) that lipids at an interface of air and water arrange themselves in a monolayer with the polar ends of the molecules directed towards the water interface and the non-polar ends at the air interface. In 1925, Gorter and Grendell found that the total amount of lipid they could extract from a red cell.

Ghost when spread in a monolayer occupied about twice the area of the membranes of the red cells. They suggested, therefore, that the lipid was arranged in the membrane as a bimolecular leaflet in which the hydrophilic polar groups were at the inner and outer surfaces and the hydrophobic carbon chains were directed towards each other in the interior of the membrane. This was a simple and attractive model. It is difficult to imagine however how such a simple structure could account for all the various functions of the cell membrane.

Dr. G. D. Robertson has described the next stages of the development of our knowledge of the cell membrane. Briefly these are as follows: Harvey and Shapiro in the 1930's and also Cole during the same period studied the surface tension of starfish eggs and came to the conclusion that, at the ranges of pH's which apparently occurred inside cells, most pure lipoid substances gave surface tension values of the order of 5 dynes/cm.^{2} or more. But intracellular oil drops were found to have a surface tension of only 0.2 dyne/cm.^{2} The surface tension of whole starfish eggs as estimated by the force required to compress them, is even lower than this.

Thus, if only a purely lipid/water interface is concerned, these low values for the living cell cannot be explained. Danielli and Harvey demonstrated that there was some substance on the cell membrane which was responsible for such low values. This substance was a surface active agent of a cytoplasmic nature and appeared to be protein. Apparently in natural membranes all lipid polar substances were covered with at least a monolayer of protein. Danielli and Davson eventually proposed that cell membranes in general had the same sort of structure and that this consisted of one or more bimolecular leaflets of lipid with each polar surface having on it a monolayer of protein.

In 1936, Schmitt, using polarization microscopy, concluded that at least in red cell membranes the lipid molecules which were present had their carbon chains radially oriented. Thus we can picture the cell membrane as being composed of two layers of lipid molecules radially arranged (at right angles to the surface of the membrane) and a monolayer of protein applied to both the inner and outer surfaces of the membrane, with the long axes of the molecules lying parallel to the surface.

The cell membrane contained either pure lipid or pure protein, with pores of dimensions about the size of a large molecule, or that it was in the form of a mosaic containing either pure lipid or pure protein in various areas. However, this seems unlikely now except in some special cases. It is of interest that a common structure for the cell membrane exists in cells as diverse as erythrocytes, axons of nerve cells, muscle fibres, leucocytes of the blood, yeast cells, algal cells, the cells of higher plants and the ova of echinoderms, e.g., sea urchins and starfish. It seems very likely, however, that there are variations to some extent between the membranes of the various cells; nevertheless it is highly probable that a general structural pattern does exist for all of them.

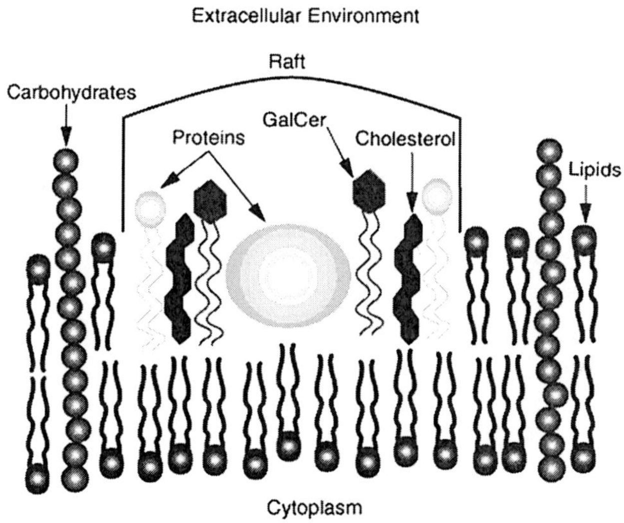

Fig. The Structure of the Cell Membrane

A typical example of variation was demonstrated by Mudd and Mudd who showed that erythrocytes are preferentially wetted by oil and leucocytes by water; this indicates that molecules at the surface of the membranes differ.

Studies with the electron-microscope have demonstrated that those membranes which have been studied have a thickness of the order of about 80 A (10,000 A = lā) so that they are about thick. Dr. Keith Porter has pointed out that the first electron-microscope photographs demonstrated the cell membrane as a solid structure, but that, as the technique improved and better resolution was obtained, the membrane was seen to be composed of two thin, very dense lines with a space between them. The lines measured about 25 A each and the space between them measured about 30 A across. It is tempting to suggest that the two outer lines represent the protein parts of the membrane and the central space the phospholipid portion, but this has yet to be proven. This electron-microscope pattern of structure and size is general in cell membranes—it is present, for example, in the nuclear membrane, the membranes of the endoplasmic reticulum and those of the mitochondria. In due course no doubt the electron-microscope will shed further light on the nature of the membranes and possibly the differences between the many types of membranes in various cells, but there are limitations to what it can do at the present moment.

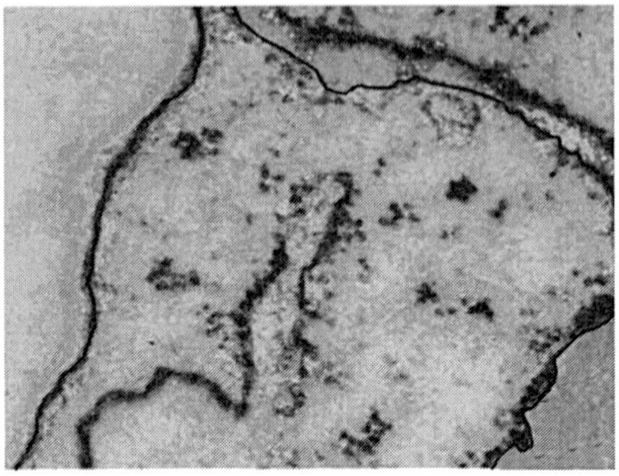

Fig. The Cell Membrane as Demonstrated in the Electron Microscope.

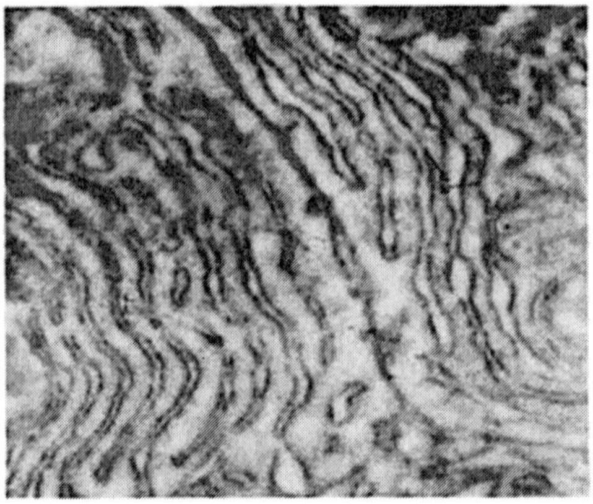

Fig. The Cell Membranes at the Junction of Two Acinar Pancreatic Cells

Some of the ideas of the nature of the cell membrane have been derived from a study of its fundamental properties. First of all it is known that lipid-soluble substances have a preferential permeability across cell membranes and this suggests that there must be in the membrane a continuous layer which is composed of phosphatides, steroids or fats or combinations of those substances. Also cell membranes have a high electrical resistance which, Danielli points out, is further evidence that there must be a continuous layer of lipids. Another property is the existence of a low surface tension at the surface of the cell membranes.

We have mentioned earlier how Danielli has suggested that this property indicates that on the surfaces of the membrane protein layers are adsorbed. Cell membranes are disrupted by digitonin which has a special ability to form a complex with cholesterol and this suggests that this latter compound may play a part in the structure of the membrane.

Analysis of red cell ghosts has in fact demonstrated that a relatively large amount of cholesterol is present in the membrane. An interesting theory concerning the structure of the red cell membrane was proposed by Winkler and Bungenburg de Jong in 1941.

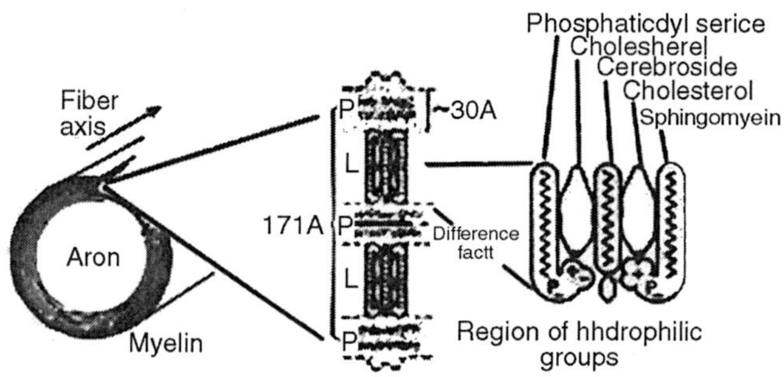

Fig. Molecular Structure of the Myelin Sheath

This theory gives the free cholesterol the role of stabilizer for the charged phospholipid molecules thus enabling them to give form to the membrane. This concept has been elaborated by Frey-Wyssling in 1953, whose studies on submicroscopic morphology and its relation to the chemical constituents of the cell are of fundamental importance. It is of interest to notice that in the two issues where cholesterol has been assigned the major role of structural component, it is present mainly if not entirely in the free unesterified form. There is one striking difference however in the nature of the cholesterol in the myelin sheath and the red cell.

In the myelin sheath it is apparently stable and once deposited under normal conditions remains static for life. In the red blood cell there is a steady turnover and exchange with plasma cholesterol. In general, in tissues where there is active metabolism of cholesterol, esterified cholesterol is found, the amount varying from tissue to tissue.

One should, perhaps, at this point say a word about the composition of the myelin sheath since this is in a sense an extended membrane. Finean, from his studies on the x-ray diffraction pattern of myelin, suggested that it was probably composed of alternate layers of lipid and non-lipid, the non-lipid probably being protein. The lipid molecules appear to be curled or tilted so that their full length is not extended and there is probably a stable complex formed between the cholesterol molecule and the larger phospholipid molecules. It is postulated that the free hydroxyl group of the cholesterol is important in binding the molecule by associating with the polar end of the lipid chain which it curls round. The hydrocarbon part of the cholesterol molecule is bound to the phospholipid by Van der Waal's forces. Cholesterol has a high dielectric constant and because of this it might act as an insulating agent.

THE PERMEABILITY OF CELL MEMBRANES

Danielli has discussed on several occasions the various problems associated with the permeability of natural membranes and the account which follows is taken largely from his writings.

PENETRATION OF COMPOUNDS

Danielli points out that there are three sites of resistance to free diffusion in cell membranes:

- The membrane/water interface for diffusion into the membrane;
- The membrane/water interface for diffusion from the membrane into the water;
- The interior of the membrane.

Every molecule which requires to pass into the interior of the cell has to get through these three sites of resistance. A molecule which has polar groups (-OH groups), as opposed, for instance, to non-polar groups such as methylene, forms at least one hydrogen bond with water for each polar group; all these hydrogen bonds must be fractured simultaneously if the molecule is to penetrate even into the lipoid membrane. Glycerol, having three OH groups, must acquire sufficient kinetic energy to break three hydrogen bonds simultaneously before it can penetrate into the membrane. This involves a large amount of energy, consequently resistance 1 (diffusion from water to membrane) is so high for glycerol that resistances 2 and 3 are reduced to insignificance.

On the other hand, a molecule such as methyl alcohol which has only one OH group and one CH_3 group penetrates easily into the membrane and can also pass easily out of the lipoid layer into water and presumably ethyl alcohol (C_2H_5OH) can function similarly. Thus we find that for such molecules the rate of diffusion across the interfaces is very large compared with rate of diffusion across the membrane. The interior of the membrane in the case of these compounds is the most important factor controlling penetration. Differing from these examples are molecules which are predominantly hydrocarbon in nature such as carotene ($C_{40}H_{56}$).

With a molecule such as this, resistance I is insignificant; the molecule has no polar groups and thus even resistance 3 is not enormous; however, resistance 2 is very large indeed, because the hydrocarbon groups are hydrophobic and a considerable amount of kinetic energy is required to transfer CH_2 groups from lipid into water. When many such groups are present they must all be transferred simultaneously from the lipoid layer into water, since otherwise the molecule remains substantially part of the lipoid layer and cannot diffuse away into the aqueous phase.With these examples in mind, Danielli classified penetrating molecules into four groups.

- Molecules with few polar and few non-polar groups —for these resistance 3 is most important and they can penetrate comparatively rapidly e.g., oxygen and methyl alcohol.
- Molecules having a predominantly polar character-for these resistance 1 is most important and penetration is slow, included are glycerols, sugar and glycogen.
- Molecules having few polar and many non-polar groups—resistance 2 is most important, penetration is slow e.g., carotene, vitamin A and fat.
- Molecules having many polar and many non-polar groups—resistances 1 anti 2 are both important and penetration is slow e.g., polyhydroxylic bile acids, the glucuronide of oestrin and proteins.

Danielli has pointed out that these principles of membrane permeability have a distinct physiological significance. For instance, oxygen is required in large amounts by the cell and carbon dioxide must be disposed of rapidly and both of these substances can penetrate the cell membrane rapidly. On the other hand, the first products of glucose utilization, for example, by muscle cells are glycerol derivatives, these are valuable and the cell membrane does not let them get through too easily and so they escape only slowly. During sustained work lactic acid is formed—this would be toxic if it accumulated. The cell membrane is relatively permeable to lactic acid which thus can escape into the blood and this permits violent exercise to be maintained for a much longer period than would be the case if the cell membrane was impermeable to lactic acid.

Amino acids penetrate all membranes moderately well but protein penetrates badly and, therefore, amino acids are stored as protein. Fatty acids will penetrate moderately well, but neutral fat very poorly, therefore fatty acids are stored as neutral fat. Glucose which can get in and out of cells fairly rapidly is polymerized in liver cells to form glycogen which passes the cell membrane only with great difficulty. Each of these three latter products, to which the membranes are impermeable, is the results of polymerization of simpler compounds.

It is of interest that, when compounds which are diffusing through a cell membrane from outside to inside are being incorporated into some compound within the cell, they appear to penetrate the membrane more easily. A typical example of this can be drawn from amino acids. If they are being actively incorporated into proteins, the rate of diffusion through the membrane will be much more rapid than if they were not being incorporated. Presumably, if they are being built into proteins, they cease to accumulate and so cease to build up a concentration gradient against the passage of further amino acids. Detoxification mechanisms also take advantage of the principles of the cell membrane permeability. For instance, toxic substances which might penetrate the cell membrane become conjugated with amino acids, sulfuric acid, or glucuronic acid. Thus toxic cell-penetrating substances such as bromobenzene or menthol are converted into new molecules which penetrate the cells with great difficulty and once in the bloodstream tend to be filtered off by the glomeruli of the kidneys and cannot be reabsorbed from the urine and returned to the bloodstream by the kidney tubules.

Penetration of Ions Through Cell Membranes

Penetration of ions through cell membranes is of special interest because for instance, Na^+ ions are more concentrated outside the cell (in the body fluids) and K+ions are more concentrated inside the cell, yet we know that K ions will pass into the cell and Na ions will pass out. Both these passages thus take place against a concentration gradient and this can only occur by the application of energy. It is claimed that oxidative enzymes and phosphatases play a part in the movement of ions in and out of membranes; a process which is known as "active transport."

Substances which interfere with the activity of these enzymes have been shown in many cases to interfere with the movement in or out of the cell of both sodium and potassium. There are certain specialized cell membranes, e.g., the cell membrane of the intestinal cells, in which the distal edge of the cell is folded to produce a larger number of extremely small fingerlike processes called microvilli; these microvilli are covered with a typical cell membrane. There are a large number of dephosphorylating enzymes localized around these microvilli. Their precise function in this position is not known, but it is almost certain, however, that they play a part in active transport by utilizing the energy derived from the hydrolysis of high-energy phosphates.

It is obvious from what we have said, therefore, that the cell membrane, in general, is a very selective structure anti exercises a good deal of control over the cell by deciding What goes in or out and by enabling the cell to be isolated from its environment. In this way it permits labour to go on in the cell which may be quite different from that going on in a neighbouring cell, or in the body fluids surrounding the cell. Here then is the first division of labour in the cell to be described by us.

However the story is not yet ended. In 1952-54, Danielli proposed the structure of the cell membrane which explained the way in which large protein molecules in the cell membrane could permit the passage of ions through the membrane. Danielli conceived of a long protein molecule extending right through the membrane and passing outside it. This end became attached to an ion and then by contraction pulled the latter through into the interior of the membrane. According to Lundgard and Hodgkin, a cytochrome oxidase energy-producing system is involved in the penetration of ions. They claim that their ion moving system requires adenosine triphosphatase (ATPase), creatine phosphatase and cytochrome oxidase. Conway has produced a redoxpump hypothesis that attempts to explain the penetration of ions through cell membranes by a mechanism which involves the successive oxidation and reduction of a compound in the membranes; possibly a phospholipid provides the energy for the movement of ions in this way. Phosphatidic acid is now thought to be concerned with ion movements through membranes.

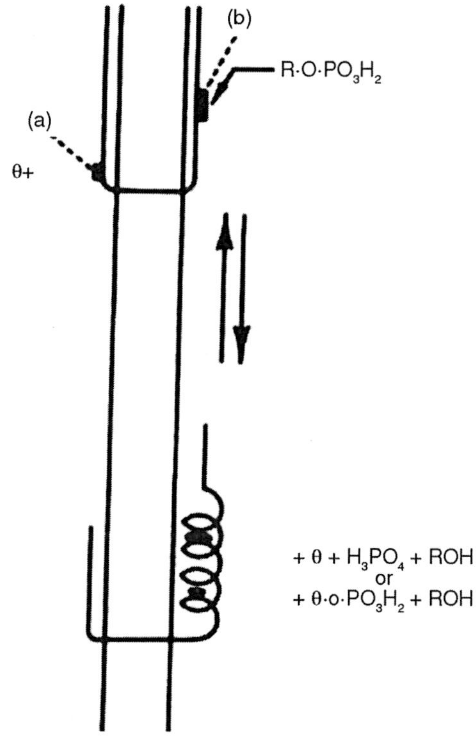

Fig. Diagram of Danielli's Theory of Secretion by a Contractile Protein

The possible relationship of phosphatases to cell membrane permeability has also been mentioned in connection with the microvilli (brush borders) of intestinal epithelium. The passage of glucose across the mebranes of gut cells could be explained by a mechanism involving these enzymes. It is of interest that in yeast cells too there is some evidence that phosphatases are localized on the membrane and that these enzymes are also involved in the permeability not only to glucose but also to phosphates. Phosphatases are also present on the brush borders of the kidney tubule cells where a good deal of absorption takes place. More recent work has indicated that ribonuclease may play a part in the penetration of cell membranes by ions. Lansing and Rosenthal and

Tenardo, for instance, showed that when ribonuclease was added from the outside to certain cells, it modified their permeability to ions; Brachet and Leduc found exactly the same thing for amphibian eggs. According to Brachet "amphibian eggs swell very much when they are immersed in a ribonuclease solution and since cell membranes often give strong cytochemical tests for ribonucleic acid it might well be that the integrity of this nucleic acid is of importance for normal permeability." Danielli made a suggestion that the cell membrane consisted, as suggested before, of bimolecular leaflets of lipoid with protein molecules stretched both inside and outside the membrane, but that it also contains a series of pores which he calls polar pores. In other words, groups of protein molecules are oriented radially with their polar groups directed towards the interior of the pore and such molecules are thus able to control the penetration of the compounds or substances through this pore according to their polar group affinities.

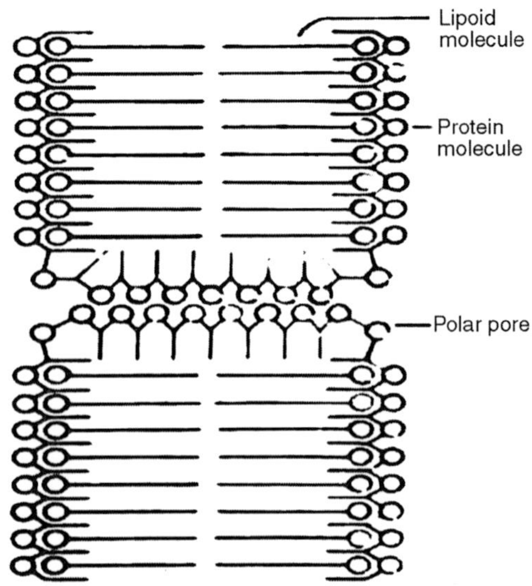

Fig. Diagram of Cell Membrane as Suggested

Danielli discusses the association with all membranes of a group of enzymes, which are called "permeases." These compounds he says "permit either facilitated diffusion or active transfer of 'signal' molecules across the membranes." It is possible that insulin may function as a permease for cells of certain organs. To quote Danielli again "Once permease becomes incorporated in the plasma membrane the situation is transformed: specific substrates can penetrate, more permease will become available by induced synthesis, induced enzyme will appear and a whole range of further induced enzymes may appear in response to the action of the induced enzymes of the inducing substrate. Thus a transient infection with permease may potentially change drastically the state of differentiation of a cell—possibly even result in carcinogenesis." It is very likely that there will be considerable further developments in this subject of permeases and their relation to cell membrane permeability and these will be awaited with interest.

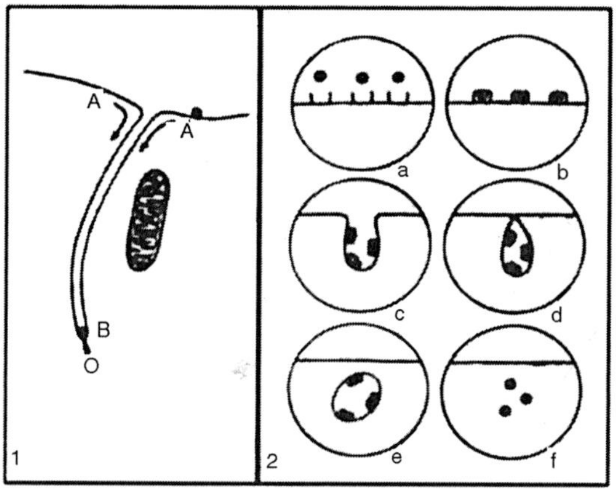

Fig. Diagrammatic Representation of Pinocytosis

A further development which should be noted when considering problems of penetration of compounds into cells is that of pinocytosis. This literally means "drinking by cells" and was first described by Warren Lewis many years ago in tissue culture cells. In this process, these cells were seen by a pseudopodial-like movement to engulf a droplet of culture fluid in which they were lying. The process was equivalent and similar to the phagocytosis of solid food by the *Amoeba*, only in the case of the tissue culture cell it is a droplet of fluid that is engulfed.

Now it seems that the cells of multicellular animals even in vivo can do the same sort of thing though there may be specialized parts of the cell membrane which carry out the process. A case in point is the kidney tubule cell. From the basal part of the cell, long internal extensions of the cell membrane pass a considerable distance up into the interior of the cytoplasm. They are present in some other cells too and are called "caveolae intracellulares." The cell appears to be able to imbibe droplets of fluid by an engulfing movement of the membranes of these caveolae. It is possible too that the cell membranes of a variety of cells may be able to carry out pinocytosis even without the presence of these intracellular membranous extensions. Thus we see that, if a cell really "wants" to take in a few macromolecules that pass its membranes with difficulty, it can always take them by "the scruff of the neck" and pull them pinocytotically into its interior.

DIFFERENTIAL CENTRIFUGATION

One of the developments in technique which has done an enormous amount to advance our studies of the cell is that of differential centrifugation and an account of this technique should be given before we pass on to the study of the cell components. The extraction of cell components from tissues dates back a very long time and studies on the chemical nature of the nuclei were carried out during the 19th century on nuclei obtained from pus cells by differential centrifugation. More recently, tissues have been homogenized and spun down at different speeds and the different portions of the cell separate out according to their density. For example, if a homogenate is taken up in 0.8-M sucrose solution which tends to preserve the morphological character of most cell structures and it is then spun at about 1000 r.p.m., the cell debris and the nuclei come down. The nuclei can be differentially centrifuged again from the cell debris so that a pure nuclear fraction suitable for chemical analysis can be obtained. At a higher speed, the mitochondria and at still higher speeds, the particles of glycogen, one part of the microsomes and other parts which are really fragmented endoplasmic reticulum aggregate at the bottom of the centrifuge tube. As a result of this technique, studies of the enzyme content of the nuclei and mitochondria have been made in great detail. This has caused some controversy because of the possibility that soluble enzymes and other substances might be leached from the nuclei during the homogenization procedures. Dilute acids, e.g., citric and acetic acids, have been used as homogenization fluids and possess certain advantages in that they make the tissue fragile and thus more easily broken. They also appear to harden the nuclear membrane and help to preserve the nuclei intact.

However, they are unsuitable for enzymatic studies and tissue homogenized in solutions of sucrose are usually used for this purpose. The latter method is claimed, because of the low ionic strength of the sucrose solution, to reduce greatly the loss of materials from the nuclei. It is said, in fact, that proteins and nucleic acids do not get through the membranes but low molecular weight substances do suffer some loss by this method.

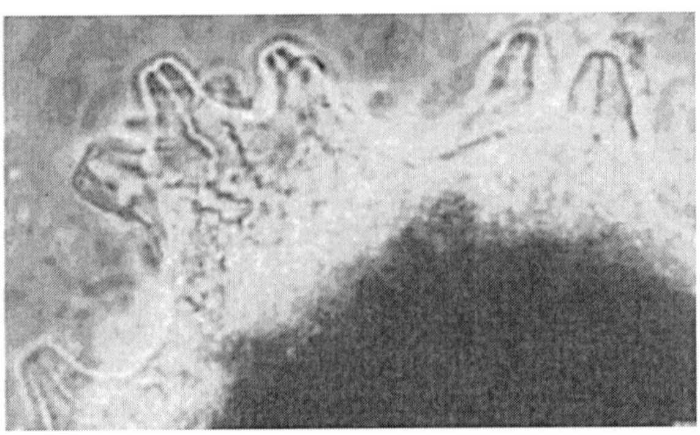

Fig. Phase Contrast Study of Pinocytosis in Amoeba

Allfrey and his colleagues at the Rockefeller Institute have tried to reduce any type of loss from the nucleus by freeze-drying the tissue and then grinding and floating it in non-aqueous media. Although in many respects this is advantageous, it appears difficult to be certain that small fragments of cytoplasm do not stick to the nuclei and contaminate the material. Extremely interesting results have been obtained by this particular technique.

Another check which was made on this differential centrifugation technique was to add various pure enzymes to the homogenate and then to calculate how much was adsorbed on the nuclei, mitochondria, microsomes and so on and how much was left in the supernatant. Although the amounts taken up with the supernatant varied with different tissues, generally speaking, the results were encouraging enough to indicate that the adsorption of enzymes from the normal supernatant was not really a significant cause of error in this technique.

Roodyn in a recently published review of the enzymic content of isolated nuclei has discussed the possibilities of contamination. He points out that we know that there is a very intimate connection between the endoplasmic reticulum and the nuclear membrane and that it is believed that the nuclear membrane is really a fold of the endoplasmic reticulum. Thus the nuclear fraction could easily be contaminated with this part of the cell.

Also there is a possibility that some of the sedimented material may contain whole cells which have escaped homogenization as well as nuclei and this would be another source of error. There is a possibility too that mitochondria may stick to the isolated nuclei and so also be contaminants. Although a certain amount of contamination is thus possible, it is only significant where a small percentage of total enzyme content of the cell is attributed to the nucleus, in this case it may be due to contamination, but where the nucleus contains an appreciable proportion of the total enzyme activity of the cell then the contamination becomes not very significant.

6

Enzymes

The use of enzymes in the diagnosis of disease is one of the important benefits derived from the intensive research in biochemistry since the 1940's. Enzymes have provided the basis for the field of clinical chemistry. It is, however, only within the recent past few decades that interest in diagnostic enzymology has multiplied. Many methods currently on record in the literature are not in wide use, and there are still large areas of medical research in which the diagnostic potential of enzyme reactions has not been explored at all. Enzymes are biological catalysts responsible for supporting almost all of the chemical reactions that maintain animal homeostasis. Because of their role in maintaining life processes, the assay and pharmacological regulation of enzymes have become key elements in clinical diagnosis and therapeutics.

The macromolecular components of almost all enzymes are composed of protein, except for a class of RNA modifying catalysts known as ribozymes. Ribozymes are molecules of ribonucleic acid that catalyze reactions on the phosphodiester bond of other RNAs. Enzymes are found in all tissues and fluids of the body. Intracellular enzymes catalyze the reactions of metabolic pathways. Plasma membrane enzymes regulate catalysis within cells in response to extracellular signals, and enzymes of the circulatory system are responsible for regulating the clotting of blood. Almost every significant life process is dependent on enzyme activity.

ENZYME CLASSIFICATIONS

Traditionally, enzymes were simply assigned names by the investigator who discovered the enzyme. As knowledge expanded, systems of enzyme classification became more comprehensive and complex. Currently enzymes are grouped into six functional classes by the International Union of Biochemists (I.U.B.).

Number	Classification	Biochemical Properties
1	Oxidoreductases	Act on many chemical groupings to add or remove hydrogen atoms.
2	Transferases	Transfer functional groups between donor and

		acceptor molecules. Kinases are specialized transferases that regulate metabolism by transferring phosphate from ATP to other molecules.
3	Hydrolases	Add water across a bond, hydrolyzing it.
4.	Lyases	Add water, ammonia or carbon dioxide across double bonds, or remove these elements to produce double bonds.
5	Isomerases	Carry out many kinds of isomerization: L to D isomerizations, mutase reactions (shifts of chemical groups) and others.
6	Ligases	Catalyze reactions in which two chemical groups are joined (or ligated) with the use of energy from ATP.

These rules give each enzyme a unique number. The I.U.B. system also specifies a textual name for each enzyme. The enzyme's name is comprised of the names of the substrate(s), the product(s) and the enzyme's functional class. Because many enzymes, such as alcohol dehydrogenase, are widely known in the scientific community by their common names, the change to I.U.B.-approved nomenclature has been slow. In everyday usage, most enzymes are still called by their common name.

Enzymes are also classified on the basis of their composition. Enzymes composed wholly of protein are known as simple enzymes in contrast to complex enzymes, which are composed of protein plus a relatively small organic molecule. Complex enzymes are also known as holoenzymes. In this terminology the protein component is known as the apoenzyme, while the non-protein component is known as the coenzyme or prosthetic group where prosthetic complex in which the small organic molecule is bound to the apoenzyme by covalent bonds; when the binding between the apoenzyme and non-protein components is non-covalent, the small organic molecule is called a coenzyme. Many prosthetic groups and coenzymes are water-soluble derivatives of vitamins. It should be noted that the main clinical symptoms of dietary vitamin insufficiency generally arise from the malfunction of enzymes, which lack sufficient cofactors derived from vitamins to maintain homeostasis.

The non-protein component of an enzyme may be as simple as a metal ion or as complex as a small non-protein organic molecule. Enzymes that require a metal in their composition are known as metalloenzymes if they bind and retain their metal atom(s) under all conditions with very high affinity. Those which have a lower affinity for metal ion, but still require the metal ion for activity, are known as metal-activated enzymes.

ENZYMES AND LIFE PROCESSES

The living cell is the site of tremendous biochemical activity called metabolism. This is the process of chemical and physical change which goes on continually in the living organism. Build-up of new tissue, replacement of old tissue, conversion of food to energy, disposal of waste materials, reproduction - all the activities that we characterize as "life." This building up and tearing down takes place in the face of an apparent paradox. The greatest majority of these biochemical reactions do not take place spontaneously. The phenomenon of catalysis makes possible biochemical reactions necessary for all life processes. Catalysis is defined as the acceleration of a chemical reaction by some substance which itself undergoes no permanent chemical change. The catalysts of biochemical reactions are enzymes and are responsible for bringing about almost all of the chemical reactions in living organisms. Without enzymes, these reactions take place at a rate far too slow for the pace of metabolism.

The oxidation of a fatty acid to carbon dioxide and water is not a gentle process in a test tube - extremes of pH, high temperatures and corrosive chemicals are required. Yet in the body, such a reaction takes place smoothly and rapidly within a narrow range of pH and temperature. In the laboratory, the average protein must be boiled for about 24 hours in a 20 per cent HCl solution to achieve a complete breakdown. In the body, the breakdown takes place in four hours or less under conditions of mild physiological temperature and pH.

CHEMICAL NATURE OF ENZYMES

All known enzymes are proteins. They are high molecular weight compounds made up principally of chains of amino acids linked together by peptide bonds.

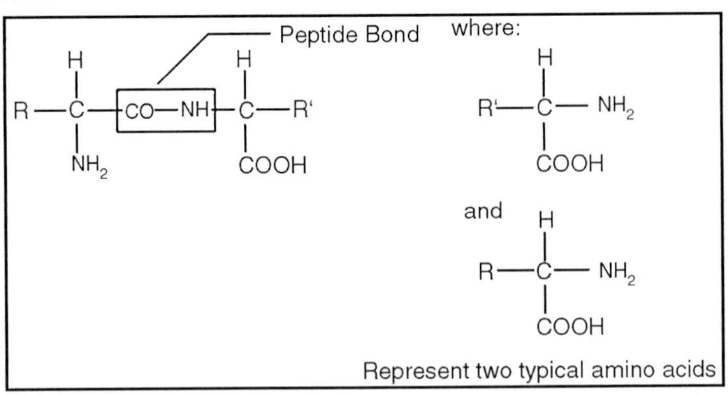

Fig. Typical Protein Structure - Two Amino Acids Joined by a Peptide Bond.

Enzymes can be denatured and precipitated with salts, solvents and other reagents. They have molecular weights ranging from 10,000 to 2,000,000. Many enzymes require the presence of other compounds - cofactors - before their catalytic activity can be exerted. This entire active complex is referred to as the holoenzyme; *i.e.*, apoenzyme (protein portion) plus the cofactor (coenzyme, prosthetic group or metal-ion-activator) is called the holoenzyme.

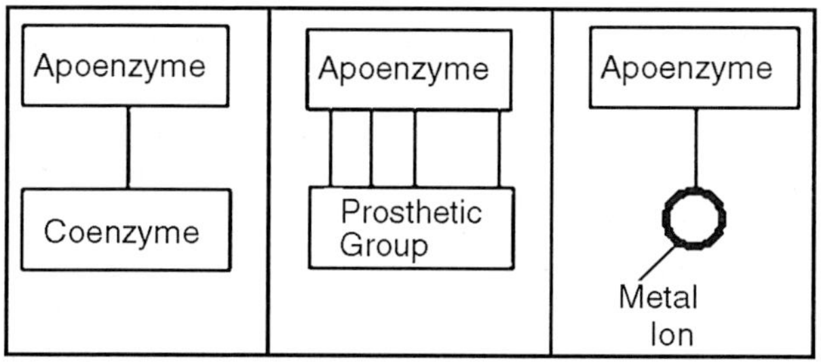

Fig. Holoenzymes- Apoenzymes Plus Various Types of Cofactors.

Apoenzyme + Cofactor = Holoenzyme

According to Holum, the cofactor may be:

- *A coenzyme*: A non-protein organic substance which is dialyzable, thermostable and loosely attached to the protein part.
- *A prosthetic group*: An organic substance which is dialyzable and thermostable which is firmly attached to the protein or apoenzyme portion.
- *A metal-ion-activator*: These include K^+, Fe^{++}, Fe^{+++}, Cu^{++}, Co^{++}, Zn^{++}, Mn^{++}, Mg^{++}, Ca^{++}, and Mo^{+++}.

Specificity of Enzymes

One of the properties of enzymes that makes them so important as diagnostic and research tools is the specificity they exhibit relative to the reactions they catalyze. A few enzymes exhibit absolute specificity; that is, they will catalyze only one particular reaction. Other enzymes will be specific for a particular type of chemical bond or functional group.

In general, there are four distinct types of specificity:

1. *Absolute specificity*: The enzyme will catalyze only one reaction.
2. *Group specificity*: The enzyme will act only on molecules that have specific functional groups, such as amino, phosphate and methyl groups.
3. *Linkage specificity*: The enzyme will act on a particular type of chemical bond regardless of the rest of the molecular structure.
4. *Stereochemical specificity*: The enzyme will act on a particular steric or optical isomer.

Though enzymes exhibit great degrees of specificity, cofactors may serve many apoenzymes. For example, nicotinamide adenine dinucleotide (NAD) is a coenzyme for a great number of dehydrogenase reactions in which it acts as a hydrogen acceptor. Among them are the alcohol dehydrogenase, malate dehydrogenase and lactate dehydrogenase reactions.

Naming and Classification

Except for some of the original enzymes such as pepsin, rennin, and trypsin, most enzyme names end in "ase". The International Union of Biochemistry

(I.U.B.) initiated standards of enzyme nomenclature which recommend that enzyme names indicate both the substrate acted upon and the type of reaction catalyzed. Under this system, the enzyme uricase is called urate: O_2 oxidoreductase, while the enzyme glutamic oxaloacetic transaminase (GOT) is called L-aspartate: 2-oxoglutarate aminotransferase.

Enzymes can be classified by the kind of chemical reaction catalyzed:

- Addition or removal of water
  - *Hydrolases*: These include esterases, carbohydrases, nucleases, deaminases, amidases, and proteases
  - Hydrases such as fumarase, enolase, aconitase and carbonic anhydrase
- Transfer of electrons
  - Oxidases
  - Dehydrogenases
- Transfer of a radical
  - *Transglycosidases*: Of monosaccharides
  - *Transphosphorylases and phosphomutases*: Of a phosphate group
  - *Transaminases*: Of amino group
  - *Transmethylases*: Of a methyl group
  - *Transacetylases*: Of an acetyl group
- Splitting or forming a C-C bond
  - Desmolases
- Changing geometry or structure of a molecule
  - Isomerases
- Joining two molecules through hydrolysis of pyrophosphate bond in ATP or other triphosphate
  - Ligases

ROLE OF COENZYMES

The functional role of coenzymes is to act as transporters of chemical groups from one reactant to another. The chemical groups carried can be as simple as the hydride ion ($H^+ + 2e^-$) carried by NAD or the mole of hydrogen carried by FAD; or they can be even more complex than the amine ($-NH_2$) carried by pyridoxal phosphate.

Since coenzymes are chemically changed as a consequence of enzyme action, it is often useful to consider coenzymes to be a special class of substrates, or second substrates, which are common to many different holoenzymes. In all cases, the coenzymes donate the carried chemical grouping to an acceptor molecule and are thus regenerated to their original form. This regeneration of coenzyme and holoenzyme fulfills the definition of an enzyme as a chemical catalyst, since (unlike the usual substrates, which are used up during the course of a reaction) coenzymes are generally regenerated.

ENZYME RELATIVE TO SUBSTRATE TYPE

Although enzymes are highly specific for the kind of reaction they catalyze, the same is not always true of substrates they attack. For example, while succinic dehydrogenase (SDH) always catalyzes an oxidation-reduction reaction and its substrate is invariably succinic acid, alcohol dehydrogenase (ADH) always catalyzes oxidation-reduction reactions but attacks a number of different alcohols, ranging from methanol to butanol. Generally, enzymes having broad substrate specificity are most active against one particular substrate. In the case of ADH, ethanol is the preferred substrate.

Enzymes also are generally specific for a particular steric configuration (optical isomer) of a substrate. Enzymes that attack D sugars will not attack the corresponding L isomer.

Enzymes that act on L amino acids will not employ the corresponding D optical isomer as a substrate. The enzymes known as racemases provide a striking exception to these generalities; in fact, the role of racemases is to convert D isomers to L isomers and vice versa. Thus racemases attack both D and L forms of their substrate.

As enzymes have a more or less broad range of substrate specificity, it follows that a given substrate may be acted on by a number of different enzymes, each of which uses the same substrate(s) and produces the same product(s). The individual members of a set of enzymes sharing such characteristics are known as isozymes. These are the products of genes that vary only slightly; often, various isozymes of a group are expressed in different tissues of the body.

The best studied set of isozymes is the lactate dehydrogenase (LDH) system. LDH is a tetrameric enzyme composed of all possible arrangements of two different protein subunits; the subunits are known as H (for heart) and M (for skeletal muscle). These subunits combine in various combinations leading to 5 distinct isozymes. The all H isozyme is characteristic of that from heart tissue, and the all M isozyme is typically found in skeletal muscle and liver. These isozymes all catalyze the same chemical reaction, but they exhibit differing degrees of efficiency. The detection of specific LDH isozymes in the blood is highly diagnostic of tissue damage such as occurs during cardiac infarct.

ENZYME-SUBSTRATE INTERACTIONS

The favoured model of enzyme substrate interaction is known as the induced fit model. This model proposes that the initial interaction between enzyme and substrate is relatively weak, but that these weak interactions rapidly induce conformational changes in the enzyme that strengthen binding and bring catalytic sites close to substrate bonds to be altered. After binding takes place, one or more mechanisms of catalysis generates transition- state complexes and reaction products.

The possible mechanisms of catalysis are four in number:

1. *Catalysis by Bond Strain*: In this form of catalysis, the induced structural rearrangements that take place with the binding of substrate and enzyme ultimately produce strained substrate bonds, which more easily attain the transition state. The new conformation often forces substrate atoms and bulky catalytic groups, such as aspartate and glutamate, into conformations that strain existing substrate bonds.
2. *Catalysis by Proximity and Orientation*: Enzyme-substrate interactions orient reactive groups and bring them into proximity with one another. In addition to inducing strain, groups such as aspartate are frequently chemically reactive as well, and their proximity and orientation towards the substrate thus favours their participation in catalysis.
3. *Catalysis Involving Proton Donors (Acids) and Acceptors (Bases)*: Other mechanisms also contribute significantly to the completion of catalytic events initiated by a strain mechanism, for example, the use of glutamate as a general acid catalyst (proton donor).
4. *Covalent Catalysis*: In catalysis that takes place by covalent mechanisms, the substrate is oriented to active sites on the enzymes in such a way that a covalent intermediate forms between the enzyme or coenzyme and the substrate. One of the best-known examples of this mechanism is that involving proteolysis by serine proteases, which include both digestive enzymes (trypsin, chymotrypsin, and elastase) and several enzymes of the blood clotting cascade. These proteases contain an active site serine whose R group hydroxyl forms a covalent bond with a carbonyl carbon of a peptide bond, thereby causing hydrolysis of the peptide bond.

Chemical Reactions and Rates

According to the conventions of biochemistry, the rate of a chemical reaction by the number of molecules of reactant(s) that are converted into product(s) in a specified time period. Reaction rate is always dependent on the concentration of the chemicals involved in the process and on rate constants that are characteristic of the reaction.

For example, the reaction in which A is converted to B is written as follows:

$$A \rightarrow B$$

The rate of this reaction is expressed algebraically as either a decrease in the concentration of reactant A:

$$-[A] = k[B]$$

or an increase in the concentration of product B:

$$[B] = k[A]$$

In the second equation (of the 3 above) the negative sign signifies a decrease in concentration of A as the reaction progresses, brackets define concentration in molarity and the *k* is known as a rate constant. Rate constants are simply proportionality constants that provide a quantitative connection between chemical concentrations and reaction rates.

Each chemical reaction has characteristic values for its rate constants; these in turn directly relate to the equilibrium constant for that reaction. Thus, reaction can be rewritten as an equilibrium expression in order to show the relationship between reaction rates, rate constants and the equilibrium constant for this simple case. The rate constant for the forward reaction is defined as k_{+1} and the reverse as k_{-1}.

At equilibrium the rate (v) of the forward reaction (A → B) is, by definition, equal to that of the reverse or back reaction (B → A), a relationship which is algebraically symbolized as:

$$v_{forward} = v_{reverse}$$

where, for the forward reaction:

$$v_{forward} = k_{+1}[A]$$

and for the reverse reaction:

$$v_{reverse} = k_{-1}[B]$$

In the above equations, k_{+1} and k_{-1} represent rate constants for the forward and reverse reactions, respectively. The negative subscript refers only to a reverse reaction, not to an actual negative value for the constant. To put the relationships of the two equations into words, we state that the rate of the forward reaction [$v_{forward}$] is equal to the product of the forward rate constant k_{+1} and the molar concentration of A. The rate of the reverse reaction is equal to the product of the reverse rate constant k_{-1} and the molar concentration of B. At equilibrium, the rate of the forward reaction is equal to the rate of the reverse reaction leading to the equilibrium constant of the reaction and is expressed by:

$$K_{eq} = \frac{K_{+1}}{K_{-1}} = \frac{[B]}{[A]}$$

This equation demonstrates that the equilibrium constant for a chemical reaction is not only equal to the equilibrium ratio of product and reactant concentrations, but is also equal to the ratio of the characteristic rate constants of the reaction.

Chemical Reaction Order

Reaction order refers to the number of molecules involved in forming a reaction complex that is competent to proceed to product(s). Empirically, order is easily determined by summing the exponents of each concentration term in the rate equation for a reaction. A reaction characterized by the conversion of one molecule of A to one molecule of B with no influence from any other reactant or solvent is a first-order reaction. The exponent on the substrate concentration in the rate equation for this type of reaction is 1. A reaction with two substrates forming two products would a second-order reaction. However, the reactants in second– and higher– order reactions need not be different chemical species. An example of a second order reaction is the formation of ATP through the condensation of ADP with orthophosphate:

$$ADP + H_2PO_4^- \leftrightarrow ATP + H_2O$$

For this reaction the forward reaction rate would be written as:

$$v_{forward} = k_1[ADP][H_2PO_4^-]$$

FACTORS AFFECTING ENZYME ACTIVITY

Knowledge of basic enzyme kinetic theory is important in enzyme analysis in order both to understand the basic enzymatic mechanism and to select a method for enzyme analysis. The conditions selected to measure the activity of an enzyme would not be the same as those selected to measure the concentration of its substrate. Several factors affect the rate at which enzymatic reactions proceed - temperature, pH, enzyme concentration, substrate concentration, and the presence of any inhibitors or activators.

Enzyme Concentration

In order to study the effect of increasing the enzyme concentration upon the reaction rate, the substrate must be present in an excess amount; *i.e.*, the reaction must be independent of the substrate concentration. Any change in the amount of product formed over a specified period of time will be dependent upon the level of enzyme present. Graphically this can be represented as:

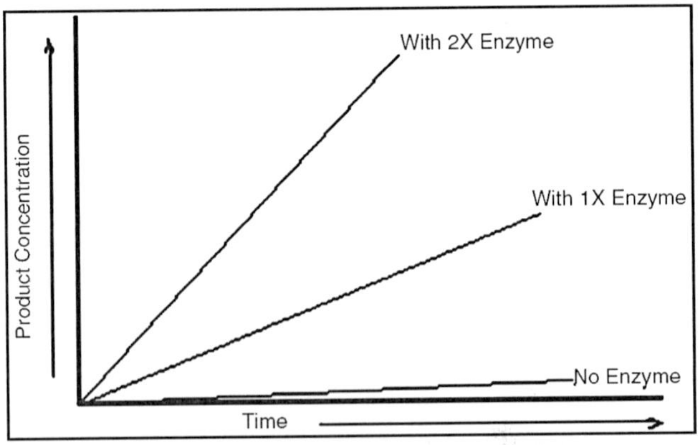

Fig. "Zero Order" Reaction rate is Independent of Substrate Concentration.

These reactions are said to be "zero order" because the rates are independent of substrate concentration, and are equal to some constant k. The formation of product proceeds at a rate which is linear with time. The addition of more substrate does not serve to increase the rate. In zero order kinetics, allowing the assay to run for double time results in double the amount of product.

Table. Reaction Orders with Respect to Substrate Concentration

Order	Rate Equation	Comments
Zero	rate = k	Rate is independent of substrate concentration
First	rate = k[S]	Rate is proportional to the first power of substrate concentration
Second	rate = $k[S][S]=k[S]^2$	Rate is proportional to the square of the substrate concentration
Second	rate = $k[S_1][S_2]$	Rate is proportional to the first power of each of two reactants

The amount of enzyme present in a reaction is measured by the activity it catalyzes. The relationship between activity and concentration is affected by many factors such as temperature, pH, etc. An enzyme assay must be designed so that the observed activity is proportional to the amount of enzyme present in order that the enzyme concentration is the only limiting factor. It is satisfied only when the reaction is zero order.

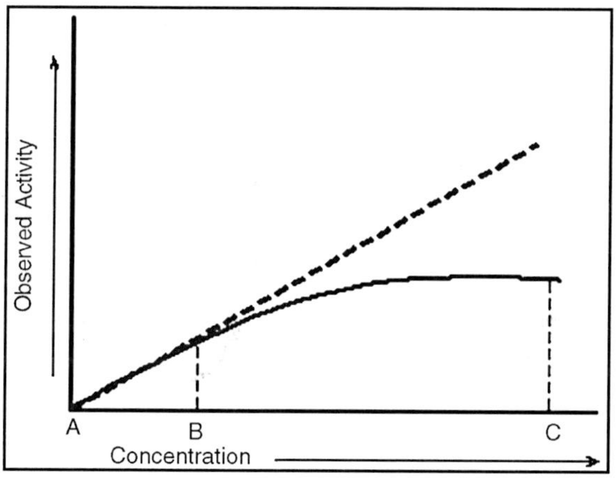

Fig. Activity vs. Concentration.

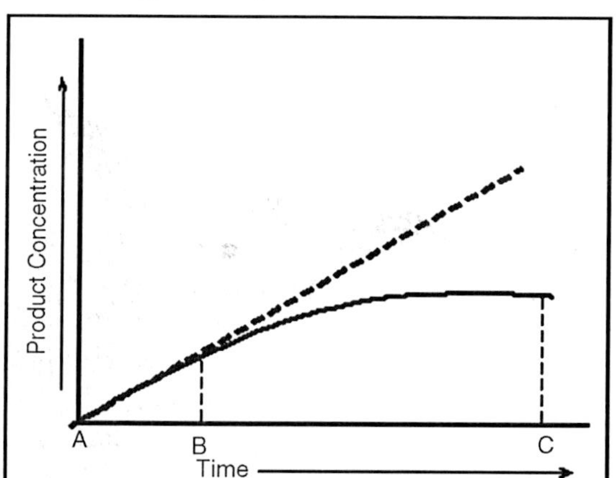

Fig. Reaction Rate Limited by Substrate Concentration.

When the concentration of the product of an enzymatic reaction is plotted against time, a similar curve results.

Between A and B, the curve represents a zero order reaction; that is, one in which the rate is constant with time. As substrate is used up, the enzyme's active sites are no longer saturated, substrate concentration becomes rate limiting, and the reaction becomes first order between B and C. To measure enzyme activity ideally, the measurements must be made in that portion of the curve where the reaction is zero order. A reaction is most likely to be zero order initially since substrate concentration is then highest. To be certain that a reaction is zero order, multiple measurements of product (or substrate) concentration must be made. Three types of reactions which might be encountered in enzyme assays and shows the problems which might be enountered if only single measurements are made.

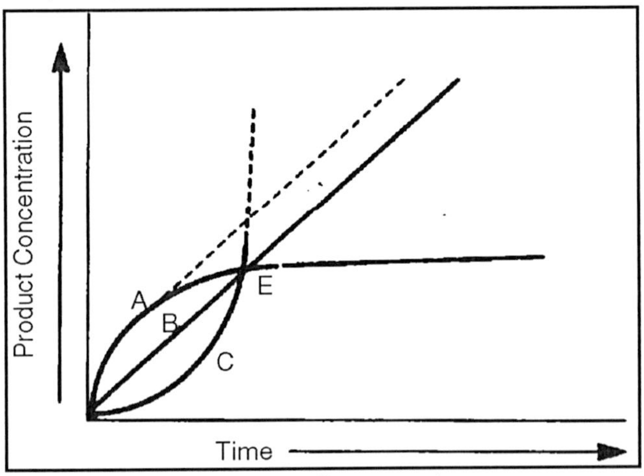

Fig. Time vs Concentration.

B is a straight line representing a zero order reaction which permits accurate determination of enzyme activity for part or all of the reaction time. This reaction is zero order initially and then slows, presumably due to substrate exhaustion or product inhibition. This type of reaction is sometimes referred to as a "leading" reaction.

True "potential" activity is represented by the dotted line. Curve C represents a reaction with an initial "lag" phase. Again the dotted line represents the potentially measurable activity. Multiple determinations of product concentration enable each curve to be plotted and true activity determined. A single end point determination at E would lead to the false conclusion that all three samples had identical enzyme concentration.

Substrate Concentration

It has been shown experimentally that if the amount of the enzyme is kept constant and the substrate concentration is then gradually increased, the reaction velocity will increase until it reaches a maximum. After this point, increases in substrate concentration will not increase the velocity ($\Delta A/\Delta T$).

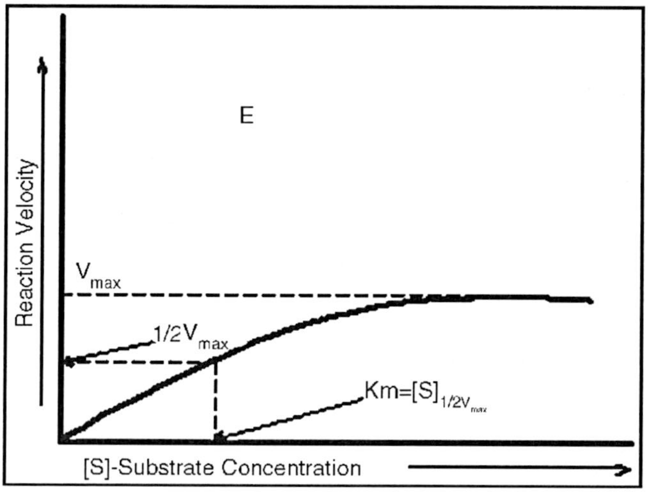

Fig. Effect of Substrate Concentration.

It is theorized that when this maximum velocity had been reached, all of the available enzyme has been converted to ES, the enzyme substrate complex. This point on the graph is designated Vmax. Using this maximum velocity and equation, Michaelis developed a set of mathematical expressions to calculate enzyme activity in terms of reaction speed from measurable laboratory data.

$$E + S \underset{K-1}{\overset{K+1}{\rightleftarrows}} ES \underset{K-2}{\overset{K+2}{\rightleftarrows}} P + E$$

The Michaelis constant Km is defined as the substrate concentration at 1/2 the maximum velocity. Using this constant and the fact that Km can also be defined as:

$$K_m = K_{-1} + K_{2/}K_{+1}$$

K^{+1}, K^{-1} and K^{+2} being the rate constants from equation. Michaelis developed the following,

$$V_1 = \frac{V_{max}[s]}{K_m + [S]}$$

where,

$V_{\uparrow}$ = the velocity at any time

[s] = the substrate concentration at this time

V_{max} = the highest under this set of experimental conditions (pH, temperature, etc.)

K_m = the Michaelis constant for the particular enzyme being investigated

Michaelis constants have been determined for many of the commonly used enzymes.

The size of K_{m3} tells us several things about a particular enzyme:

- A small K_{m3} indicates that the enzyme requires only a small amount of substrate to become saturated. Hence, the maximum velocity is reached at relatively low substrate concentrations.
- A large K_{m3} indicates the need for high substrate concentrations to achieve maximum reaction velocity.
- The substrate with the lowest K_{m3} upon which the enzyme acts as a catalyst is frequently assumed to be enzyme's natural substrate, though this is not true for all enzymes.

ENZYMES AS BIOLOGICAL CATALYSTS

In cells and organisms most reactions are catalyzed by enzymes, which are regenerated during the course of a reaction. These biological catalysts are physiologically important because they speed up the rates of reactions that would otherwise be too slow to support life. Enzymes increase reaction rates, sometimes by as much as one million-fold, but more typically by about one thousand fold. Catalysts speed up the forward and reverse reactions proportionately so that, although the magnitude of the rate constants of the forward and reverse reactions is are increased, the ratio of the rate constants remains the same in the presence or absence of enzyme. Since the equilibrium constant is equal to a ratio of rate constants, it is apparent that enzymes and other catalysts have no effect on the equilibrium constant of the reactions they catalyze.

Enzymes increase reaction rates by decreasing the amount of energy required to form a complex of reactants that is competent to produce reaction products. This complex is known as the activated state or transition state complex for the reaction. Enzymes and other catalysts accelerate reactions by lowering the energy of the transition state. The free energy required to form an activated complex is much lower in the catalyzed reaction. The amount of energy required to achieve the transition state is lowered; consequently, at any instant a greater proportion of the molecules in the population can achieve the transition state. The result is that the reaction rate is increased.

Michaelis-Menten Kinetics

In typical enzyme-catalyzed reactions, reactant and product concentrations are usually hundreds or thousands of times greater than the enzyme concentration. Consequently, each enzyme molecule catalyzes the conversion to product of many reactant molecules. In biochemical reactions, reactants are commonly known as substrates. The catalytic event that converts substrate to product involves the formation of a transition state, and it occurs most easily at a specific binding site on the enzyme. This site, called the catalytic site of the enzyme, has been evolutionarily structured to provide specific, high-affinity binding of substrate(s) and to provide an environment that favours the catalytic events. The complex that forms, when substrate(s) and enzyme combine, is called the enzyme substrate (ES) complex. Reaction products arise when the ES complex breaks down releasing free enzyme.

Between the binding of substrate to enzyme, and the reappearance of free enzyme and product, a series of complex events must take place. At a minimum an ES complex must be formed; this complex must pass to the transition state (ES*); and the transition state complex must advance to an enzyme product complex (EP). The latter is finally competent to dissociate to product and free enzyme.

The series of events can be shown thus:

$$E + S \leftrightarrow ES \leftrightarrow ES^* \leftrightarrow EP \leftrightarrow E + P$$

The kinetics of simple reactions like that above were first characterized by biochemists Michaelis and Menten. The concepts underlying their analysis of enzyme kinetics continue to provide the cornerstone for understanding metabolism today, and for the development and clinical use of drugs aimed at selectively altering rate constants and interfering with the progress of disease states. The Michaelis-Menten equation is a quantitative description of the relationship among the rate of an enzyme- catalyzed reaction [v_1], the concentration of substrate [S] and two constants, V_{max} and K_m (which are set by the particular equation). The symbols used in the Michaelis-Menten equation refer to the reaction rate [v_1], maximum reaction rate (V_{max}), substrate concentration [S] and the Michaelis-Menten constant (K_m).

$$V_1 = \frac{V_{max}[s]}{\{K_m + [S]\}}$$

The Michaelis-Menten equation can be used to demonstrate that at the substrate concentration that produces exactly half of the maximum reaction rate, *i.e.* ½ V_{max}, the substrate concentration is numerically equal to K_m. This fact provides a simple yet powerful bioanalytical tool that has been used to characterize both normal and altered enzymes, such as those that produce the symptoms of genetic diseases. Rearranging the Michaelis-Menten equation leads to:

$$K_m = [S]\left\{\left[\frac{V_{max}}{V_1}\right] - 1\right\}$$

From this equation it should be apparent that when the substrate concentration is half that required to support the maximum rate of reaction, the observed rate, v_1, will, be equal to V_{max} divided by 2; in other words, v_1 = [V_{max}/2]. At this substrate concentration V_{max}/v_1 will be exactly equal to 2, with the result that:

$$[S](1) = K_m$$

The latter is an algebraic statement of the fact that, for enzymes of the Michaelis-Menten type, when the observed reaction rate is half of the maximum possible reaction rate, the substrate concentration is numerically equal to the Michaelis-Menten constant. In this derivation, the units of K_m are those used to specify the concentration of S, usually Molarity.

The Michaelis-Menten equation has the same form as the equation for a rectangular hyperbola; graphical analysis of reaction rate (v) versus substrate concentration [S] produces a hyperbolic rate plot.

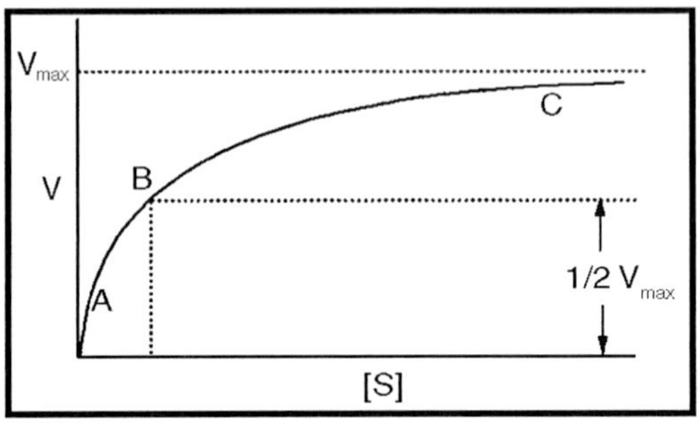

Fig. Plot of Substrate Concentration Versus Reaction Velocity

The key features of the plot are marked by points A, B and C. At high substrate concentrations the rate represented by point C the rate of the reaction is almost equal to V_{max}, and the difference in rate at nearby concentrations of substrate is almost negligible. If the Michaelis-Menten plot is extrapolated to infinitely high substrate concentrations, the extrapolated rate is equal to V_{max}. When the reaction rate becomes independent of substrate concentration, or nearly so, the rate is said to be zero order. (Note that the reaction is zero order only with respect to this substrate. If the reaction has two substrates, it may or may not be zero order with respect to the second substrate). The very small differences in reaction velocity at substrate concentrations around point C (near V_{max}) reflect the fact that at these concentrations almost all of the enzyme molecules are bound to substrate and the rate is virtually independent of substrate, hence zero order. At lower substrate concentrations, such as at points A and B, the lower reaction velocities indicate that at any moment only a portion of the enzyme molecules are bound to the substrate. In fact, at the substrate concentration denoted by point B, exactly half the enzyme molecules are in an ES complex at any instant and the rate is exactly one half of V_{max}. At substrate concentrations near point A the rate appears to be directly proportional to substrate concentration, and the reaction rate is said to be first order.

Inhibition of Enzyme Catalyzed Reactions

To avoid dealing with curvilinear plots of enzyme catalyzed reactions, biochemists Lineweaver and Burk introduced an analysis of enzyme kinetics based on the following rearrangement of the Michaelis-Menten equation:

$$\frac{1}{V} = \left[\frac{K_m (1)}{V_{max} (S)} + \frac{1}{V_{max}} \right]$$

Plots of 1/v versus 1/[S] yield straight lines having a slope of K_m/V_{max} and an intercept on the ordinate at $1/V_{max}$.

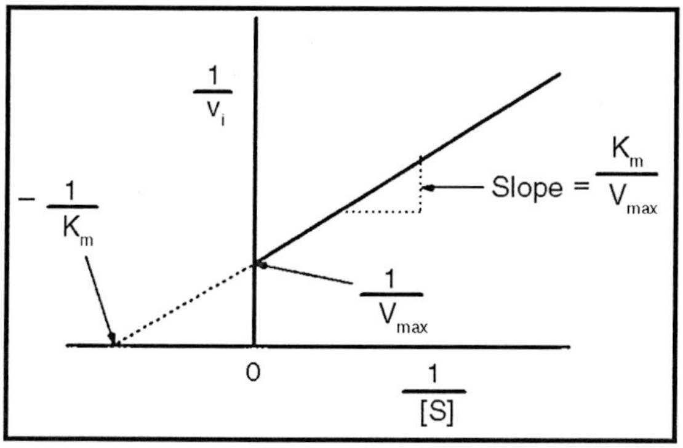

Fig. A Lineweaver-Burk Plot

An alternative linear transformation of the Michaelis-Menten equation is the Eadie-Hofstee transformation:

$$\frac{V}{[S]} = -V\left[\frac{1}{K_m} + \right] + \left[\frac{V_{max}}{K_m} \right]$$

and when v/[S] is plotted on the y-axis versus v on the x-axis, the result is a linear plot with a slope of $-1/K_m$ and the value V_{max}/K_m as the intercept on the y-axis and V_{max} as the intercept on the x-axis.

Both the Lineweaver-Burk and Eadie-Hofstee transformation of the Michaelis-Menton equation are useful in the analysis of enzyme inhibition. Well-known examples of such therapy include the use of methotrexate in cancer chemotherapy to semi-selectively inhibit DNA synthesis of malignant cells, the use of aspirin to inhibit the synthesis of prostaglandins which are at least partly responsible for the aches and pains of arthritis, and the use of sulfa drugs to inhibit the folic acid synthesis that is essential for the metabolism and growth of disease-causing bacteria. In addition, many poisons, such as cyanide, carbon monoxide and polychlorinated biphenols (PCBs) produce their life-threatening effects by means of enzyme inhibition.

Enzyme inhibitors fall into two broad classes: those causing irreversible inactivation of enzymes and those whose inhibitory effects can be reversed. Inhibitors of the first class usually cause an inactivating, covalent modification of enzyme structure. Cyanide is a classic example of an irreversible enzyme inhibitor: by covalently binding mitochondrial cytochrome oxidase, it inhibits all the reactions associated with electron transport. The kinetic effect of irreversible inhibitors is to decrease the concentration of active enzyme, thus decreasing the maximum possible concentration of ES complex. Since the limiting enzyme reaction rate is often k_2[ES], it is clear that under these circumstances the reduction of enzyme concentration will lead to decreased reaction rates. Note that when enzymes in cells are only partially inhibited by irreversible inhibitors, the remaining unmodified enzyme molecules are not distinguishable from those in untreated cells; in particular, they have the same turnover number and the same K_m. Turnover number, related to V_{max}, is defined as the maximum number of moles of substrate that can be converted to product per mole of catalytic site per second. Irreversible inhibitors are usually considered to be poisons and are generally unsuitable for therapeutic purposes.

Reversible inhibitors can be divided into two main categories; competitive inhibitors and noncompetitive inhibitors, with a third category, uncompetitive inhibitors, rarely encountered.

Inhibitor Type	Binding Site on Enzyme	Kinetic effect
Competitive Inhibitor	Specifically at the catalytic site, where it competes with substrate for binding in a dynamic equilibrium-like process. Inhibition is reversible by substrate.	V_{max} is unchanged; K_m, as defined by [S] required for ½ maximal activity, is increased.
Noncompetitive Inhibitor	Binds E or ES complex other than at the catalytic site. Substrate binding unaltered, but ESI complex cannot form products. Inhibition cannot be reversed by substrate. concentration.	K_m appears unaltered; V_{max} is decreased proportionately to inhibitor
Uncompetitive Inhibitor	Binds only to ES complexes at locations other than the catalytic site. Substrate binding modifies enzyme structure, making inhibitor- binding site available. Inhibition cannot be reversed by substrate.	Apparent V_{max} decreased; K_m, as defined by [S] required for ½ maximal activity, is decreased.

The hallmark of all the reversible inhibitors is that when the inhibitor concentration drops, enzyme activity is regenerated. Usually these inhibitors bind to enzymes by non-covalent forces and the inhibitor maintains a reversible equilibrium with the enzyme. The equilibrium constant for the dissociation of enzyme inhibitor complexes is known as K_i:

$$K_i = \frac{[E][I]}{[EI]}$$

The importance of K_I is that in all enzyme reactions where substrate, inhibitor and enzyme interact, the normal K_m and or V_{max} for substrate enzyme interaction appear to be altered. These changes are a consequence of the influence of K_i on the overall rate equation for the reaction. The effects of K_i

are best observed in Lineweaver-Burk plots. Probably the best known reversible inhibitors are competitive inhibitors, which always bind at the catalytic or active site of the enzyme. Most drugs that alter enzyme activity are of this type. Competitive inhibitors are especially attractive as clinical modulators of enzyme activity because they offer two routes for the reversal of enzyme inhibition, while other reversible inhibitors offer only one. First, as with all kinds of reversible inhibitors, a decreasing concentration of the inhibitor reverses the equilibrium regenerating active free enzyme. Second, since substrate and competitive inhibitors both bind at the same site they compete with one another for binding

Raising the concentration of substrate (S), while holding the concentration of inhibitor constant, provides the second route for reversal of competitive inhibition. The greater the proportion of substrate, the greater the proportion of enzyme present in competent ES complexes.

As noted earlier, high concentrations of substrate can displace virtually all competitive inhibitor bound to active sites. Thus, it is apparent that V_{max} should be unchanged by competitive inhibitors. This characteristic of competitive inhibitors is reflected in the identical vertical-axis intercepts of Lineweaver-Burk plots, with and without inhibitor.

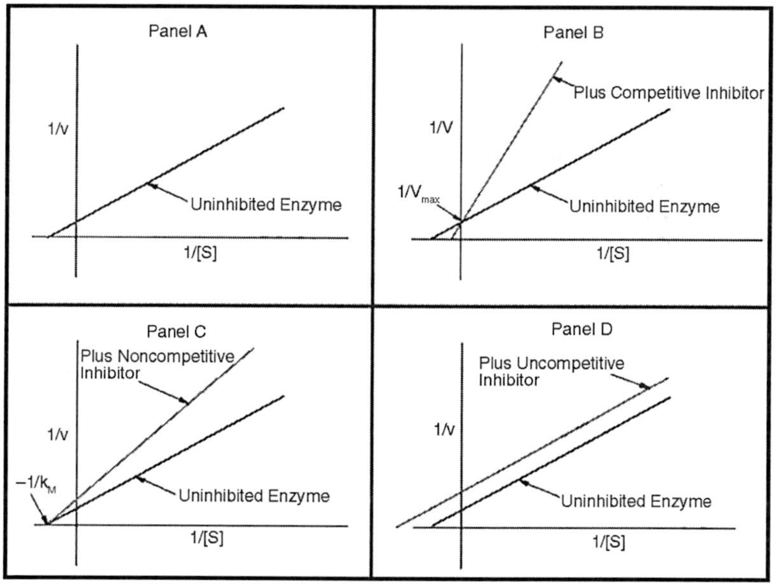

Fig. Lineweaver-Burk Plots of Inhibited Enzymes

Since attaining V_{max} requires appreciably higher substrate concentrations in the presence of competitive inhibitor, K_m (the substrate concentration at half maximal velocity) is also higher, as demonstrated by the differing negative intercepts on the horizontal axis in panel B. Analogously, panel C illustrates that noncompetitive inhibitors appear to have no effect on the intercept at the x-axis implying that noncompetitive inhibitors have no effect on the K_m of the enzymes they inhibit. Since non-competitive inhibitors do not interfere in the equilibration of enzyme, substrate and ES complexes, the K_m's of Michaelis-Menten type enzymes are not expected to be affected by non-competitive inhibitors, as demonstrated by x-axis intercepts in panel C. However, because complexes that contain inhibitor (ESI) are incapable of progressing to reaction products, the effect of a non-competitive inhibitor is to reduce the concentration of ES complexes that can advance to product. Since $V_{max} = k_2[E_{total}]$, and the concentration of competent E_{total} is diminished by the amount of ESI formed, non-competitive inhibitors are expected to decrease V_{max}, as illustrated by the y-axis intercepts in panel C.

A corresponding analysis of uncompetitive inhibition leads to the expectation that these inhibitors should change the apparent values of K_m as well as V_{max}. Changing both constants leads to double reciprocal plots, in which intercepts on the x and y axes are proportionately changed; this leads to the production of parallel lines in inhibited and uninhibited reactions.

REGULATION OF ENZYME ACTIVITY

While it is clear that enzymes are responsible for the catalysis of almost all biochemical reactions, it is important to also recognize that rarely, if ever, do enzymatic reactions proceed in isolation. The most common scenario is that enzymes catalyze individual steps of multi-step metabolic pathways, as is the case with glycolysis, gluconeogenesis or the synthesis of fatty acids. As a consequence of these lock- step sequences of reactions, any given enzyme is dependent on the activity of preceding reaction steps for its substrate.

In humans, substrate concentration is dependent on food supply and is not usually a physiologically important mechanism for the routine regulation of enzyme activity. Enzyme concentration, by contrast, is continually modulated in response to physiological needs.

Three principal mechanisms are known to regulate the concentration of active enzyme in tissues:

1. Regulation of gene expression controls the quantity and rate of enzyme synthesis.
2. Proteolytic enzyme activity determines the rate of enzyme degradation.
3. Covalent modification of pre-existing pools of inactive proenzymes produces active enzymes.

Enzyme synthesis and proteolytic degradation are comparatively slow mechanisms for regulating enzyme concentration, with response times of hours, days or even weeks. Proenzyme activation is a more rapid method of increasing enzyme activity but, as a regulatory mechanism, it has the disadvantage of not being a reversible process. Proenzymes are generally synthesized in abundance, stored in secretory granules and covalently activated upon release from their storage sites. Examples of important proenzymes include pepsinogen, trypsinogen and chymotrypsinogen, which give rise to the proteolytic digestive enzymes. Likewise, many of the proteins involved in the cascade of chemical reactions responsible for blood clotting are synthesized as proenzymes. Other important proteins, such as peptide hormones and collagen, are also derived by covalent modification of precursors.

Another mechanism of regulating enzyme activity is to sequester enzymes in compartments where access to their substrates is limited. For example, the proteolysis of cell proteins and glycolipids by enzymes responsible for their degradation is controlled by sequestering these enzymes within the lysosome.

In contrast to regulatory mechanisms that alter enzyme concentration, there is an important group of regulatory mechanisms that do not affect enzyme concentration, are reversible and rapid in action, and actually carry out most of the moment to moment physiological regulation of enzyme activity. These mechanisms include allosteric regulation, regulation by reversible covalent modification and regulation by control proteins such as calmodulin. Reversible covalent modification is a major mechanism for the rapid and transient regulation of enzyme activity.

The best examples, again, come from studies on the regulation of glycogen metabolism where phosphorylation of glycogen synthase and glycogen phosphorylase kinase results in the stimulation of glycogen degradation while glycogen synthesis is coordinately inhibited. Numerous other enzymes of intermediary metabolism are affected by phosphorylation, either positively or negatively.

These covalent phosphorylations can be reversed by a separate sub-subclass of enzymes known as phosphatases. Recent research has indicated that the aberrant phosphorylation of growth factor and hormone receptors, as well as of proteins that regulate cell division, often leads to unregulated cell growth or cancer. The usual sites for phosphate addition to proteins are the serine, threonine and tyrosine R group hydroxyl residues.

ENZYME NOMENCLATURE

All enzymes contain a protein backbone. In some enzymes this is the only component in the structure. However there are additional non-protein moieties usually present which may or may not participate in the catalytic activity of the enzyme. Covalently attached carbohydrate groups are commonly encountered structural features which often have no direct bearing on the catalytic activity, although they may well effect an enzyme's stability and solubility. Other factors often found are metal ions (cofactors) and low molecular weight organic molecules (coenzymes). These may be loosely or tightly bound by non-covalent or covalent forces. They are often important constituents contributing to both the activity and stability of the enzymes. This requirement for cofactors and coenzymes must be recognised if the enzymes are to be used efficiently and is particularly relevant in continuous processes where there may be a tendency for them to become separated from an enzyme's protein moiety.

Enzymes are classified according the report of a Nomenclature Committee appointed by the International Union of Biochemistry. This enzyme commission assigned each enzyme a recommended name and a 4-part distinguishing number. It should be appreciated that some alternative names remain in such common usage that they will be used, where appropriate, in this text. The enzyme commission (EC) numbers divide enzymes into six main groups according to the type of reaction catalysed:

- Oxidoreductases which involve redox reactions in which hydrogen or oxygen atoms or electrons are transferred between molecules. This extensive class includes the dehydrogenases (hydride transfer), oxidases (electron transfer to molecular oxygen), oxygenases (oxygen transfer from molecular oxygen) and peroxidases (electron transfer to peroxide). For example: glucose oxidase.

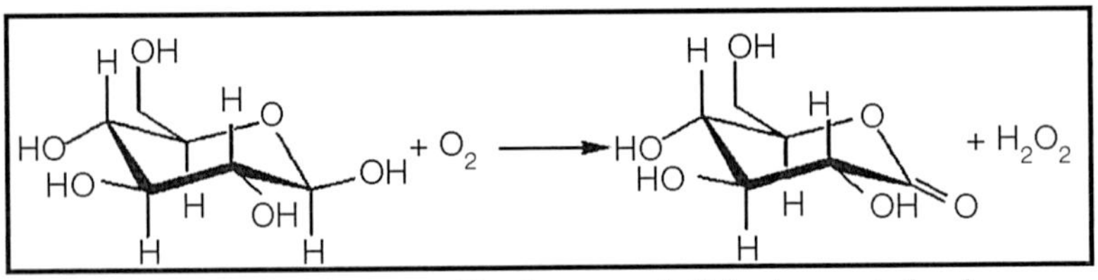

β–D–glucose + oxygen→D-glucono–1,5-lactone + hydrogen peroxide

- Transferases which catalyse the transfer of an atom or group of atoms (e.g. acyl-, alkyl- and glycosyl-), between two molecules, but excluding such transfers as are classified in the other groups (e.g. oxidoreductases and hydrolases). For example: aspartate aminotransferase.

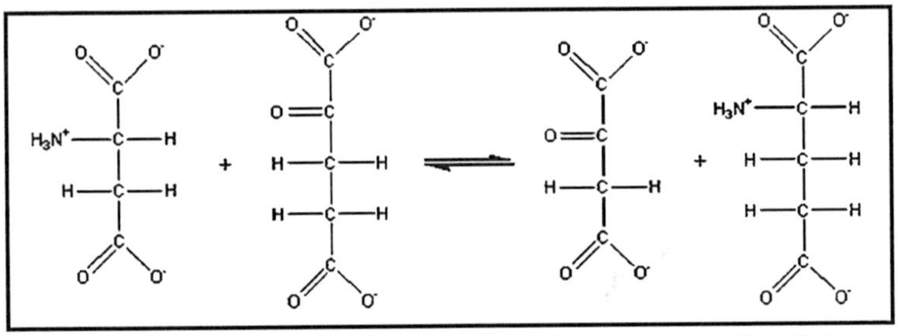

L-aspartate + 2–oxoglutarate ƒoxaloacetate + L–glutamate

- Hydrolases which involve hydrolytic reactions and their reversal. This is presently the most commonly encountered class of enzymes within the field of enzyme technology and includes the esterases, glycosidases, lipases and proteases. For example: chymosin.

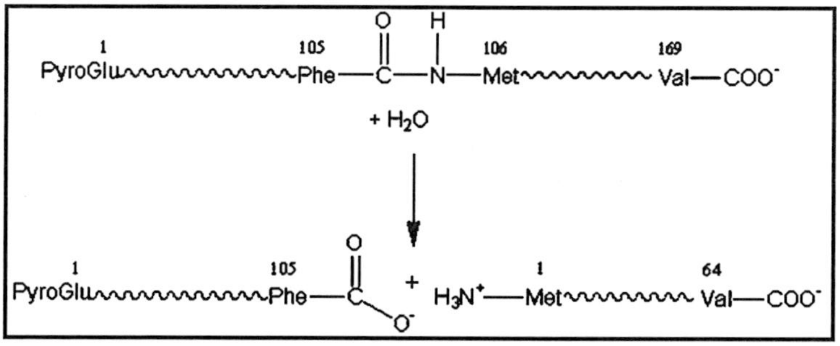

k-casein + water → *para*-k-casein + caseino macropeptide

- Isomerases which catalyse molecular isomerisations and includes the epimerases, racemases and intramolecular transferases. For example: xylose isomerase.

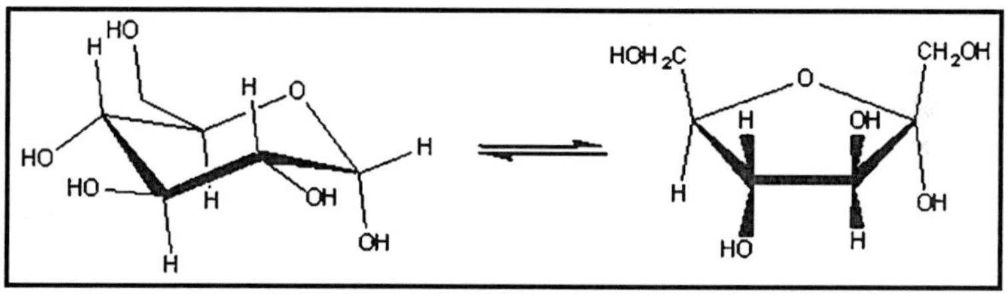

α-D-glucopyranose ƒα-D-fructofuranose

- Lyases which involve elimination reactions in which a group of atoms is removed from the substrate. This includes the aldolases, decarboxylases, dehydratases and some pectinases but does not include hydrolases. For example: histidine ammonia-lyase.

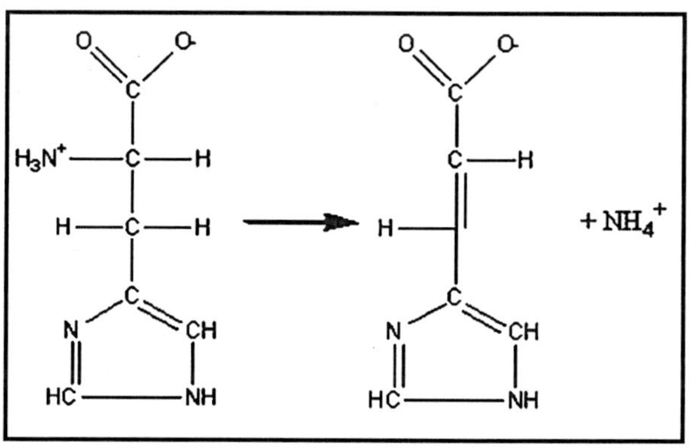

L-histidine → urocanate + ammonia

- Ligases, also known as synthetases, form a relatively small group of enzymes which involve the formation of a covalent bond joining two molecules together, coupled with the hydrolysis of a nucleoside triphosphate. For example: glutathione synthase (EC 6.3.2.3, systematic name,γ-L-glutamyl-L-cysteine:glycine ligase (ADP-forming); also called glutathione synthetase).

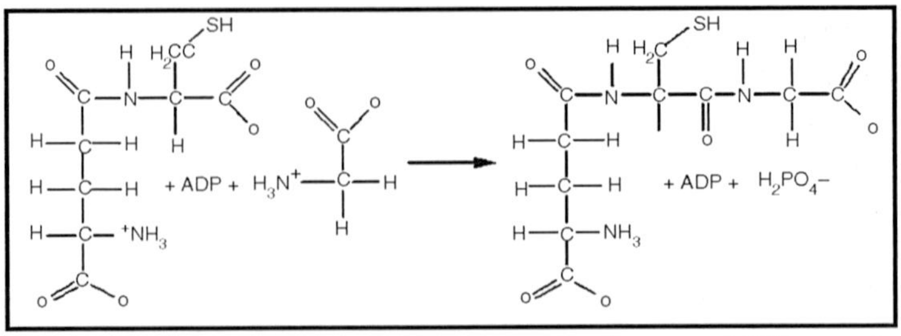

ATP + γ-L-glutamyl-L-cysteine + glycine → ADP + phosphate + glutathione

ENZYME UNITS

The amount of enzyme present or used in a process is difficult to determine in absolute terms (e.g. grams), as its purity is often low and a proportion may be in an inactive, or partially active, state. More relevant parameters are the activity of the enzyme preparation and the activities of any contaminating enzymes. These activities are usually measured in terms of the activity unit (U) which is defined as the amount which will catalyse the transformation of 1 micromole of the substrate per minute under standard conditions.

Typically, this represents 10^{-6} - 10^{-11} Kg for pure enzymes and 10^{-4} - 10^{-7} Kg for industrial enzyme preparations. Another unit of enzyme activity has been recommended. This is the katal (kat) which is defined as the amount which will catalyse the transformation of one mole of substance per second (1 kat = 60 000 000 U). It is an impracticable unit and has not yet received widespread acceptance. Sometimes non-standard activity units are used, such as Soxhet, Anson and Kilo Novo units, which are based on physical changes such as lowering viscosity and supposedly better understood by industry. Rightfully, such units are gradually falling into disuse. The activity is a measure of enzyme content that is clearly of major interest when the enzyme is to be used in a process.

For this reason, enzymes are usually marketed in terms of activity rather than weight. The specific activity (e.g. U Kg^{-1}) is a parameter of interest, some utility as an index of purity but lesser importance. There is a major problem with these definitions of activity; the rather vague notion of "standard conditions". These are meant to refer to optimal conditions, especially with regard to pH, ionic strength, temperature, substrate concentration and the presence and concentration of cofactors and coenzymes.

However, these so-termed optimal conditions vary both between laboratories and between suppliers. They also depend on the particular application in which the enzyme is to be used. Additionally, preparations of the same notional specific activity may differ with respect to stability and be capable of very different total catalytic productivity (this is the total substrate converted to product during the lifetime of the catalyst, under specified conditions).

Conditions for maximum initial activity are not necessarily those for maximum stability. Great care has to be taken over the consideration of these factors when the most efficient catalyst for a particular purpose is to be chosen.

THE MECHANISM OF ENZYME CATALYSIS

In order for a reaction to occur, reactant molecules must contain sufficient energy to cross a potential energy barrier, the activation energy. All molecules possess varying amounts of energy depending, for example, on their recent collision history but, generally, only a few have sufficient energy for reaction. The lower the potential energy barrier to reaction, the more reactants have sufficient energy and, hence, the faster the reaction will occur.

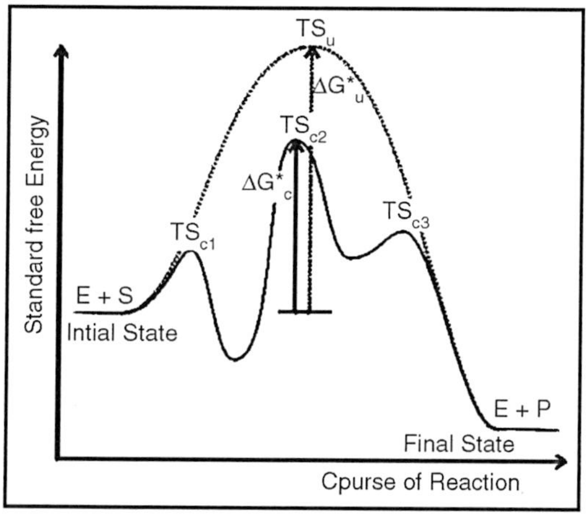

Fig. A Schematic Diagram Showing the free Energy Profile of the Course of an Enzyme Catalysed Reaction Involving the Formation of Enzyme-substrate (ES) and Enzyme-product (EP) Complexes, i.e.

$$E+S \rightleftharpoons ES \rightleftharpoons EP \rightleftharpoons E+P$$

All catalysts, including enzymes, function by forming a transition state, with the reactants, of lower free energy than would be found in the uncatalysed reaction. Even quite modest reductions in this potential energy barrier may produce large increases in the rate of reaction (e.g. the activation energy for the uncatalysed breakdown of hydrogen peroxide to oxygen and water is 76 kJ M^{-1} whereas, in the presence of the enzyme catalase, this is reduced to 30 kJ M^{-1} and the rate of reaction is increased by a factor of 10^8, sufficient to convert a reaction time measured in years into one measured in seconds).

The catalysed reaction pathway goes through the transition states TS_{c1}, TS_{c2} and TS_{c3}, with standard free energy of activation G_c^*, whereas the uncatalysed reaction goes through the transition state TS_u with standard free energy of activation G_u^*. In this example the rate limiting step would be the conversion of ES into EP. Reactions involving several substrates and products, or more intermediates, are even more complicated.

The Michaelis-Menten reaction scheme would give a similar profile but without the EP-complex free energy trough. The schematic profile for the uncatalysed reaction is shown as the dashed line. It should be noted that the catalytic effect only concerns the lowering of the standard free energy of activation from G_u^* to G_c^* and has no effect on the overall free energy change (i.e.. the difference between the initial and final states) or the related equilibrium constant.

There are a number of mechanisms by which this activation energy decrease may be achieved. The most important of these involves the enzyme initially binding the substrate(s), in the correct orientation to react, close to the catalytic groups on the active enzyme complex and any other substrates. In this way the binding energy is used partially in order to reduce the contribution of the considerable activation entropy, due to the loss of the reactants' (and catalytic groups') translational and rotational entropy, towards the total activation energy. Other contributing factors are the introduction of strain into the reactants (allowing more binding energy to be available for the transition state), provision of an alternative reactive pathway and the desolvation of reacting and catalysing ionic groups.

The energies available to enzymes for binding their substrates are determined primarily by the complementarity of structures (i.e. a good 3-dimensional fit plus optimal non-covalent ionic and/or hydrogen bonding forces). The specificity depends upon minimal steric repulsion, the absence of unsolvated or unpaired charges, and the presence of sufficient hydrogen bonds. These binding energies are capable of being quite large. As examples, antibody-antigen dissociation constants are characteristically near 10^{-8} M (free energy of binding is 46 kJ M^{-1}), ATP binds to myosin with a dissociation constant of 10^{-13} M (free energy of binding is 75 kJ M^{-1}) and biotin binds to avidin, a protein found in egg white, with a dissociation constant of 10^{-15} M (free energy of binding is 86 kJ M^{-1}). However, enzymes do not use this potential binding energy simply in order to bind the substrate(s) and form stable long-lasting complexes. If this were to be the case, the formation of the transition state between ES and EP would involve an extremely large free energy change due to the breaking of these strong binding forces, and the rate of formation of products would be very slow. They must use this binding energy for reducing the free energy of the transition state. This is generally achieved by increasing the binding to the transition state rather than the reactants and, in the process, introducing an energetic strain into the system and allowing more favourable interactions between the enzyme's catalytic groups and the reactants.

SIMPLE KINETICS OF ENZYME ACTION

It is established that enzymes form a bound complex to their reactants (i.e. substrates) during the course of their catalysis and prior to the release of products. This can be simply illustrated, using the mechanism based on that of Michaelis and Menten for a one-substrate reaction, by the reaction sequence:

Enzyme + Substrate ⇌ (Enzyme-substrate complex) → Enzyme + Product

$$E + S \underset{k-1}{\overset{k+1}{\rightleftharpoons}} ES \xrightarrow{K+2} P + E$$

where k_{+1}, k_{-1} and k_{+2} are the respective rate constants, typically having values of $10^5 - 10^8\ M^{-1}\ s^{-1}$, $1 - 10^4\ s^{-1}$ and $1 - 10^5\ s^{-1}$ respectively; the sign of the subscripts indicating the direction in which the rate constant is acting. For the sake of simplicity the reverse reaction concerning the conversion of product to substrate is not included in this scheme. This is allowable (1) at the beginning of the reaction when there is no, or little, product present, or (2) when the reaction is effectively irreversible. The rate of reaction (v) is the rate at which the product is formed.

$$V = \frac{d[P]}{df} = k_{+2}[ES]$$

where indicates the molar concentration of the material enclosed (i.e. [ES] is the concentration of the enzyme-substrate complex). The rate of change of the concentration of the enzyme-substrate complex equals the rate of its formation minus the rate of its breakdown, forwards to give product or backwards to regenerate substrate.

Therefore:

$$\frac{d[ES]}{df} = k_{+1}[E][S] - (k_{-1} + k_{+2})[ES]$$

During the course of the reaction, the total enzyme at the beginning of the reaction ($[E]_0$, at zero time) is present either as the free enzyme ([E]) or the ES complex ([ES]).

$$\text{i.e. } [E]_0 = [E] + [ES]$$

Therefore:

$$\frac{d[ES]}{dt} = k_{+1}([E]_0 - [ES])[S] - (k_{-1} + k_{+2})[ES]$$

Gathering terms together,

$$\frac{d[ES]}{df} = k_{+1}([E]_0[S] - k_{+1}[ES][S] - (k_{-1} + k_{+2})[ES]$$

$$\frac{d[ES]}{dt} = k_{+1}([E]_0[S] - (k_{+1}[S] + k_{-1} + k_{+2})[ES]$$

This gives:

$$\frac{\frac{d[ES]}{df}}{k_{+1}[S] + k_{-1} + k_{+2})} + [ES] = \frac{k_{+1}[E]_0[S]}{k_{+1}[S] + k_{-1} + k_{+2}}$$

The differential equation 1.5 is difficult to handle, but may be greatly simplified if it can be assumed that the left hand side is equal to [ES] alone. This assumption is valid under the sufficient but unnecessarily restrictive steady state approximation that the rate of formation of ES equals its rate of disappearance by product formation and reversion to substrate (i.e. d[ES]/dt is zero). It is additionally valid when the condition:

$$\frac{\frac{d[ES]}{df}}{k_{+1}[S]+k_{-1}+k_{+2})} << [ES]$$

is valid. This occurs during a substantial part of the reaction time-course over a wide range of kinetic rate constants and substrate concentrations and at low to moderate enzyme concentrations. The variation in [ES], d[ES]/dt, [S] and [P] with the time-course of the reaction, where it may be seen that the simplified equation is valid throughout most of the reaction.

$$\frac{\frac{d[ES]}{df}}{k_{+1}[S]+k_{-1}+k_{+2})}$$

is much less than [ES] during both of the latter two phases.

The Michaelis-Menten equation is simply derived from equations, by substituting K_m for $\frac{k_{-1}+k_{+2}}{k_{+1}}$. K_m is known as the Michaelis constant with a value typically in the range 10^{-1} – 10^{-5} M. When $k_{+2} << k_{-1}$, K_m equals the dissociation constant (k_{-1}/k_{+1}) of the enzyme substrate complex.

$$V = k_{+2}[ES] = \frac{k_{+2}[E]_0[S]}{[S]+K_m}$$

or, more simply

$$V = \frac{V_{max}[S]}{[S]+K_m}$$

where V_{max} is the maximum rate of reaction, which occurs when the enzyme is completely saturated with substrate (i.e. when [S] is very much greater than K_m, V_{max} equals $k_{+2}[E]_0$, as the maximum value [ES] can have is $[E]_0$ when $[E]_0$ is less than $[S]_0$). Equation 1.8 may be rearranged to show the dependence of the rate of reaction on the ratio of [S] to K_m,

$$V = \frac{V_{max}}{1+\frac{K_m}{[S]}}$$

and the rectangular hyperbolic nature of the relationship, having asymptotes

at v = V_{max} and [S] = $-K_m$,

$$(V_{max}-v)(K_m+[S])=V_{max}K_m$$

The substrate concentration in these equations is the actual concentration at the time and, in a closed system, will only be approximately equal to the initial substrate concentration ($[S]_0$) during the early phase of the reaction. Hence, it is usual to use these equations to relate the initial rate of reaction to the initial, and easily predetermined, substrate concentration. This also avoids any problem that may occur through product inhibition or reaction reversibility.

It has been established that few enzymes follow the Michaelis-Menten equation over a wide range of experimental conditions. However, it remains by far the most generally applicable equation for describing enzymic reactions. Indeed it can be realistically applied to a number of reactions which have a far more complex mechanism than the one described here. In these cases K_m remains an important quantity, characteristic of the enzyme and substrate, corresponding to the substrate concentration needed for half the enzyme molecules to bind to the substrate but the precise kinetic meaning derived earlier may not hold and may be misleading.

In these cases the K_m is likely to equal a much more complex relationship between the many rate constants involved in the reaction scheme. It remains independent of the enzyme and substrate concentrations and indicates the extent of binding between the enzyme and its substrate for a given substrate concentration, a lower K_m indicating a greater extent of binding. V_{max} clearly depends on the enzyme concentration and for some, but not all, enzymes may be largely independent of the specific substrate used. K_m and V_{max} may both be influenced by the charge and conformation of the protein and substrate(s) which are determined by pH, temperature, ionic strength and other factors.

It is often preferable to substitute k_{cat} for k_{+2}, where $V_{max} = k_{cat}[E]_0$, as the precise meaning of k_{+2}, above, may also be misleading. k_{cat} is also known as the turnover number as it represents the maximum number of substrate molecules that the enzyme can 'turn over' to product in a set time (e.g. the turnover numbers of α-amylase, glucoamylase and glucose isomerase are 500 s^{-1}, 160 s^{-1} and 3 s^{-1} respectively; an enzyme with a relative molecular mass of 60000 and specific activity 1 U mg^{-1} has a turnover number of 1 s^{-1}). The ratio k_{cat}/K_m determines the relative rate of reaction at low substrate concentrations, and is known as the specificity constant. It is also the apparent 2nd order rate constant at low substrate concentrations, where

$$V=\frac{k_{cat}}{k_m}+[E]_0[S]$$

Many applications of enzymes involve open systems, where the substrate concentration remains constant, due to replenishment, throughout the reaction time-course. This is, of course, the situation that often prevails *in vivo*. Under these circumstances, the Michaelis-Menten equation is obeyed over an even wider range of enzyme concentrations than allowed in closed systems, and is commonly used to model immobilised enzyme kinetic systems.

Enzymes have evolved by maximising k_{cat}/K_m (i.e. the specificity constant for the substrate) whilst keeping K_m approximately identical to the naturally encountered substrate concentration. This allows the enzyme to operate efficiently and yet exercise some control over the rate of reaction.

The specificity constant is limited by the rate at which the reactants encounter one another under the influence of diffusion. For a single-substrate reaction the rate of encounter between the substrate and enzyme is about 10^8 - 10^9 M^{-1} s^{-1}. The specificity constant of some enzymes approach this value although the range of determined values is very broad (e.g. k_{cat}/K_m for catalase is 4 x 10^7 M^{-1} s^{-1}, whereas it is 25 M^{-1} s^{-1} for glucose isomerase, and for other enzymes varies from less than 1 M^{-1} s^{-1} to greater than 10^8 M^{-1} s^{-1}).

EFFECT OF PH AND IONIC STRENGTH

Enzymes are amphoteric molecules containing a large number of acid and basic groups, mainly situated on their surface. The charges on these groups will vary, according to their acid dissociation constants, with the pH of their environment. This will effect the total net charge of the enzymes and the distribution of charge on their exterior surfaces, in addition to the reactivity of the catalytically active groups. These effects are especially important in the neighbourhood of the active sites. Taken together, the changes in charges with pH affect the activity, structural stability and solubility of the enzyme.

Table. pK_as and Heats of Ionisation of the Ionising Groups Commonly Found in Enzymes.

Group	Usual pK_a range	Approximate charge at pH 7	Heats of ionisation (kJ $mole^{-1}$)
Carboxyl (C-terminal, glutamic acid, aspartic acid)	3 - 6	-1.0	± 5
Ammonio (N-terminal)	7 - 9	+1.0	+45
(lysine)	9 - 11	+1.0	+45
Imidazolyl (histidine)	5 - 8	+0.5	+30
Guanidyl (arginine)	11 - 13	+1.0	+50
Phenolic (tyrosine)	9 - 12	0.0	+25
Thiol (cysteine)	8 - 11	0.0	+25

[a] The pK_a (defined as $-Log_{10}(K_a)$) is the pH at which half the groups are ionised. Note the similarity between the K_a of an acid and the K_m of an enzyme, which is the substrate concentration at which half the enzyme molecules have bound substrate.

[b] By convention, the heat (enthalpy) of ionisation is positive when heat is withdrawn from the surrounding solution (i.e. the reaction is endothermic) by the dissociation of the hydrogen ions.

There will be a pH, characteristic of each enzyme, at which the net charge on the molecule is zero. This is called the isoelectric point (pI), at which the enzyme generally has minimum solubility in aqueous solutions. In a similar manner to the effect on enzymes, the charge and charge distribution on the substrate(s), product(s) and coenzymes (where applicable) will also be affected by pH changes. Increasing hydrogen ion concentration will, additionally, increase the successful competition of hydrogen ions for any metal cationic binding sites on the enzyme, reducing the bound metal cation concentration.

Decreasing hydrogen ion concentration, on the other hand, leads to increasing hydroxyl ion concentration which compete against the enzymes' ligands for divalent and trivalent cations causing their conversion to hydroxides and, at high hydroxyl concentrations, their complete removal from the enzyme. The temperature also has a marked effect on ionisations, the extent of which depends on the heats of ionisation of the particular groups concerned. The relationship between the change in the pK_a and the change in temperature is given by a derivative of the Gibbs-Helmholtz equation:

$$\frac{d(pK_a)_{cat}}{dt} = \frac{-\Delta H}{2.303RT^2}$$

where T is the absolute temperature (K), R is the gas law constant (8.314 J M^{-1} K^{-1}), ?H is the heat of ionisation and the numeric constant (2.303) is the natural logarithm of 10, as pK_a's are based on logarithms with base 10. This variation is sufficient to shift the pI of enzymes by up to one unit towards lower pH on increasing the temperature by 50°C.

These charge variations, plus any consequent structural alterations, may be reflected in changes in the binding of the substrate, the catalytic efficiency and the amount of active enzyme. Both V_{max} and K_m will be affected due to the resultant modifications to the kinetic rate constants k_{+1}, k_{-1} and k_{cat} (k_{+2} in the Michaelis-Menten mechanism), and the variation in the concentration of active enzyme. The effect of pH on the V_{max} of an enzyme catalysed reaction may be explained using the, generally true, assumption that only one charged form of the enzyme is optimally catalytic and therefore the maximum concentration of the enzyme-substrate intermediate cannot be greater than the concentration of this species. In simple terms, assume EH^- is the only active form of the enzyme,

$$EH^- + S \underset{k_{-1}}{\overset{k_{+1}}{\rightleftharpoons}} EH^- \xrightarrow{k_{+2}} P$$

The concentration of EH^- is determined by the two dissociations

$$EH_2 \overset{K_{a1}}{\rightleftharpoons} EH^- + H^+$$

$$EH^- \overset{K_{a2}}{\rightleftharpoons} E^{2-} + H^+$$

with

$$K_{a1} = \frac{[EH^-]_0[H^+]}{[EH_2]_0}$$

and

$$K_{a2} = \frac{[EH^{2-}]_0[H^+]}{[EH^-]_0}$$

However,

$$[E]_0 = [EH_2]_0 + [EH^-]_0 + [E^{2-}]_0$$

Therefore:

$$[E]_0 = [EH^-]_0\left(\frac{[H^+]}{K_{a1}} + 1 + \frac{K_{a2}}{[H^+]}\right)$$

As the rate of reaction is given by $k_{+2}[EH^-S]$ and this is maximal when $[EH^-S]$ is maximal (i.e.. when $[EH^-S] = [EH^-]_0$):

$$V_{max} = k_{+2}[EH] = \frac{K_{+2}[E]_0}{\left(\frac{[H^+]}{K_{a1}} + 1 + \frac{K_{a2}}{[H^+]}\right)}$$

The V_{max} will be greatest when

$$\frac{[H^+]}{K_{a1}} + 1 + \frac{K_{a2}}{[H^+]}$$

Therefore:

$$pH_{optimum} = \frac{pK_{a1} + pk_{a2}}{2}$$

This derivation has involved a number of simplifications on the real situation; it ignores the effect of the ionisation of substrates, products and enzyme-substrate complexes and it assumes EH^- is a single ionised species when it may contain a mixture of differently ionised groups but with identical overall charge, although the process of binding substrate will tend to fix the required ionic species.

It does, however, produce a variation of maximum rate with pH which gives the commonly encountered 'bell-shaped' curve. Where the actual reaction scheme is more complex, there may be a more complex relationship between

V_{max} and pH. In particular, there may be a change in the rate determining step with pH.

It should be recognised that K_m may change with pH in an independent manner to the V_{max} as it usually involves other, or additional, ionisable groups. It is clear that at lower non-saturating substrate concentrations the activity changes with pH may or may not reflect the changes in V_{max}. It should also be noted from the foregoing that the variation of activity with pH depends on the reaction direction under consideration.

The $pH_{optimum}$ may well be different in the forward direction from that shown by the reverse reaction. This is particularly noticeable when reactions which liberate or utilise protons are considered (e.g. dehydrogenases) where there may well be greater than 2 pH units difference between the $pH_{optimum}$ shown by the rates of forward and reverse reactions.

The variation of activity with pH, within a range of 2-3 units each side of the pI, is normally a reversible process. Extremes of pH will, however, cause a time- and temperature-dependent, essentially irreversible, denaturation. In alkaline solution ($pH > 8$), there may be partial destruction of cystine residues due to base catalysed b-elimination reactions whereas, in acid solutions ($pH < 4$), hydrolysis of the labile peptide bonds, sometimes found next to aspartic acid residues, may occur. The importance of the knowledge concerning the variation of activity with pH cannot be over-emphasised.

However, a number of other factors may mean that the optimum pH in the V_{max}– pH diagram may not be the pH of choice in a technological process involving enzymes. These include the variation of solubility of substrate and product, changes in the position of equilibrium for a reaction, suppression of the ionisation of a product to facilitate its partition and recovery into an organic solvent, and the reduction in susceptibility to oxidation or microbial contamination. The major such factor is the effect of pH on enzyme stability.

This relationship is further complicated by the variation in the effect of the pH with both the duration of the process and the temperature or temperature-time profile. The important parameter derived from these influences is the productivity of the enzyme (i.e. how much substrate it is capable of converting to product).

The variation of productivity with pH may be similar to that of the V_{max}-pH relationship but changes in the substrate stream composition and contact time may also make some contribution. Generally, the variation must be determined under the industrial process conditions. It is possible to alter the pH-activity profiles of enzymes. The ionisation of the carboxylic acids involves the separation of the released groups of opposite charge. This process is encouraged within solutions of higher polarity and reduced by less polar solutions. Thus, reducing the dielectric constant of an aqueous solution by the addition of a co-solvent of low polarity (e.g. dioxan, ethanol), or by immobilisation, increases the pK_a of carboxylic acid groups. This method is sometimes useful but not generally applicable to enzyme catalysed reactions as it may cause a drastic change on an enzyme's productivity due to denaturation.

The pK_a of basic groups are not similarly affected as there is no separation of charges when basic groups ionise. However, protonated basic groups which are stabilised by neighbouring negatively charged groups will be stabilised (i.e. have lowered pK_a) by solutions of lower polarity. Changes in the *ionic strength* (I) of the solution may also have some effect. The ionic strength is defined as half of the total sum of the concentration (c_i) of every ionic species (i) in the solution times the square of its charge (z_i); i.e.. $I = 0.5 \Sigma (c_i z_i^2)$.

For example, the ionic strength of a 0.1 M solution of $CaCl_2$ is $0.5 \times (0.1 \times 2^2 + 0.2 \times 1^2) = 0.3$ M.

At higher solution ionic strength, charge separation is encouraged with a concomitant lowering of the carboxylic acid pK_as. These changes, extensive as they may be, have little effect on the overall charge on the enzyme molecule at neutral pH and are, therefore, only likely to exert a small influence on the enzyme's isoelectric point. Chemical derivatisation methods are available for converting surface charges from positive to negative and vice-versa. It is found that a single change in charge has little effect on the pH-activity profile, unless it is at the active site. However if all lysines are converted to carboxylates (e.g. by reaction with succinic anhydride) or if all the carboxylates are converted to amines the profile can be shifted about a pH unit towards higher or lower pH, respectively. The cause of these shifts is primarily the stabilisation or destabilisation of the charges at the active site during the reaction, and the effects are most noticeable at low ionic strength. Some, more powerful, methods for shifting the pH-activity profile are specific to immobilised enzymes.

The ionic strength of the solution is an important parameter affecting enzyme activity. This is especially noticeable where catalysis depends on the movement of charged molecules relative to each other. Thus both the binding of charged substrates to enzymes and the movement of charged groups within the catalytic 'active' site will be influenced by the ionic composition of the medium. If the rate of the reaction depends upon the approach of charged moieties the following approximate relationship may hold,

$$\log(k) = \log(k_0) + Z_A Z_B \sqrt{I}$$

where k is the actual rate constant, k_0 is the rate constant at zero ionic strength, z_A and z_B are the electrostatic charges on the reacting species, and I?is the ionic strength of the solution. If the charges are opposite then there is a decrease in the reaction rate with increasing ionic strength whereas if the charges are identical, an increase in the reaction rate will occur (e.g. the rate controlling step in the catalytic mechanism of chymotrypsin involves the approach of two positively charged groups, ^{57}histidine$^+$ and ^{145}arginine$^+$ causing a significant increase in k_{cat} on increasing the ionic strength of the solution). Even if a more complex relationship between the rate constants and the ionic strength holds, it is clearly important to control the ionic strength of solutions in parallel with the control of pH.

EFFECT OF TEMPERATURE AND PRESSURE

Rates of all reactions, including those catalysed by enzymes, rise with increase in temperature in accordance with the Arrhenius equation.

$$k = Ae^{-\Delta G^*/RT}$$

where k is the kinetic rate constant for the reaction, A is the Arrhenius constant, also known as the frequency factor, ΔG^* is the standard free energy of activation (kJ M^{-1}) which depends on entropic and enthalpic factors, R is the gas law constant and T is the absolute temperature. Typical standard free energies of activation (15 - 70 kJ M^{-1}) give rise to increases in rate by factors between 1.2 and 2.5 for every 10°C rise in temperature.

This factor for the increase in the rate of reaction for every 10°C rise in temperature is commonly denoted by the term Q_{10} (i.e. in this case, Q_{10} is within the range 1.2 - 2.5). All the rate constants contributing to the catalytic mechanism will vary independently, causing changes in both K_m and V_{max}. It follows that, in an exothermic reaction, the reverse reaction (having a higher activation energy) increases more rapidly with temperature than the forward reaction.

This, not only alters the equilibrium constant, but also reduces the optimum temperature for maximum conversion as the reaction progresses. The reverse holds for endothermic reactions such as that of glucose isomerase where the ratio of fructose to glucose, at equilibrium, increases from 1.00 at 55°C to 1.17 at 80°C.

In general, it would be preferable to use enzymes at high temperatures in order to make use of this increased rate of reaction plus the protection it affords against microbial contamination. Enzymes, however, are proteins and undergo essentially irreversible denaturation (i.e.. conformational alteration entailing a loss of biological activity) at temperatures above those to which they are ordinarily exposed in their natural environment. These denaturing reactions have standard free energies of activation of about 200 - 300 kJ mole^{-1} (Q_{10} in the range 6 - 36) which means that, above a critical temperature, there is a rapid rate of loss of activity. The actual loss of activity is the product of this rate and the duration of incubation. It may be due to covalent changes such as the deamination of asparagine residues or non-covalent changes such as the rearrangement of the protein chain. Inactivation by heat denaturation has a profound effect on the enzymes productivity.

The thermal denaturation of an enzyme may be modelled by the following serial deactivation scheme:

$$E \xrightarrow{K_{d1}} E_1 \xrightarrow{K_{d2}} E_2$$

where k_{d1} and k_{d2} are the first-order deactivation rate coefficients, E is the native enzyme which may, or may not, be an equilibrium mixture of a number of species, distinct in structure or activity, and E_1 and E_2 are enzyme molecules of average specific activity relative to E of A_1 and A_2. A_1 may be greater or less than unity (i.e. E_1 may have higher or lower activity than E) whereas A_2 is normally very small or zero.

This model allows for the rare cases involving free enzyme (e.g. tyrosinase) and the somewhat commoner cases involving immobilised enzyme where there is a small initial activation or period of grace involving negligible discernible loss of activity during short incubation periods but prior to later deactivation. Assuming, at the beginning of the reaction:

$$[E] = [E]_0$$

and:

$$[E_1] = [E_2] = 0$$

At time t,

$$[E] + [E_1] + [E_2] = [E]_0$$

It follows from the reaction scheme,

$$-\frac{d[E]}{df} = k_{d1}[E]$$

Integrating equation 1.25 using the boundary condition in equation gives:

$$[E] = [E]_0 e^{(-kdt)}$$

From the reaction scheme,

$$-\frac{d[E_1]}{df} = k_{d2}[E_1] - k_{d1}[E]$$

Substituting for [E] from equation,

$$-\frac{d[E_1]}{df} = k_{d2}[E_1] - k_{d1}[E]_0 e(-kdt)$$

Integrating equation 1.27 using the boundary condition in equation gives:

$$[E_1] = \frac{k_{d1}[E]_0}{k_{d2} - k_{d1}}\left(e^{(-kdt)} - e^{(-kdz)}\right)$$

If the term 'fractional activity' (A^f) is introduced where,

$$A^t = \frac{[E] + A_1[E_1] + A_2[E_2]}{[E]_0}$$

then, substituting for $[E_2]$ from equation, gives:

$$A^t = \frac{[E] + A_1[E_1] + A_2[E]_0 - [E] - [E_1]}{[E]_0}$$

therefore:

$$A^t = A_2 + \left(a + \frac{A_1 k_{d1} - A_2 k_{d2}}{k_{d2} - k_{d1}} \right)$$

$$e^{(-kdt)} - \frac{(A_1 - A_2)k_{d1}}{k_{d2} - k_{d1}} e^{(-Kdz)}$$

When both A_1 and A_2 are zero, the simple first order deactivation rate expression results

$$A^t = e^{(-kdt)}$$

The half-life ($t_{1/2}$) of an enzyme is the time it takes for the activity to reduce to a half of the original activity (i,e. $A^f = 0.5$). If the enzyme inactivation obeys equation 1.33, the half-life may be simply derived,

$$\text{Ln}\,(1/2) = -k_{d1} f_{1/2}$$

therefore:

$$f_{1/2} = \frac{0.693}{k_{d1}}$$

In this simple case, the half-life of the enzyme is shown to be inversely proportional to the rate of denaturation.

Many enzyme preparations, both free and immobilised, appear to follow this series-type deactivation scheme. However because reliable and reproducible data is difficult to obtain, inactivation data should, in general, be assumed to be rather error-prone. It is not surprising that such data can be made to fit a model involving four determined parameters (A_1, A_2, k_{d1} and k_{d2}). Despite this possible reservation, equations remain quite useful and the theory possesses the definite advantage of simplicity. In some cases the series-type deactivation may be due to structural microheterogeneity, where the enzyme preparation consists of a heterogeneous mixture of a large number of closely related structural forms. These may have been formed during the past history of the enzyme during preparation and storage due to a number of minor reactions, such as deamidation of one or two asparagine or glutamine residues, limited proteolysis or disulphide interchange.

Alternatively it may be due to quaternary structure equilibria or the presence of distinct genetic variants. In any case, the larger the variability the more apparent will be the series-type inactivation kinetics. The practical effect of this is that usually k_{d1} is apparently much larger than k_{d2} and A_1 is less than unity.

In order to minimise loss of activity on storage, even moderate temperatures should be avoided. Most enzymes are stable for months if refrigerated (0 - 4°C). Cooling below 0°C, in the presence of additives (e.g. glycerol) which prevent freezing, can generally increase this storage stability even further. Freezing enzyme solutions is best avoided as it often causes denaturation due to the stress and pH variation caused by ice-crystal formation. The first order deactivation constants are often significantly lower in the case of enzyme-substrate, enzyme-inhibitor and enzyme-product complexes which helps to explain the substantial stabilising effects of suitable ligands, especially at concentrations where little free enzyme exists (e.g. [S] >> K_m). Other factors, such as the presence of thiol anti-oxidants, may improve the thermal stability in particular cases.

It has been found that the heat denaturation of enzymes is primarily due to the proteins' interactions with the aqueous environment. They are generally more stable in concentrated, rather than dilute, solutions. In a dry or predominantly dehydrated state, they remain active for considerable periods even at temperatures above 100°C. This property has great technological significance and is currently being exploited by the use of organic solvents.

Pressure changes will also affect enzyme catalysed reactions. Clearly any reaction involving dissolved gases (e.g. oxygenases and decarboxylases) will be particularly affected by the increased gas solubility at high pressures. The equilibrium position of the reaction will also be shifted due to any difference in molar volumes between the reactants and products. However an additional, if rather small, influence is due to the volume changes which occur during enzymic binding and catalysis.

Some enzyme-reactant mixtures may undergo reductions in volume amounting to up to 50 ml $mole^{-1}$ during reaction due to conformational restrictions and changes in their hydration. This, in turn, may lead to a doubling of the k_{cat}, and/or a halving in the K_m for a 1000 fold increase in pressure. The relative effects on k_{cat} and K_m depend upon the relative volume changes during binding and the formation of the reaction transition states.

REVERSIBLE REACTIONS

A reversible enzymic reaction (e.g. the conversion of glucose to fructose, catalysed by glucose isomerase) may be represented by the following scheme where the reaction goes through the reversible stages of enzyme-substrate (ES) complex formation, conversion to enzyme-product (EP) complex and finally desorption of the product. No step is completely rate controlling.

$$E + S \underset{K_{-1}}{\overset{K_{+1}}{\rightleftharpoons}} ES \underset{K_{-2}}{\overset{K_{+2}}{\rightleftharpoons}} EP \underset{K_{-3}}{\overset{K_{+3}}{\rightleftharpoons}} E + P$$

Pairs of symmetrical equations may be obtained for the change in the concentration of the intermediates with time:

$$\frac{d[ES]}{df} = K_{+1}[E][S] + K_{-2}[EP] - (k_{-1} + k_{+2})[ES]$$

$$\frac{d[EP]}{df} = K_{-3}[E][P] + K_{+2}[ES] - (k_{+3} + k_{-2})[EP]$$

Assuming that there is no denaturation, the total enzyme concentration must remain constant and:

$$[E]+[ES]+[EP]=[E]_0$$

therefore:

$$\frac{d[ES]}{df}=K_{+1}([E]_0-[ES]-[EP])$$
$$[S]+k_{-2}[EP]-(k_{-1}+k_{+2})[ES]$$

gathering terms in [ES]

$$\frac{d[ES]}{df}=K_{+1}([E]_0-[EP])[S]+k_{-2}[EP]-$$
$$(k_{+1}[S]+k_{-1}+k_{+2})[ES]$$

$$\frac{\frac{d[ES]}{df}}{k_{+1}[S]+k_{-1}+k_{+2}}+[ES]=\frac{K_{+1}([E]_0-[EP])[S]+K_{-2}[EP]}{k_{+1}[S]+k_{-1}+k_{+2}}$$

and,

$$\frac{d[EP]}{df}=K_{-3}[([E]_0-[EP]-[ES])[P]+k_{+2}[ES]-$$
$$(k_{+3}+k_{-2})[EP]$$

gathering terms in [EP]

$$\frac{d[EP]}{df}=K_{-3}[([E]_0-[ES])[P]+k_{+2}[ES]-$$
$$(k_{-3}[P]+k_{+3}+k_{-2})[EP]$$

$$\frac{\frac{d[EP]}{df}}{k_{-3}[P]+k_{+3}+k_{-2}}+[EP]=\frac{K_{-3}([E]_0-[ES])[P]+K_{+2}[ES]}{k_{-3}[P]+k_{+1}+k_{-2}}$$

Under similar conditions to those for the Michaelis-Menten mechanism (e.g. under the steady-state assumptions when both d[ES]/dt and d[EP]/dt are zero, or more exactly when

$$\frac{\frac{d[ES]}{df}}{K_{+1}[S]+K_{-1}+K_{+2}}<<[ES]$$

and

$$\frac{\frac{d[EP]}{df}}{K_{-3}[P]+K_{+3}+K_{-2}}<<[EP]$$

are both true. The following equations may be derived from equation using the approximation, given by equations and collecting terms.

$$k_{+1}([E]_0-[EP])[S]+k_{-2}[EP]=(K_{+1}[S]+k_{-1}+k_{+2})[ES]$$

$$-k_{+1}[EP][S]+k_{-2}[EP]=(K_{+1}[S]+k_{-1}+k_{+2})[ES]-k_{+1}[E]_0[S]$$

$$[EP]=\frac{(k_{+1}[s]+K_{-1}+K_{+2})[ES]-K_{+1}[E]_0[S]}{K_{-2}-K_{+1}[S]}$$

Also, the following equations (symmetrical to the above) may be derived from equation by using the approximation, given by equation, and collecting terms.

$$k_{-3}([E]_0-[ES])[P]+k_{+2}[ES]=(k_{-3}[P]+k_{+3}+k_{-2})[EP]$$

$$[ES]=\frac{(k_{-3}[P]+K_{+3}+K_{-2})[EP]-k_{-3}[E]_0[P]}{K_{+2}-K_{-3}[P]}$$

Substituting for [ES] from equation into

$$k_{+1}([E]_0-[EP])[S]+k_{-2}[EP]=$$

$$\frac{(K_{+1}[S]+K_{-1}+K_{+2})\{(K_{-3}[S]+K_{+3}+K_{-2})[EP]-K_{-3}[E]_0[P]\}}{K_{+2}-K_{-3}[P]}$$

$$(K_{+2}-K_{-3}[P]\{K_{+1}([E]_0-[EP])\}=K_{+1}[S]+K_{-1}+K_{+2})\{(K_{-3}$$
$$[P]+K_{+3}+K_{-2}[EP]-K_{-3}[E]_0[P]\}$$

$$(K_{+1}K_{+2}[E]_0[S]-K_{+1}K_{+2}([EP][S]+K_{+2}K_{-2}[EP]-K_{+1}K_{-3})$$
$$[E]_0[S][P]-k_{-2}K_{-3}[EP][P]$$

$$=(K_{+1}K_{-3}[EP][S][P]+K_{+1}K_{+3}[EP][S]+K_{+2}K_{-2}$$
$$[EP][S]-K_{+1}K_{-3}[E]_0[S][P]+k_{-1}K_{+3}[EP]$$
$$+(K_{-1}K_{-2}[EP]-K_{-1}K_{-3}[E]_0[P]+K_{+2}K_{-3}[P][EP]$$
$$+K_{+2}K_{-2}[EP]-k_{+2}K_{-3}[E]_0[P]$$

Removing identical terms from both sides of the equation:

$$(K_{+1}K_{+2}[E]_0[S]-K_{+1}K_{+2}([EP][S]+K_{+2}K_{-2}$$
$$[EP]-K_{-2}K_{-3}[EP][P]$$
$$=K_{+1}K_{+3}[EP][S]+k_{+1}K_{-2}[EP]$$
$$[S]+K_{-1}K_{-3}[P][EP]+k_{-1}k_{+3}[EP]$$
$$+K_{-1}K_{-2}[EP]-K_{-1}K_{-3}[E]_0[P]+K_{+2}K_{-3}$$
$$[P][EP]+K_{+2}K_{+3}[EP]+k_{+2}k_{-2}[EP]-K_{+2}K_{-3}[E]_0[P]$$

Gathering all the terms in [EP]:

$$K_{+1}K_{+2}[EP][S]-K_{+2}K_{-2}[EP]+K_{-2}K_{-3}[EP][P]+$$
$$K_{+1}K_{+3}[EP][S]+K_{-1}K_{-3}[P][EP]$$
$$+K_{-1}K_{+3}[EP]+K_{-1}K_{-2}[EP]+K_{+2}K_{-3}$$
$$[P][EP]+K_{+2}K_{+3}[EP]+K_{+2}K_{-2}[EP]$$
$$=K_{-1}K_{-3}[E]_0[P]+K_{+1}K_{+2}[E]_0[S]$$

$$[EP] = \frac{K_{+1}K_{+2}[E]_0[S] + (K_{-1} + K_{+2})K_{-3}[E]_0[P]}{(K_{-1}K_{+3} + K_{-1}K_{-2} + K_{+2}K_{+3}) + (K_{+2} + K_{-2} + k_{+3})}$$
$$K_{+1}[S] + (K_{+2}K_{-2} + K_{-1})K_{-3}[P]$$

Also, substituting for [EP] from equation into

$$K_{-3}([E]_0 - [ES])[P] + k_{+2}[ES] =$$
$$\frac{(k_{-3}[P] + k_{+3} + k_{-2})\{(k_{+1}[S] + k_{-1} + k_{+2})[ES] - k_{+1}[E]_0[S]\}}{k_{-2} - k_{+1}[S]}$$

$$K_{-2} - k_{+1}([S]\{k_{-3}([E]_0 - [ES])[P] + k_{+2}[ES] =$$
$$(k_{-3}[P] + k_{+3} + k_{-2})\{(k_{+1}[S] + k_{-1} + k_{+2})[ES] - k_{+1}[E]_0[S]\}$$

$$K_{-3} - k_{-2}[E]_0[P] - k_{-3}k_{-2}[ES][P] + K_{+2}k_{-2}[ES] -$$
$$k_{+1}k_{-3}[E]_0[S][P] - k_{+1}k_{-3}[E]_0[S][P] - K_{+1}K_{+2}[ES][S] =$$
$$k_{+1}k_{-3}[EP][S][P] + k_{-1} + k_{-3}[ES][P] + k_{+2}k_{-3}[ES][P] -$$
$$k_{+1}k_{-3}[E]_0[S][P] + k_{+1} + k_{+3}[S][ES] + k_{-1}k_{+3}[ES]$$
$$+ K_{+2}K_{+3} - K_{+1}K_{+3}[E]_0[S] + K_{-2}K_{+3}[S][ES] +$$
$$K_{-1}K_{-2}[ES] + K_{+2}K_{-2}[ES] - K_{+1}K_{-2}[E]_0[S]$$

Removing identical terms from both sides of the equation:

$$K_{-3} - k_{-2}[E]_0[P] - k_{-3}k_{-2}[ES][P] + K_{+2}k_{-2}[ES] -$$
$$k_{+1}k_{-3}[E]_0[S][P] - k_{+1}k_{-3}[E]_0[S][P] - K_{+1}K_{+2}[ES][S] =$$
$$k_{+1}k_{-3}[EP][S][P] + k_{-1} + k_{-3}[ES][P] + k_{+2}k_{-3}[ES][P] -$$
$$k_{+1}k_{-3}[E]_0[S][P] + k_{+1} + k_{+3}[S][ES] + k_{-1}k_{+3}[ES]$$
$$+ K_{+2}K_{+3} - K_{+1}K_{+3}[E]_0[S] + K_{-2}K_{+3}[S][ES] +$$
$$K_{-1}K_{-2}[ES] + K_{+2}K_{-2}[ES] - K_{+1}K_{-2}[E]_0[S]$$

Gathering all the terms in [ES]:

$$K_{-3}K_{-2}[ES][P] - K_{-2}K_{+2}[ES] + K_{+2}K_{+1}[ES][S] +$$
$$K_{-3}K_{-1}[ES][P] + K_{-3}K_{+2}[ES][P]$$
$$+ K_{+3}K_{+1}[S][ES] + K_{+3}K_{-1}[ES] + K_{+3}K_{+2}$$
$$[ES] + K_{-2}K_{+1}[S][ES] + K_{-2}K_{-1}[ES] + K_{-2}K_{+2}[ES]$$
$$= K_{+3}K_{+1}[E]_0[S] + K_{-2}K_{+1}[E]_0[S] + K_{-3}K_{-2}[E]_0[P]$$

$$[ES] = \frac{K_{-3}K_{-2}[E]_0[P] + (K_{+3} + K_{-2})K_{+1}[E]_0[S]}{(K_{-1}K_{+3} + K_{-1}K_{-2} + K_{+2}K_{+3}) + (K_{+2} + K_{-2} + k_{+3})}$$
$$K_{+1}[S] + (K_{+2}K_{-2} + K_{-1})K_{-3}[P]$$

The net rate of reaction (i.e.. rate at which substrate is converted to product less the rate at which product is converted to substrate) may be denoted by *v* where,

$$V = k_{+2}[ES] - k_{-2}[EP]$$

Substituting from equations

$$V = \frac{k_{+2}\{k_{-3}k_{-2}[E]_0[P] + (k_{+3} + k_{-2})k_{+1}[E]_0[S]\} - k_{-2}\{k_{+1}k_{+2}[E]_0[S] + (k_{-1} + k_{+2})k_{-3}[E]_0[P]\}}{(k_{-1}k_{+3} + k_{-1}k_{-2} + k_{+2}k_{+3}) (k_{+2} + k_{-2} + k_{+3}) k_{+1}[S] + (k_{+2} + k_{-2} + k_{-1}) k_{-3}[P]}$$

Simplifying:

$$V = \frac{k_{+1}k_{+2}k_{+3}[E]_0[S] - k_{-1}k_{-2}k_{-3}[E]_0[P]\}}{(k_{-1}k_{+3} + k_{-1}k_{-2} + k_{+2}k_{+3}) (k_{+2} + k_{-2} + k_{+3}) k_{+1}[S] + (k_{+2} + k_{-2} + k_{-1}) k_{-3}[P]}$$

Therefore,

$$\frac{\frac{V^t[S]}{K_m^S} - \frac{V^r[P]}{K_m^P}}{1 + \frac{[S]}{K_m^S} + \frac{[P]}{K_m^P}}$$

where:

$$V^t = \frac{k_{+2}k_{+3}[E]_0}{k_{+2} + k_{-2} + k_{+3}}$$

$$V^r = \frac{k_{-2}k_{-1}[E]_0}{k_{-2} + k_{+2} + k_{-1}}$$

$$K_m^S = \frac{k_{-1}k_{-2} + k_{-1}k_{+3} + k_{+2}k_{+3}}{k_{+1}(k_{+2} + k_{-2} + k_{+3})}$$

$$K_m^P = \frac{k_{+3}k_{+2} + k_{+3}k_{-1} + k_{-2}k_{-1}}{k_{-3}(k_{-2} + k_{+2} + k_{-1})}$$

At equilibrium:

$$V = 0$$

and, because the numerator of equation must equal zero,

$$\frac{V^t[S]_\infty}{K_m^S} = \frac{V^r[P]_\infty}{K_m^P}$$

where $[S]_\infty$ and $[P]_\infty$ are the equilibrium concentrations of substrate and product (at infinite time). But by definition,

$$K_{eq} = \frac{[P]\infty}{[S]\infty}$$

Substituting from equation

$$K_{eq} = \frac{V^t K_m^P}{V^r K_m^S}$$

This is the Haldane relationship.

Therefore:

$$V = \frac{\dfrac{V^t}{K_m^S}\left([S] - \dfrac{P}{K_{eq}}\right)}{1 + \dfrac{[S]}{K_m^S} + \dfrac{[P]}{K_m^P}}$$

If $K_m{}^S$ and $K_m{}^P$ are approximately equal (e.g. the commercial immobilised glucose isomerase, Sweetase, has K_m(glucose) of 840 mM and K_m(fructose) of 830 mM at 70°C), and noting that the total amount of substrate and product at any time must equal the sum of the substrate and product at the start of the reaction:

$$[S] + [P] = [S]_0 + [P]_0$$

Therefore:

$$V = \frac{V^t\left([S] - \dfrac{P}{K_{eq}}\right)}{K_m^S + [S]_0 + [P]_0}$$

Therefore:

$$V = K'\left([S] - \frac{P}{K_{eq}}\right)$$

where:

$$K' = \frac{V^t}{K_m^S + [S]_0 + [P]_0}$$

K' is not a true kinetic constant as it is only constant if the initial substrate plus product concentration is kept constant.

Also,

$$[S] + [P] = [S]_\infty + [P]_\infty$$

Substituting from equation, $[P] = k_{eq}[S]_\infty + [S]_\infty - [S]$

Let $[S^\#]$ equal the concentration difference between the actual concentration of substrate and the equilibrium concentration.

$$[S^\#] = [S] - [S]_\infty$$

Therefore:

$$[P] = K_{eq}[S]_\infty[S]_\infty - [S^\#]$$

Substituting in equation

$$V = \frac{\frac{V^t[S^\#]}{K_m^S} + \frac{V^t[S]_\infty}{K_m^S} - \frac{V^tK_{eq}[S]_\infty}{K_m^P} + \frac{V^t[S^\#]}{K_m^P}}{1 + \frac{[S^\#]}{K_m^S} + \frac{[S]_\infty}{K_m^S} + \frac{K_{eq}[S]_\infty}{K_m^P} - \frac{[S^\#]}{K_m^P}}$$

Rearranging equation,

$$\frac{V^t}{K_m^S} = \frac{V^rK_{eq}}{K_m^P}$$

Therefore:

$$V = \frac{\left(\frac{V^t}{K_m^S} = \frac{V^r}{K_m^P}\right)[S^\#]}{1 + \frac{[S]_\infty}{K_m^S} + \frac{K_{eq}[S]_\infty}{K_m^P} + \frac{\left(K_m^P - K_m^S\right)[S^\#]}{K_m^S K_m^P}}$$

Therefore:

$$V = \frac{V[S^\#]}{K + [S^\#]}$$

Where

$$V = \frac{K_m^P V^t + K_m^S V^t}{K_m^P - K_m^S}$$

Therefore:

$$V = \frac{K_m^P V^t + \frac{K_m^P V^t}{K_{eq}}}{K_m^P - K_m^S}$$

Therefore:

$$V = \left(\frac{K_{eq} + 1}{Keq}\right) \times \left(\frac{K_m^P}{K_m^P - K_m^S}\right) \times V^t$$

and:

$$K = \frac{K_m^S K_m^P + (K_m^P + K_m^S K_{eq})[S]_\infty}{K_m^P - K_m^S}$$

As in the case of K' in equation, *K* is not a true kinetic constant as it varies with $[S]_\infty$ and hence the sum of $[S]_0$ and $[P]_0$. It is only constant if the initial substrate plus product concentration is kept constant. By a similar but symmetrical argument, the net reverse rate of reaction,

$$V_{rev} = \frac{V_{rev}[P^\#]}{K_{rev} + [P^\#]}$$

with constants defined as above but by symmetrically exchanging $K_m{}^P$ with $K_m{}^S$, and V^r with V^f.

Both equations are useful when modelling reversible reactions, particularly the technologically important reaction catalysed by glucose isomerase. They may be developed further to give productivity-time estimates and for use in the comparison of different reactor configurations.

Although an enzyme can never change the equilibrium position of a catalysed reaction, as it has no effect on the standard free energy change involved, it can favour reaction in one direction rather than its reverse. It achieves this by binding strongly, as enzyme-reactant complexes, the reactants in this preferred direction but only binding the product(s) weakly. The enzyme is bound up with the reactant(s), encouraging their reaction, leaving little free to catalyse the reaction in the reverse direction. It is unlikely, therefore, that the same enzyme preparation would be optimum for catalysing a reversible reaction in both directions.

ENZYME INHIBITION

A number of substances may cause a reduction in the rate of an enzyme catalysed reaction. Some of these (e.g. urea) are non-specific protein denaturants. Others, which generally act in a fairly specific manner, are known as inhibitors. Loss of activity may be either reversible, where activity may be restored by the removal of the inhibitor, or irreversible, where the loss of activity is time dependent and cannot be recovered during the timescale of interest. If the inhibited enzyme is totally inactive, irreversible inhibition behaves as a time-dependent loss of enzyme concentration (i.e.. lower V_{max}), in other cases, involving incomplete inactivation, there may be time-dependent changes in both K_m and V_{max}. Heavy metal ions (e.g. mercury and lead) should generally be prevented from coming into contact with enzymes as they usually cause such irreversible inhibition by binding strongly to the amino acid backbone. More important for most enzyme-catalysed processes is the effect of reversible inhibitors.

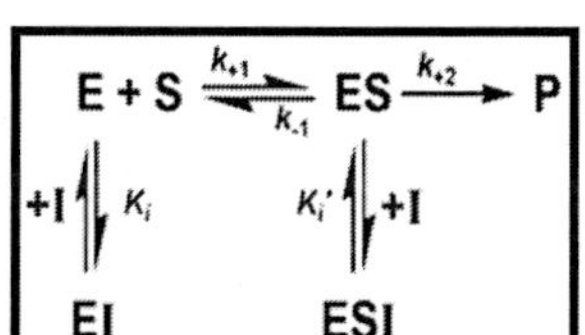

where I represents the reversible inhibitor and the inhibitory (dissociation) constants K_i and K_i' are given by

$$K_i = \frac{[E][I]}{[I]}$$

and,

$$K_i' = \frac{[E][I]}{[ESI]}$$

For the present purposes, it is assumed that neither EI nor ESI may react to form product. Equilibrium between EI and ESI is allowed, but makes no net contribution to the rate equation as it must be equivalent to the equilibrium established through:

$$EI + S \rightleftharpoons EI + S + I \rightleftharpoons ES + I \rightleftharpoons ESI$$

Binding of inhibitors may change with the pH of the solution, for substrate binding, and result in the independent variation of both K_i and K_i' with pH.

In order to simplify the analysis substantially, it is necessary that the rate of product formation (k_{+2}) is slow relative to the establishment of the equilibria between the species.

Therefore:

$$K_m = \frac{K_{-1}}{K_{+1}} + \frac{[E][S]}{[ES]}$$

also:

$$\frac{V}{V_{max}} = \frac{[ES]}{[E]_0}$$

where:

$$[E]_0 = [E] + [EI] + [ES] + [ESI]$$

therefore:

$$\frac{V}{V_{max}} = \frac{[ES]}{[E] + [EI] + [ES] + ESI}$$

Substituting from equations followed by simplification, gives:

$$\frac{V}{V_{max}} = \frac{1}{\frac{K_m}{[S]} + \frac{K_m[I]}{[S]k_i} + 1 + \frac{[I]}{K_i}}$$

therefore:

$$V = \frac{V_{max}[S]}{K_m\left(1 + \frac{[I]}{K_i}\right) + [S]\left(1 + \frac{[I]}{K_i}\right)}$$

If the total enzyme concentration is much less than the total inhibitor concentration (i.e. $[E]_0 << [I]_0$), then:

$$V = \frac{V_{max}[S]}{K_m\left(1+\frac{[I]}{K_i}\right)+[S]\left(1+\frac{[I]}{K_i}\right)}$$

This is the equation used generally for mixed inhibition involving both EI and ESI complexes. A number of simplified cases exist.

COMPETITIVE INHIBITION

K_i' is much greater than the total inhibitor concentration and the ESI complex is not formed. This occurs when both the substrate and inhibitor compete for binding to the active site of the enzyme. The inhibition is most noticeable at low substrate concentrations but can be overcome at sufficiently high substrate concentrations as the V_{max} remains unaffected. The rate equation is given by:

$$V = \frac{V_{max}[S]}{K_m^{app}+[S]}$$

where K_m^{app} is the apparent K_m for the reaction, and is given by:

$$K_m^{app} = K_m\left(1+\frac{[I]}{K_i}\right)$$

Normally the competitive inhibitor bears some structural similarity to the substrate, and often is a reaction product (product inhibition, e.g. inhibition of lactase by galactose), which may cause a substantial loss of productivity when high degrees of conversion are required.

The rate equation for product inhibition is derived from equations.

$$V = \frac{V_{max}[S]}{K_m\left(1+\frac{[P]}{K_P}\right)+[S]}$$

A similar effect is observed with competing substrates, quite a common state of affairs in industrial conversions, and especially relevant to macromolecular hydrolyses where a number of different substrates may coexist, all with different kinetic parameters. The reaction involving two co-substrates may be modelled by the scheme.

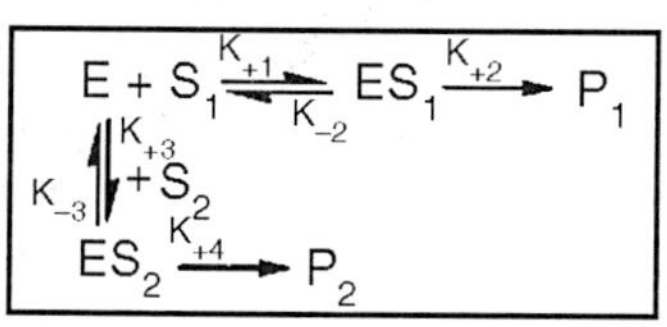

Both substrates compete for the same catalytic site and, therefore, their binding is mutually exclusive and they behave as competitive inhibitors of each others reactions. If the rates of product formation are much slower than attainment of the equilibria (i.e. k_{+2} and k_{+4} are very much less than k_{-1} and k_{-3} respectively), the rate of formation of P_1 is given by:

$$V_1 = \frac{V^1{}_{max}[S_1]}{K^1_m\left(1+\frac{[S_2]}{K^2_m}\right)+[S_1]}$$

and the rate of formation of P_2 is given by

$$V_2 = \frac{V^2{}_{max}[S_2]}{K^2_m\left(1+\frac{[S_1]}{K^1_m}\right)+[S_2]}$$

If the substrate concentrations are both small relative to their K_m values:

$$\frac{V_1}{V_2} = \frac{V^1_{max}/K^1_m}{V^2_{max}/K^2_m} \times \frac{[S_1]}{[S_2]}$$

Therefore, in a competitive situation using the same enzyme and with both substrates at the same concentration:

$$\frac{V_1}{V_2} = \frac{K^1_{cat}/K^1_m}{V^2_{cat}/K^2_m}$$

where $K^1_{cat} = K_{+2}$ and $K^2_{cat} = K_{+4}$ > in this simplified case. The relative rates of reaction are in the ratio of their specificity constants. If both reactions produce the same product (e.g. some hydrolyses):

$$V_1 = \frac{V^1_{max}[S_1]}{K^1_m\left(1+\frac{[S_1]}{K^1_m}+\frac{[S_2]}{K^2_m}\right)} + \frac{V^2_{max}[S_2]}{K^2_m\left(1+\frac{[S_1]}{K^1_m}+\frac{[S_2]}{K^2_m}\right)}$$

therefore:

$$V_1 = \frac{\frac{V^1_{max}[S_1]}{K^1_m}+\frac{V^2_{max}[S_2]}{K^2_m}}{1+\frac{[S_1]}{K^1_m}+\frac{[S_2]}{K^2_m}}$$

Uncompetitive Inhibition

K_i is much greater than the total inhibitor concentration and the EI complex is not formed. This occurs when the inhibitor binds to a site which only becomes available after the substrate (S_1) has bound to the active site of the enzyme. This inhibition is most commonly encountered in multi-substrate reactions where the inhibitor is competitive with respect to one substrate (e.g. S_2) but uncompetitive with respect to another (e.g. S_1), where the reaction scheme may be represented by

$$E + S \rightleftharpoons ES_1 + S_2 \rightleftharpoons ES_1S_2 \longrightarrow \text{product}$$

$$E + S_1 \rightleftharpoons ES_1 + I \rightleftharpoons ES_1I \not\longrightarrow s$$

The inhibition is most noticeable at high substrate concentrations (i.e. S_1 in the scheme above) and cannot be overcome as both the V_{max} and K_m are equally reduced. The rate equation is:

$$V = \frac{V^{app}_{max}[S]}{V^{app}_{max} + [S]}$$

where $V_{max}{}^{app}$ and $K_m{}^{app}$ are the apparent V_{max} and K_m given by:

$$V^{app}_{max} = \frac{V_{max}}{1 + \frac{[I]}{K_i}}$$

and

$$K^{app}_{m} = \frac{K_m}{1 + \frac{[I]}{K_i}}$$

In this case the specificity constant remains unaffected by the inhibition. Normally the uncompetitive inhibitor also bears some structural similarity to one of the substrates and, again, is often a reaction product.

A special case of uncompetitive inhibition is substrate inhibition which occurs at high substrate concentrations in about 20 per cent of all known enzymes (e.g. invertase is inhibited by sucrose). It is primarily caused by more than one substrate molecule binding to an active site meant for just one, often by different parts of the substrate molecules binding to different subsites within the substrate binding site. If the resultant complex is inactive this type of inhibition causes a reduction in the rate of reaction, at high substrate concentrations. It may be modelled by the following scheme

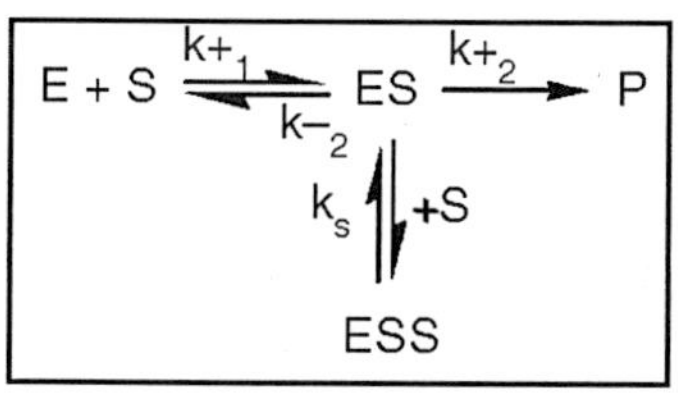

where:

$$K_s = \frac{[ES][S]}{[ESS]}$$

The assumption is made that ESS may not react to form product. It follows from equation that:

$$V = \frac{V_{max}[S]}{K_m + [S]\left(1 + \frac{[S]}{K_s}\right)}$$

Even quite high values for K_S lead to a levelling off of the rate of reaction at high substrate concentrations, and lower K_S values cause substantial inhibition.

Noncompetitive Inhibition

Both the EI and ESI complexes are formed equally well (i.e. K_i equals K_i'). This occurs when the inhibitor binds at a site away from the substrate binding site, causing a reduction in the catalytic rate.

It is quite rarely found as a special case of mixed inhibition. The fractional inhibition is identical at all substrate concentrations and cannot be overcome by increasing substrate concentration due to the reduction in V_{max}. The rate equation is given by:

$$V = \frac{V_{max}^{app}[S]}{K_m + [S]}$$

where $V_{max}{}^{app}$ is given by:

$$V_{max}^{app} = \frac{V_{max}}{1 + \frac{[I]}{K_i}}$$

The diminution in the rate of reaction with pH, described earlier, may be considered as a special case of non-competitive inhibition; the inhibitor being the hydrogen ion on the acid side of the optimum or the hydroxide ion on the alkaline side.

Determination of V_{max} and K_m

It is important to have as thorough knowledge as is possible of the performance characteristics of enzymes, if they are to be used most efficiently. The kinetic parameters V_{max}, K_m and k_{cat}/K_m should, therefore, be determined. There are two approaches to this problem using either the reaction progress curve (integral method) or the initial rates of reaction (differential method).

Use of either method depends on prior knowledge of the mechanism for the reaction and, at least approximately, the optimum conditions for the reaction. If the mechanism is known and complex then the data must be reconciled to the appropriate model (hypothesis), usually by use of a computer-aided analysis involving a weighted least-squares fit. Many such computer programmes are currently available and, if not, the programming skill involved is usually fairly low. If the mechanism is not known, initial attempts are usually made to fit the data to the Michaelis-Menten kinetic model. Combining equations,

$$\frac{d[P]}{df} = V = \frac{V_{max}[S]}{K_m + [S]}$$

which, on integration, using the boundary condition that the product is absent at time zero and by substituting [S] by ($[S]_0$ - [P]), becomes

$$t = \frac{[P]}{V_{max}} - \frac{K_m}{V_{max}} Ln \frac{[S]_0 - [P]}{[S]_0}$$

If the fractional conversion (*X*) is introduced, where

$$X = \frac{[S]_0 - [P]}{[S]_0}$$

then equation may be simplified to give:

$$t = \frac{X[S]_0 - [P]}{V_{max}} - \frac{K_m Ln(1 - X)}{V_{max}}$$

Use of equation involves the determination of the initial rate of reaction over a wide range of substrate concentrations. The initial rates are used, so that $[S] = [S]_0$, the predetermined and accurately known substrate concentration at the start of the reaction. Its use also ensures that there is no effect of reaction reversibility or product inhibition which may affect the integral method based on equation. Equation can be utilised directly using a computer programme, involving a weighted least-squares fit, where the parameters for determining the hyperbolic relationship between the initial rate of reaction and initial substrate concentration (i.e.. K_m and V_{max}) are chosen in order to minimise the errors between the data and the model, and the assumption is made that the errors inherent in the practically determined data are normally distributed about their mean (error-free) value. Alternatively the direct linear plot may be used. This is a powerful non-parametric statistical method which depends upon the assumption that any errors in the experimentally derived data are as likely to be positive as negative. It is common practice to show the data obtained by the above statistical methods on one of three linearised plots, derived from equation. Of these, the double reciprocal plot is preferred to test for the qualitative correctness of a proposed mechanism, and the Eadie-Hofstee plot is preferred for discovering deviations from linearity.

Relation Between Hyperbolic and the Initial Rate of Reaction and the Initial Substrate Concentration

Three ways in which the hyperbolic relationship between the initial rate of reaction and the initial substrate concentration

$$V = \frac{V_{max}[S]_0}{K_m + [S]_0}$$

can be rearranged to give linear plots. The examples are drawn using K_m = 2 mM and V_{max} = 4 mM min^{-1}.

(a) Lineweaver-Burk (double-reciprocal) plot of 1/v against $1/[S]_0$ giving intercepts at $1/V_{max}$ and $-1/K_m$

$$\frac{1}{V}=\frac{K_m}{V_{max}}\times\frac{1}{[s]_0}+\frac{1}{V_{max}}$$

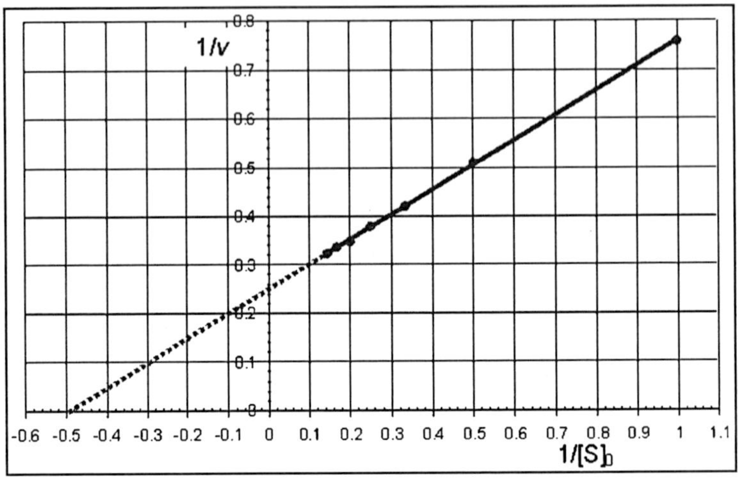

(b) Eadie-Hofstee plot of v against $v/[S]_0$ giving intercepts at V_{max} and V_{max}/K_m

$$V=-K_m\times\frac{V}{[S]_0}+V_{max}$$

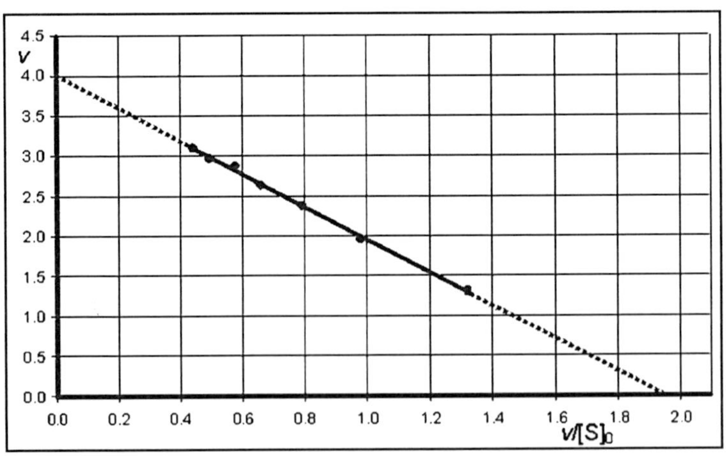

c) Hanes-Woolf (half-reciprocal) plot of $[S]_0/v$ against $[S]_0$ giving intercepts at K_m/V_{max} and K_m.

$$\frac{[S]_0}{V}=\frac{1}{V_{max}}\times[S]_0+\frac{K_m}{V_{max}}$$

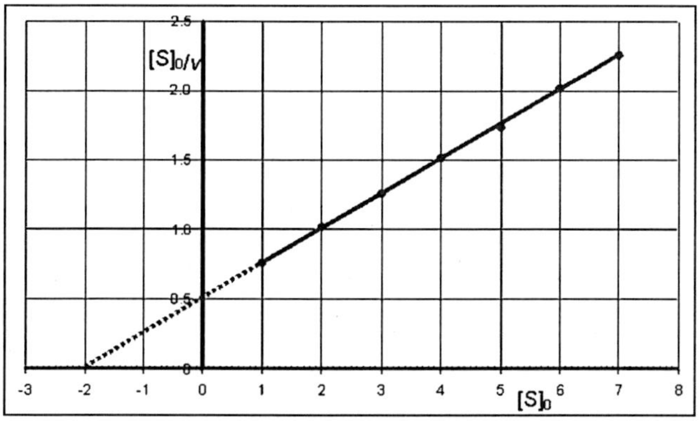

The progress curve of the reaction can be used to determine the specificity constant (k_{cat}/K_m) by making use of the relationship between time of reaction and fractional conversion. This has the advantage over the use of the initial rates in that fewer determinations need to be made, possibly only one progress curve is necessary, and sometimes the initial rate of reaction is rather difficult to determine due to its rapid decline. If only the early part of the progress curve, or its derivative, is utilised in the analysis, this procedure may even be used in cases where there is competitive inhibition by the product, or where the reaction is reversible.

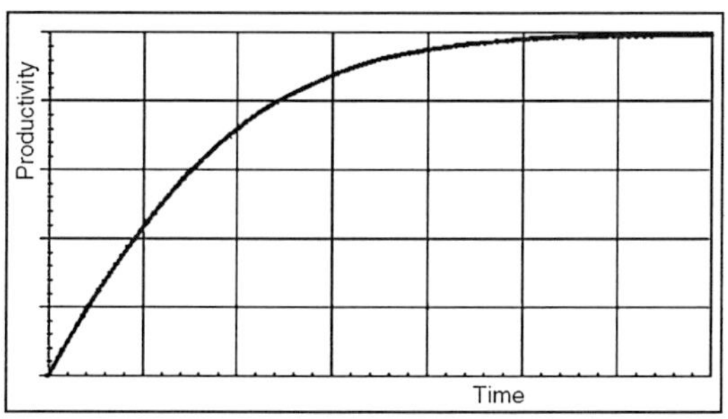

Fig. A Schematic Plot Showing the Amount of Product Formed (productivity).

Against the time of Reaction, in a Closed System. The Specificity Constant may be Determined by a Weighted Least-squared fit of the data to the Relationship given by Equation. The type of inhibition and the inhibition constants may be determined from the effect of differing concentrations of inhibitor on the apparent K_m, V_{max} and k_{cat}/K_m, although some more specialised plots do exist.

7

Photosynthetic Energy Transformation

ROLE OF PHOTOSYNTHESIS

Photosynthesis is the physico-chemical process by which plants, algae and photosynthetic bacteria use light energy to drive the synthesis of organic compounds. In plants, algae and certain types of bacteria, the photosynthetic process results in the release of molecular oxygen and the removal of carbon dioxide from the atmosphere that is used to synthesize carbohydrates (oxygenic photosynthesis).

Other types of bacteria use light energy to create organic compounds but do not produce oxygen (anoxygenic photosynthesis). Photosynthesis provides the energy and reduced carbon required for the survival of virtually all life on our planet, as well as the molecular oxygen necessary for the survival of oxygen consuming organisms1.In addition, the fossil fuels currently being burned to provide energy for human activity were produced by ancient photosynthetic organisms. Although photosynthesis occurs in cells or organelles that are typically only a few microns across, the process has a profound impact on the earth's atmosphere and climate.

Each year more than 10 per cent of the total atmospheric carbon dioxide is reduced to carbohydrate by photosynthetic organisms. Most, if not all, of the reduced carbon is returned to the atmosphere as carbon dioxide by microbial, plant and animal metabolism, and by biomass combustion. In turn, the performance of photosynthetic organisms depends on the earth's atmosphere and climate.

Over the next century, the large increase in the amount of atmospheric carbon dioxide created by human activity is certain to have a profound impact on the performance and competition of photosynthetic organisms. Knowledge of the physico-chemical process of photosynthesis is essential for understanding the relationship between living organisms and the atmosphere and the balance of life on earth. Several books on photosynthesis are available for the uninitiated or advanced student. Taiz and Zeiger place the photosynthetic process in the context of over all plant physiology, and Cramer and Knaff describe the bioenergetic foundation of photosynthesis. The overall equation for photosynthesis is deceptively simple. In fact, a complex set of physical and chemical reactions must occur in a coordinated manner for the synthesis of carbohydrates. To produce a sugar molecule such as sucrose, plants require nearly 30 distinct proteins that work within a complicated membrane structure.

Research into the mechanism of photosynthesis centers on understanding the structure of the photosynthetic components and the molecular processes that use radiant energy to drive carbohydrate synthesis. The research involves several disciplines, including physics, biophysics, chemistry, structural biology, biochemistry, molecular biology and physiology, and serves as an outstanding example of the success of multidisciplinary research. As such, photosynthesis presents a special challenge in understanding several interrelated molecular processes.

BRIEF HISTORY

In the 1770s Joseph Priestley, an English chemist and clergyman, performed experiments showing that plants release a type of air that allows combustion. He demonstrated this by burning a candle in a closed vessel until the flame went out. He placed a sprig of mint in the chamber and after several days showed that the candle could burn again. Although Priestley did not know about molecular oxygen, his work showed that plants release oxygen into the atmosphere. It is noteworthy that over 200 years later, investigating the mechanism by which plants produce oxygen is one of the most active areas of photosynthetic research.

Building on the work of Priestley, Jan Ingenhousz, a Dutch physician, demonstrated that sunlight was necessary for photosynthesis and that only the green parts of plants could release oxygen. During this period Jean Senebier, a Swiss botanist and naturalist, discovered that CO_2 is required for photosynthetic growth and Nicolas- The odoure de Saussure, a Swiss chemist and plant physiologist, showed that water is required. It was not until 1845 that Julius Robert von Mayer, a German physician and physicist, proposed that photosynthetic organisms convert light energy into chemical free energy. An interesting time line of the history of photosynthesis has been presented by Huzisige and Ke. By the middle of the nineteenth century the key features of plant photosynthesis were known, namely, that plants could use light energy to make carbohydrates from CO_2 and water. The empirical equation representing the net reaction of photosynthesis for oxygen evolving organisms is :

$$CO_2 + 2H_2O + \text{Light Energy} \rightarrow [CH_2O] + O_2 + H_2O, \quad (1)$$

where [CH2O] represents a carbohydrate (e.g., glucose, a six-carbon sugar). The synthesis of carbohydrate from carbon and water requires a large input of light energy. The standard free energy for the reduction of one mole of CO_2 to the level of glucose is +478 kJ/mol. Because glucose, a six carbon sugar, is often an intermediate product of photosynthesis, the net equation of photosynthesis is frequently written as:

$$6CO_2 + 12H_2O + \text{Light Energy} \rightarrow C_6H_{12}O_6 + 6O_2 + 6H_2O. \quad (2)$$

The standard free energy for the synthesis of glucose is +2,870 kJ/mol. Not surprisingly, early scientists studying photosynthesis concluded that the O_2 released by plants came from CO_2, which was thought to be split by light energy. In the 1930s comparison of bacterial and plant photosynthesis lead Cornelis van Niel to propose the general equation of photosynthesis that applies to plants, algae and photosynthetic bacteria. Van Niel was aware that some photosynthetic bacteria could use hydrogen sulfide (H_2S) instead of water for photosynthesis and that these organisms released sulfur instead of oxygen. Van Niel, among others, concluded that photosynthesis depends on electron donation and acceptor reactions and that the O_2 released during photosynthesis comes from the oxidation of water. Van Niel's generalized equation is :

$$CO_2 + 2H_2A + \text{Light Energy} \rightarrow [CH_2O] + 2A + H_2O. \quad (3)$$

In oxygenic photosynthesis, 2A is O_2, whereas in anoxygenic photosynthesis, which occurs in some photosynthetic bacteria, the electron donor can be an inorganic hydrogen donor, such as H2S (in which case A is elemental sulfur) or an organic hydrogen donor such as succinate (in which case, A is fumarate). Experimental evidence that molecular oxygen came from water was provided by Hill and Scarisbrick (1940) who demonstrated oxygen evolution in the absence of CO_2 in illuminated chloroplasts and by Ruben et al. (1941) who used 18O enriched water.

The biochemical conversion of CO_2 to carbohydrate is a reduction reaction that involves the rearrangement of covalent bonds between carbon, hydrogen and oxygen. The energy for the reduction of carbon is provided by energy rich molecules that are produced by the light driven electron transfer reactions. Carbon reduction can occur in the dark and involves a series of biochemical reactions that were elucidated by Melvin Calvin, Andrew Benson and James Bassham in the late 1940s and 1950s. Using the radioisotope 14C, most of the intermediate steps that result in the production of carbohydrate were identified. Calvin was awarded the Nobel Prize for Chemistry in 1961 for this work.

In 1954 Daniel Arnon and coworkers discovered that plants, and A. Frenkel discovered that photosynthetic bacteria, use light energy to produce ATP, an organic molecule that serves as an energy source for many biochemical reactions. During the same period L.N.M. Duysens showed that the primary photochemical reaction of photosynthesis is an oxidation/reduction reaction that occurs in a protein complex (the reaction centre). Over the next few years the work of several groups, including those of Robert Emerson, Bessel Kok, L.N.M. Duysens, Robert Hill and Horst Witt, combined to prove that plants, algae and cyanobacteria require two reaction centers, photosystem II and photosystem

I, operating in series. In 1961 Peter Mitchell suggested that cells can store energy by creating an electric field or a proton gradient across a membrane. Mitchell's proposal that energy is stored as an electrochemical gradient across a vesicular membrane opened the door for understanding energy transformation by membrane systems. He was awarded the Nobel Prize in Chemistry in 1978 for his theory of chemiosmotic energy transduction.

Most of the proteins required for the conversion of light energy and electron transfer reactions of photosynthesis are located in membranes. Despite decades of work, efforts to determine the structure of membrane bound proteins had little success. This changed in the 1980s when Johann Deisenhofer, Hartmut Michel, Robert Huber and co-workers determined the structure of the reaction centre of the purple bacterium Rhodospeudomonas viridis.. They were awarded the Nobel Prize for Chemistry in 1988 for their work, which has provided insight into the relationship between structure and function in membrane-bound proteins.

A key element in photosynthetic energy conversion is electron transfer within and between protein complexes and simple organic molecules. The electron transfer reactions are rapid (as fast as a few picoseconds) and highly specific. Much of our current understanding of the physical principles that guide electron transfer is based on the pioneering work of Rudolph A. Marcus, who received the Nobel Prize in Chemistry in 1992 for his contributions to the theory of electron transfer reaction in chemical systems.

CLASSIFICATION OF PHOTOSYNTHETIC ORGANISMS

All life can be divided into three domains, Archaea, Bacteria and Eucarya, which originated from a common ancestor. Historically, the term photosynthesis has been applied to organisms that depend on chlorophyll (or bacteriochlorophyll) for the conversion of light energy into chemical free energy. These include organisms in the domains Bacteria (photosynthetic bacteria) and Eucarya (algae and higher plants). The most primitive domain, Archaea, includes organisms known as halobacteria, that convert light energy into chemical free energy. However, the mechanism by which halobacteria convert light is fundamentally different from that of higher organisms because there is no oxidation/reduction chemistry and halobacteria cannot use CO_2 as their carbon source. Consequently some biologists do not consider halobacteria as photosynthetic. This chapter will follow the historical definition of photosynthesis and omit halobacteria.

Oxygenic Photosynthetic Organisms

The photosynthetic process in all plants and algae as well as in certain types of photosynthetic bacteria involves the reduction of CO_2 to carbohydrate and removal of electrons from H_20, which results in the release of O_2. In this process, known as oxygenic photosynthesis, water is oxidized by the photosystem II reaction centre, a multisubunit protein located in the photosynthetic membrane.

Years of research have shown that the structure and function of photosystem II is similar in plants, algae and certain bacteria, so that knowledge gained in one species can be applied to others. This homology is a common feature of proteins that perform the same reaction in different species. This homology at the molecular level is important because there are estimated to be 300,000-500,000 species of plants. If different species had evolved diverse mechanisms for oxidizing water, research aimed at a general understanding of photosynthetic water oxidation would be hopeless.

Anoxygenic Photosynthetic Organisms

Some photosynthetic bacteria can use light energy to extract electrons from molecules other than water. These organisms are of ancient origin, presumed to have evolved before oxygenic photosynthetic organisms. Anoxygenic photosynthetic organisms occur in the domain Bacteria and have representatives in four phyla - Purple Bacteria, Green Sulfur Bacteria, Green Gliding Bacteria, and Gram Positive Bacteria.

PRINCIPLES OF PHOTOSYNTHETIC ENERGY TRANSFORMATION

The energy that drives photosynthesis originates in the centre of the sun, where mass is converted to heat by the fusion of hydrogen. Over time, the heat energy reaches the sun's surface, where some of it is converted to light by black body radiation that reaches the earth. A small fraction of the visible light incident on the earth is absorbed by plants. Through a series of energy transducing reactions, photosynthetic organisms are able to transform light energy into chemical free energy in a stable form that can last for hundreds of millions of years (e.g., fossil fuels). A simplified scheme describing how energy is transformed in the photosynthetic process is presented in this section. The focus is on the structural and functional features essential for the energy transforming reactions. For clarity, mechanistic and structural details are omitted. A more highly resolved description of oxygenic and anoxygenic photosynthesis is given in the remaining sections.

The photosynthetic process in plants and algae occurs in small organelles known as chloroplasts that are located inside cells. The more primitive photosynthetic organisms, for example oxygenic cyanobacteria, prochlorophytes and anoxygenic photosynthetic bacteria, lack organelles. The photosynthetic reactions are traditionally divided into two stages - the "light reactions," which consist of electron and proton transfer reactions and the "dark reactions," which consist of the biosynthesis of carbohydrates from CO_2.The light reactions occur in a complex membrane system (the photosynthetic membrane) that is made up of protein complexes, electron carriers, and lipid molecules. The photosynthetic membrane is surrounded by water and can be thought of as a two-dimensional surface that defines a closed space, with an inner and outer water phase.

A molecule or ion must pass through the photosynthetic membrane to go from the inner space to the outer space. The protein complexes embedded in the photosynthetic membrane have a unique orientation with respect to the inner and outer phase. The asymmetrical arrangement of the protein complexes allows some of the energy released during electron transport to create an electrochemical gradient of protons across the photosynthetic membrane.

Photosynthetic electron transport consists of a series of individual electron transfer steps from one electron carrier to another. The electron carriers are metal ion complexes and aromatic groups. The metal ion complexes and most of the aromatic groups are bound within proteins. Most of the proteins involved in photosynthetic electron transport are composed of numerous polypeptide chains that lace through the membrane, providing a scaffolding for metal ions and aromatic groups. An electron enters a protein complex at a specific site, is transferred within the protein from one carrier to another, and exits the protein at a different site. The protein controls the pathway of electrons between the carriers by determining the location and environment of the metal ion complexes and aromatic groups. By setting the distance between electron carriers and controlling the electronic environment surrounding a metal ion complex or aromatic group, the protein controls pairwise electron transfer reactions. Between proteins, electron transfer is controlled by distance and free energy, as for intraprotein transfer, and by the probability that the two proteins are in close contact. Protein association is controlled by a number of factors, including the structure of the two proteins, their surface electrical and chemical properties and the probability that they collide with one another. Not all electron carriers are bound to proteins. The reduced forms of plastoquinone or ubiquinone and nicotinamide adenine dinucleotide phosphate (NADPH) or NADH act as mobile electron carriers operating between protein complexes. For electron transfer to occur, these small molecules must bind to special pockets in the proteins known as binding sites. The binding sites are highly specific and are a critical factor in controlling the rate and pathway of electron transfer.

The light reactions convert energy into several forms. The first step is the conversion of a photon to an excited electronic state of an antenna pigment molecule located in the antenna system. The antenna system consists of hundreds of pigment molecules (mainly chlorophyll or bacteriochlorophyll and carotenoids) that are anchored to proteins within the photosynthetic membrane and serve a specialized protein complex known as a reaction centre. The electronic excited state is transferred over the antenna molecules as an exciton.

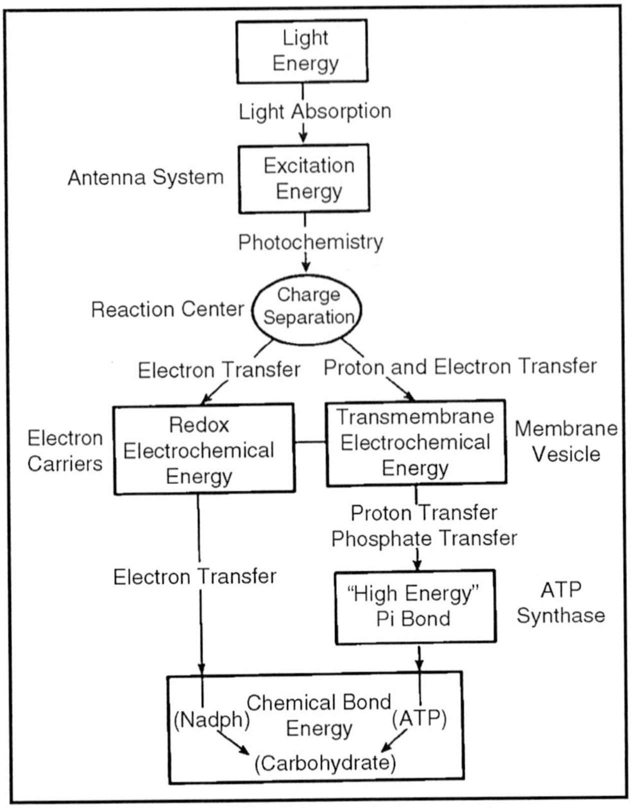

Some excitons are converted back into photons and emitted as fluorescence, some are converted to heat, and some are trapped by a reaction centre protein. Excitons trapped by a reaction centre provide the energy for the primary photochemical reaction of photosynthesis - the transfer of an electron from a donor molecule to an acceptor molecule. Both the donor and acceptor molecules are attached to the reaction centre protein complex. Once primary charge separation occurs, the subsequent electron transfer reactions are energetically downhill. In oxygenic photosynthetic organisms, two different reaction centers, known as photosystem II and photosystem I, work concurrently but in series. In the light photosystem II feeds electrons to photosystem I. The electrons are transferred from photosystem II to the photosystem I by intermediate carriers. The net reaction is the transfer of electrons from a water molecule to $NADP^+$, producing the reduced form, NADPH. In the photosynthetic process, much of the energy initially provided by light energy is stored as redox free energy (a form of chemical free energy) in NADPH, to be used later in the reduction of carbon. In addition, the electron transfer reactions concentrate protons inside the membrane vesicle and create an electric field across the photosynthetic membrane. In this process the electron transfer reactions convert redox free energy into an electrochemical potential of protons. The energy stored in the proton electrochemical potential is used by a membrane bound protein complex (ATP-Synthase) to covalently attach a phosphate group to adenosine diphosphate (ADP), forming adenosine triphosphate (ATP). Protons pass through the ATP-Synthase protein complex that transforms electrochemical free energy into a type of chemical free energy known as phosphate group-transfer potential (or a high-energy phosphate bond).

The energy stored in ATP can be transferred to another molecule by transferring the phosphate group. The net effect of the light reactions is to convert radiant energy into redox free energy in the form of NADPH and phosphate group-transfer energy in the form of ATP. In the light reactions, the transfer of a single electron from water to $NADP^+$ involves about 30 metal ions and 7 aromatic groups.

The metal ions include 19 Fe, 5 Mg, 4 Mn, and 1 Cu. The aromatics include quinones, pheophytin, NADPH, tyrosine and a flavoprotein. The NADPH and ATP formed by the light reactions provide the energy for the dark reactions of photosynthesis, known as the Calvin cycle or the photosynthetic carbon reduction cycle. The reduction of atmospheric CO_2 to carbohydrate occurs in the aqueous phase of the chloroplast and involves a series of enzymatic reactions. The first step is catalyzed by the protein Rubisco (D-ribulose 1,5-bisphosphate carboxylase/oxygenase), which attaches CO_2 to a five-carbon compound. The reaction produces two molecules of a three-carbon compound. Subsequent biochemical reactions involve several enzymes that reduce carbon by hydrogen transfer and rearrange the carbon compounds to synthesize carbohydrates. The carbon reduction cycle involves the transfer and rearrangement of chemical bond energy. In anoxygenic photosynthetic organisms water is not used as the electron donor. Electron flow is cyclic and is driven by a single photosystem, producing a proton electrochemical gradient that is used to provide energy for the reduction of NAD^+ by an external H-atom or e-donor (e.g., H_2S or an organic acid) in a process known as "reverse electron flow". Fixation of CO_2 occurs via different pathways in different organisms.

OXYGENIC PHOTOSYNTHESIS

CHLOROPLASTS - STRUCTURE AND ORGANIZATION

In plants the photosynthetic process occurs inside chloroplasts, which are organelles found in certain cells. Chloroplasts provide the energy and reduced carbon needed for plant growth and development, while the plant provides the chloroplast with CO_2, water, nitrogen, organic molecules and minerals necessary for the chloroplast biogenesis. Most chloroplasts are located in specialized leaf cells, which often contain 50 or more chloroplasts per cell. Each chloroplast is defined by an inner and an outer envelope membrane and is shaped like a meniscus convex lens that is 5-10 microns in diameter, although many different shapes and sizes can be found in plants. For details of chloroplast structure. The inner envelope membrane acts as a barrier, controlling the flux of organic and charged molecules in and out of the chloroplast. Water passes freely through the envelope membranes, as do other small neutral molecules like CO_2 and O_2. There is evidence that chloroplasts were once free living bacteria that invaded a non-photosynthetic cell long ago. They have retained some of the DNA necessary for their assembly, but much of the DNA necessary for their biosynthesis is located in the cell nucleus. This enables a cell to control the biosynthesis of chloroplasts within its domain.

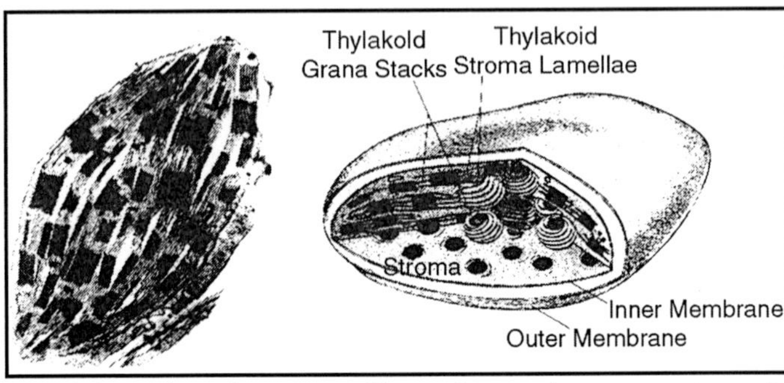

Inside the chloroplast is a complicated membrane system, known as the photosynthetic membrane (or thylakoid membrane), that contains most of the proteins required for the light reactions. The proteins required for the fixation and reduction of CO_2 are located outside the photosynthetic membrane in the surrounding aqueous phase. The photosynthetic membrane is composed mainly of glycerol lipids and protein.

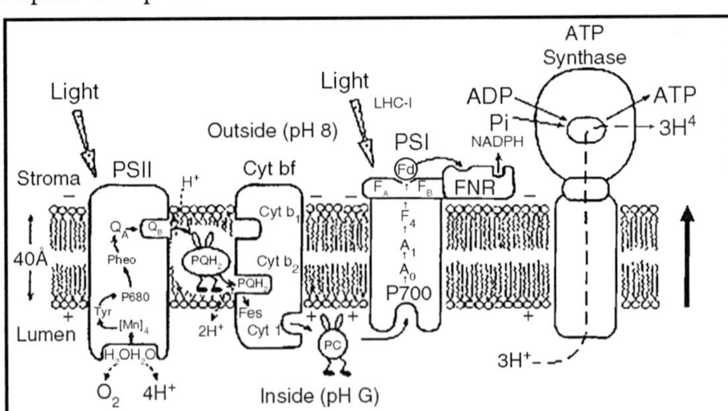

The glycerol lipids are a family of molecules characterized by a polar head group that is hydrophilic and two fatty acid side chains that are hydrophobic. In membranes, the lipid molecules arrange themselves in a bilayer, with the polar head towards the water phase and the fatty acid chains aligned inside the membrane forming a hydrophobic core. The photosynthetic membrane is vesicular, defining a closed space with an outer water space (stromal phase) and an inner water space (lumen). The organization of the photosynthetic membrane can be described as groups of stacked membranes (like stacks of pita or chapati bread with the inner pocket representing the inner aqueous space), interconnected by non-stacked membranes that protrude from the edges of the stacks. Experiments indicate that the inner aqueous space of the photosynthetic membrane is likely continuous inside of the chloroplast. It is not known why the photosynthetic membrane forms such a convoluted structure. To understand the energetics of photosynthesis the complicated structure can be ignored and the photosynthetic membrane can be viewed as a simple vesicle.

Light Absorption - The Antenna System

Plant photosynthesis is driven primarily by visible light (wavelengths from 400 to 700 nm) that is absorbed by pigment molecules (mainly chlorophyll a and b and carotenoids). The chemical structure of chlorophyll a molecule is shown in Fig..

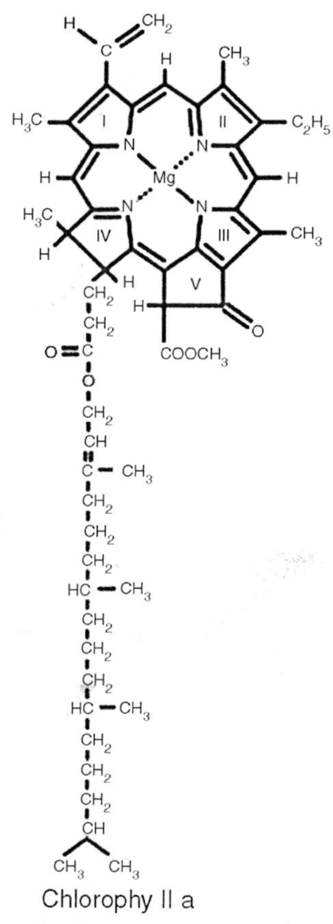

Chlorophy ll a

In chlorophyll b, CH3 in ring II is replaced by CHO group. Plants appear green because of chlorophyll, which is so plentiful that regions of the earth appear green from space. The absorption spectrum of chloroplast chlorophyll a and b and carotenoids along with the action spectrum of photosynthesis of a chloroplast is shown in Figure.

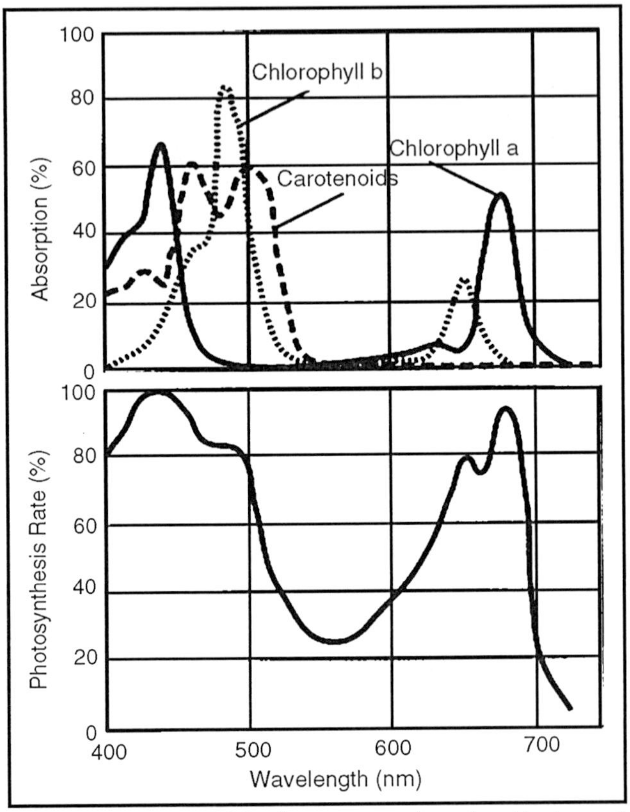

Light is collected by 200-300 pigment molecules that are bound to light-harvesting protein complexes located in the photosynthetic membrane. The light-harvesting complexes surround the reaction centers that serve as an antenna. The three-dimensional structure of the light-harvesting complex shows that the protein determines the position and orientation of the antenna pigments.

Photosynthesis is initiated by the absorption of a photon by an antenna molecule, which occurs in about a femtosecond (10-15 s) and causes a transition from the electronic ground state to an excited state. Within 10-13 s the excited state decays by vibrational relaxation to the first excited singlet state.

The fate of the excited state energy is guided by the structure of the protein. Because of the proximity of other antenna molecules with the same or similar energy states, the excited state energy has a high probability of being transferred by resonance energy transfer to a near neighbour. Exciton energy transfer between antenna molecules is due to the interaction of the transition dipole moment of the molecules. The probability of transfer is dependent on the distance between the transition dipoles of the donor and acceptor molecules (1/R6), the relative orientation of the transition dipoles, and the overlap of the emission spectrum of the donor molecule with the absorption spectrum of the acceptor molecule. Photosynthetic antenna systems are very efficient at this transfer process.

Under optimum conditions over 90 per cent of the absorbed quanta are transferred within a few hundred picoseconds from the antenna system to the reaction centre which acts as a trap for the exciton. A simple model of the antenna and its reaction centre is shown in Fig.

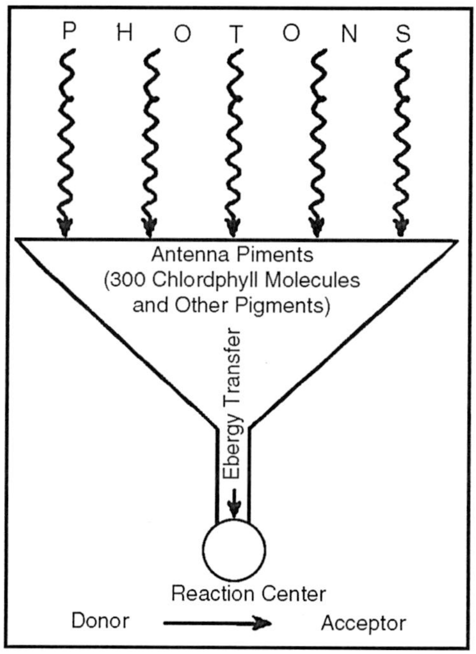

Primary Photochemistry - Photosystem II and Photosystem I Reaction Centers

Photosystem II uses light energy to drive two chemical reactions - the oxidation of water and the reduction of plastoquinone. The photosystem II complex is composed of more than fifteen polypeptides and at least nine different redox components (chlorophyll, pheophytin, plastoquinone, tyrosine, Mn, Fe, cytochrome b559, carotenoid and histidine) have been shown to undergo light-induced electron transfer.

However, only five of these redox components are known to be involved in transferring electrons from H_2O to the plastoquinone pool - the water oxidizing manganese cluster $(Mn)_4$, the amino acid tyrosine, the reaction centre chlorophyll (P680), pheophytin, and the plastoquinone molecules, QA and QB. Of these essential redox components, tyrosine, P680, pheophytin, QA and QB have been shown to be bound to two key polypeptides that form the heterodimeric reaction centre core of photosystem II (D1 and D2).

Recent work indicates that the D1 and D2 polypeptides also provide ligands for the (Mn)4 cluster. The three-dimensional structure of photosystem II is not known. Our knowledge of its structure is guided by the known structure of the reaction centre in purple bacteria and biochemical and spectroscopic data. Fig. shows a schematic view of photosystem II that is consistent with current data.

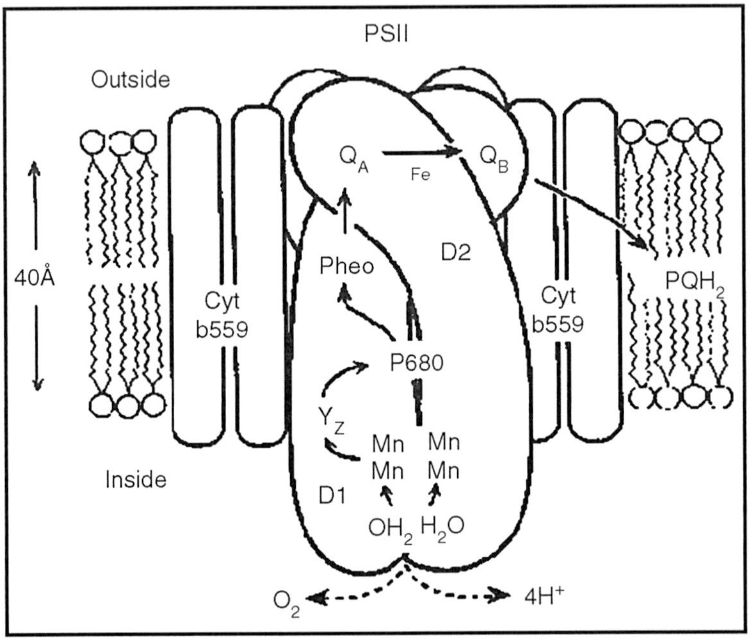

Photochemistry in photosystem II is initiated by charge separation between P680 and pheophytin, creating P680+/Pheo-. Primary charge separation takes about a few picoseconds. Subsequent electron transfer steps have been designed through evolution to prevent the primary charge separation from recombining. This is accomplished by transferring the electron within 200 picoseconds from pheophytin to a plastoquinone molecule (QA) that is permanently bound to photosystem II.

Although plastoquinone normally acts as a two-electron acceptor, it works as a one-electron acceptor at the QA-site. The electron on QA- is then transferred to another plastoquinone molecule that is loosely bound at the QB-site. Plastoquinone at the QB-site differs from QA in that it works as a two-electron acceptor, becoming fully reduced and protonated after two photochemical turnovers of the reaction centre.

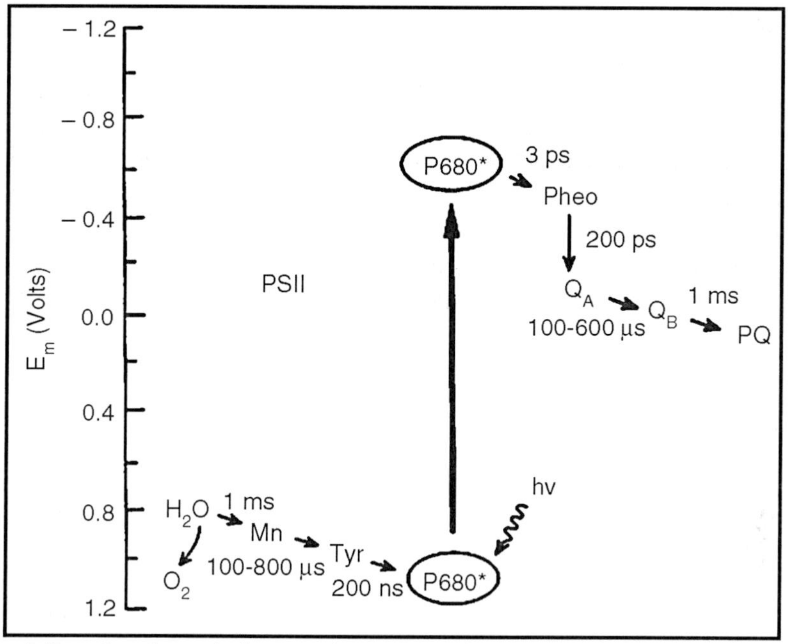

The full reduction of plastoquinone requires the addition of two electrons and two protons, i.e., the addition of two hydrogen atoms. The reduced plastoquinone then debinds from the reaction centre and diffuses into the hydrophobic core of the membrane. After which, an oxidized plastoquinone molecule finds its way to the QB-binding site and the process is repeated. Because the QB-site is near the outer aqueous phase, the protons added to plastoquinone during its reduction are taken from the outside of the membrane.

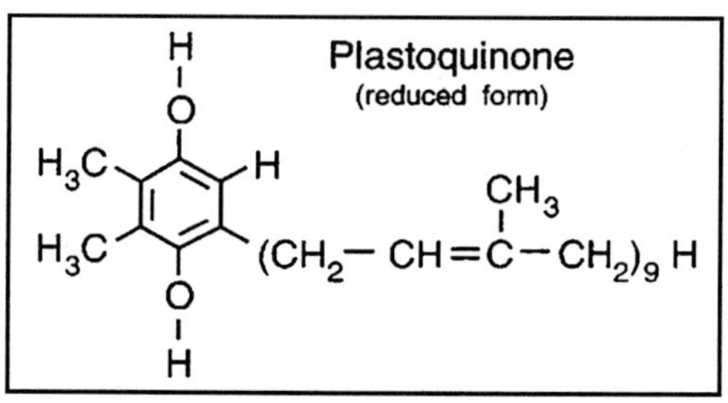

Photosystem II is the only known protein complex that can oxidize water, resulting in the release of 02 into the atmosphere. Despite years of research, little is known about the molecular events that lead to water oxidation. Energetically, water is a poor electron donor. The oxidation- reduction midpoint potential (Em,7) of water is +0.82 V (pH 7). In photosystem II this reaction is driven by the oxidized reaction centre, P680+ (the midpoint potential of P680/ P680+ is estimated to be +1.2 V at pH 7).

How electrons are transferred from water to P680+ remains a mystery. It is known that P680+ oxidizes a tyrosine on the D1 protein and that Mn plays a key role in water oxidation. Four Mn ions are present in the water oxidizing complex. X-ray absorption spectroscopy shows that Mn undergoes light-induced oxidation.

Water oxidation requires two molecules of water and involves four sequential turnovers of the reaction centre. This was shown by an experiment demonstrating that oxygen release by photosystem II occurs with a four flash dependence. Each photochemical reaction creates an oxidant that removes one electron. The net reaction results in the release of one 02 molecule, the deposition of four protons into the inner water phase, and the transfer of four electrons to the QB-site (producing two reduced plastoquinone molecules)

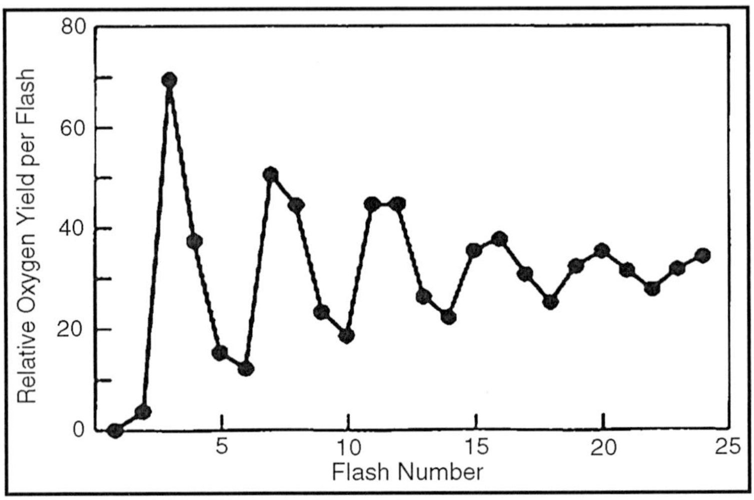

Photosystem II reaction centers contain a number of redox components with no known function. An example is cytochrome b559, a heme protein, that is an essential component of all photosystem II reaction centers. If the cytochrome is not present in the membrane, a stable PS II reaction centre cannot be formed.

Although the structure and function of Cyt b559 remain to be discovered, it is known that the cytochrome is not involved in the primary enzymatic activity of PS II, which is the transfer of electrons from water to plastoquinone. Why PS II reaction centers contain redox components that are not involved in the primary enzymatic reactions is a puzzling question. The answer may be found in the unusual chemical reactions occurring in PS II and the fact that the reaction centre operates at a very high power level.

Photosystem II is an energy transforming enzyme that must switch between various high energy states that involve the creation of the powerful oxidants required for removing electrons from water and the complex chemistry of plastoquinone reduction which is strongly influenced by protons. In saturating light a single reaction centre can have an energy throughput of 600 eV/s.

Operating at such a high power level results in damage to the reaction centre. It may be that some of the "extra" redox components in photosystem II may serve to protect the reaction centre.

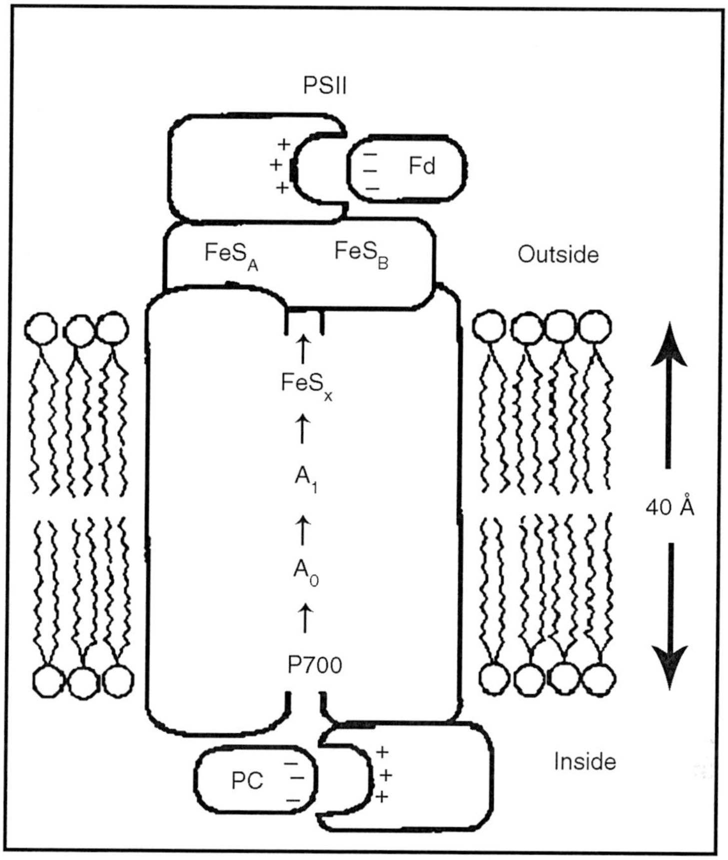

Photosystem II has another perplexing feature. Many plants and algae have been shown to have a significant number of photosystem II reaction centers that do not contribute to photosynthetic electron transport. Why plants devote resources for the synthesis of reaction centers that apparently do not contribute to energy conversion is unknown. The photosystem I complex catalyzes the oxidation of plastocyanin, a small soluble Cu- protein, and the reduction of ferredoxin, a small FeS protein. Photosystem I is composed of a heterodimer of proteins that act as ligands for most of the electron carriers.

The reaction centre is served by an antenna system that consists of about two hundred chlorophyll molecules (mainly chlorophyll a) and primary photochemistry is initiated by a chlorophyll a dimer, P700. In contrast to photosystem II, many of the antenna chlorophyll molecules in photosystem I are bound to the reaction centre proteins. Also, FeS centers serve as electron carriers in photosystem I and, so far as is known, photosystem I electron transfer is not coupled to proton translocation. Primary charge separation occurs between a primary donor, P700, a chlorophyll dimer, and a chlorophyll monomer (Ao). The subsequent electron transfer events and rates are shown in Figure.

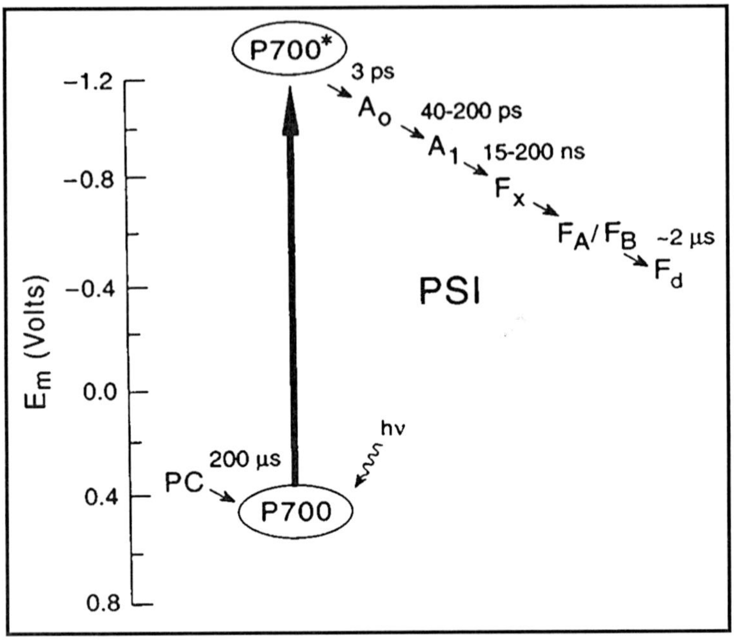

Electron Transport

Electron transport from water to $NADP^+$ requires three membrane bound protein complexes operating in series - photosystem II, the cytochrome bf complex and photosystem I. Electrons are transferred between these large protein complexes by small mobile molecules (plastoquinone and plastocyanin in plants). Because these small molecules carry electrons (or hydrogen atoms) over relatively long distances, they play a unique role in photosynthetic energy conversion. This is illustrated by plastoquinone (PQ), which serves two key functions. Plastoquinone transfers electrons from the photosystem II reaction centre to the cytochrome bf complex and carries protons across the photosynthetic membrane. It does this by shuttling hydrogen atoms across the membrane from photosystem II to the cytochrome bf complex. Because plastoquinone is hydrophobic its movement is restricted to the hydrophobic core of the photosynthetic membrane.

Plastoquinone operates by diffusing through the membrane until, due to random collisions, it becomes bound to a specific site on the photosystem II complex. The photosystem II reaction centre reduces plastoquinone at the QB-site by adding two electrons and two protons creating PQH2. The reduced plastoquinone molecule debinds from photosystem II and diffuses randomly in the photosynthetic membrane until it encounters a specific binding site on the cytochrome bf complex. The cytochrome bf complex is a membrane bound protein complex that contains four electron carriers, three cytochromes and an FeS centre.

The crystal structure has been solved for cytochrome f from turnip and the FeS centre from bovine heart mitochondria. In a complicated reaction sequence that is not fully understood, the cytochrome bf complex removes the electrons from reduced plastoquinone and facilitates the release of the protons into the inner aqueous space. The electrons are eventually transferred to the photosystem I reaction centre. The protons released into the inner aqueous space contribute to the proton chemical free energy across the membrane.

Electron transfer from the cytochrome bf complex to photosystem I is mediated by a small Cu-protein, plastocyanin (PC). Plastocyanin is water soluble and operates in the inner water space of the photosynthetic membrane. Electron transfer from photosystem I to $NADP^+$ requires ferredoxin, a small FeS protein, and ferredoxin-NADP oxidoreductase, a peripheral flavoprotein that operates on the outer surface of the photosynthetic membrane. Ferredoxin and $NADP^+$ are water soluble and are found in the outer aqueous phase.

The pathway of electrons is largely determined by the energetics of the reaction and the distance between the carriers. The electron affinity of the carriers is represented in Figure. by their midpoint potentials, which show the free energy available for electron transfer reactions under equilibrium conditions. (It should be kept in mind that reaction conditions during photosynthesis are not in equilibrium.)

Subsequent to primary charge separation, electron transport is energetically downhill (from a lower (more negative) to a higher (more positive) redox potential). It is the downhill flow of electrons that provides free energy for the creation of a proton chemical gradient.

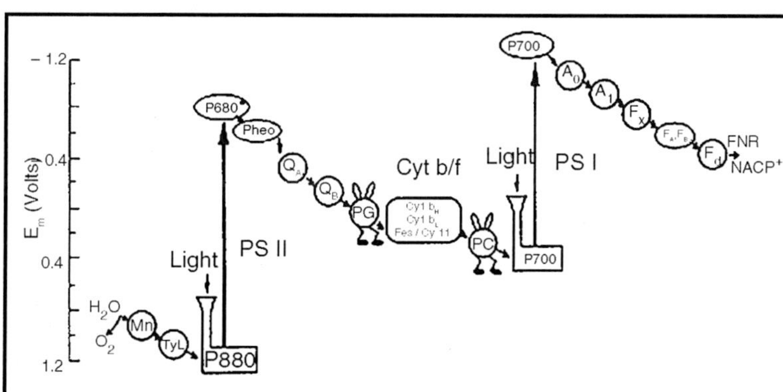

Photosynthetic membranes effectively limit electron transport to two dimensions. For mobile electron carriers, limiting diffusion to two dimensions increases the number of random encounters. Furthermore, because plastocyanin is mobile, any one cytochrome bf complex can interact with a number of photosystem I complexes. The same is true for plastoquinone, which commonly operates at a stoichiometry of about six molecules per photosystem II complex.

Creation of a Proton Electrochemical Potential

Electron transport creates the proton electrochemical potential of the photosynthetic membrane by two types of reactions. (1) The release of protons during the oxidation of water by photosystem II and the translocation of protons from the outer aqueous phase to the inner aqueous phase by the coupled reactions of photosystem II and the cytochrome bf complex in reducing and oxidizing plastoquinone on opposite sides of the membrane. This creates a concentration difference of protons across the membranes (DpH = pHin – pHout). (2) Primary charge separation at the reaction centre drives an electron across the photosynthetic membrane, which creates an electric potential across the membrane (DY = Yin – Yout). Together, these two forms of energy make up the proton electrochemical potential across the photosynthetic membrane (DmH+) which is related to the pH difference across the membrane and the electrical potential difference across the membrane by the following equation:

$$DmH+ = F\ DY - 2.3\ RT\ DpH, \quad (4)$$

where F is the Faraday constant, R is the gas constant, and T the temperature in Kelvin. Although the value of DY across the photosynthetic membrane in chloroplasts can be as large as 100 mV, under normal conditions the proton gradient dominates. For example, during photosynthesis the outer pH is typically near 8 and the inner pH is typically near 6, giving a pH difference of 2 across the membrane that is equivalent to 120 mV. Under these conditions the free energy for proton transfer from the inner to the outer aqueous phase is - 12 kJ/mol of protons.

Synthesis of ATP by the ATP Synthase Enzyme

The conversion of proton electrochemical energy into chemical free energy is accomplished by a single protein complex known as ATP synthase. This enzyme catalyzes a phosphorylation reaction, which is the formation of ATP by the addition of inorganic phosphate (Pi) to ADP

$$ADP{-}3 + Pi{-}2 + H+ \rightarrow ATP{-}4 + H_2O. \quad (5)$$

The reaction is energetically uphill (DG = +32 kJ/mol) and is driven by proton transfer through the ATP synthase protein. The ATP Synthase complex is composed of two major subunits, CF0 and CF1.

The CF0 subunit spans the photosynthetic membrane and forms a proton channel through the membrane. The CF1 subunit is attached to the top of the CF0 on the outside of the membrane and is located in the aqueous space. CF1 is composed of several different protein subunits, referred to as a, b, g, d and e.

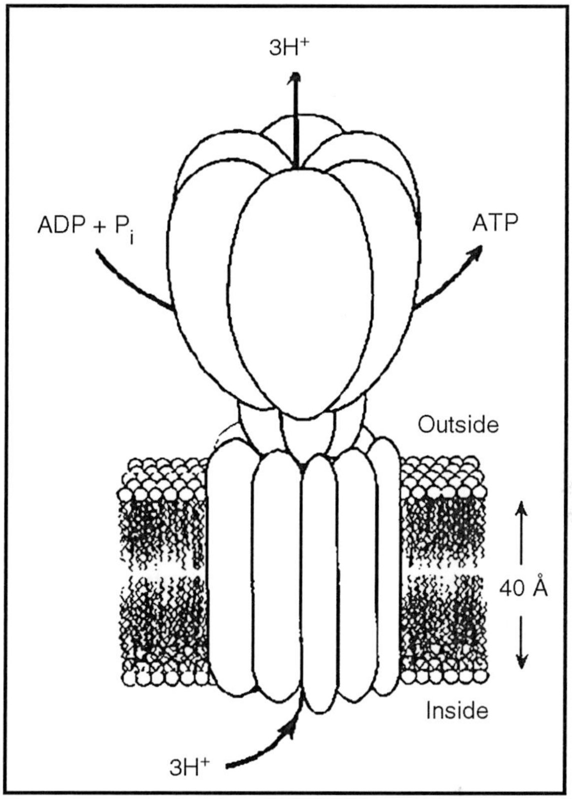

The top portion of the CF1 subunit is composed of three ab-dimers that contain the catalytic sites for ATP synthesis. A recent major breakthrough has been the elucidation of the structure of ATPase of beef heart mitochondria by Abrahams et al. (1994).

The molecular processes that couple proton transfer through the protein to the chemical addition of phosphate to ADP are poorly understood. It is known that phosphorylation can be driven by a pH gradient, a transmembrane electric field, or a combination of the two. Experiments indicate that three protons must pass through the ATP synthase complex for the synthesis of one molecule of ATP. However, the protons are not involved in the chemistry of adding phosphate to ADP. Paul Boyer and coworkers have proposed an alternating binding site mechanism for ATP synthesis. One model based on their proposal is that there are three catalytic sites on each CF1 that cycle among three different states.

The states differ in their affinity for ADP, Pi and ATP. At any one time, each site is in a different state. This model is supported by the structure of ATPase elucidated by Abrahams et al.. Initially, one catalytic site on CF1 binds one ADP and one inorganic phosphate molecule relatively loosely.

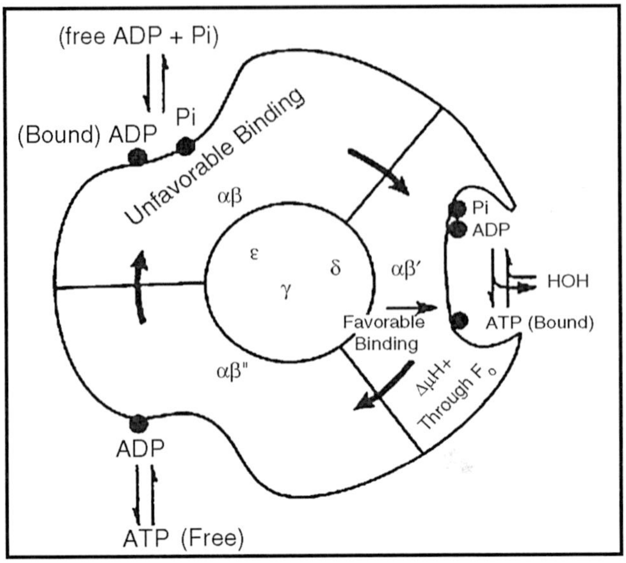

Due to a conformational change of the protein, the site becomes a tight binding site, that stabilizes ATP. Next, proton transfer induces an alteration in protein conformation that causes the site to release the ATP molecule into the aqueous phase. In this model, the energy from the proton electrochemical gradient is used to lower the affinity of the site for ATP, allowing its release to the water phase.

The three sites on CF1 act cooperatively, i.e., the conformational states of the sites are linked. It has been proposed that protons affect the conformational change by driving the rotation of the top part (the three ab-dimers) of CF1. Such a rotating model has recently been supported by recording of a rotation of the gamma subunit relative to the alpha-beta subunits by Sabbert et al.. This revolving site mechanism would require rates as high as 100 revolutions per second. It is worth noting that flagella that propel some bacteria are driven by a proton pump and can rotate at 60 revolutions per second.

Synthesis of Carbohydrates

All plants and algae remove CO2 from the environment and reduce it to carbohydrate by the Calvin cycle. The process is a sequence of biochemical reactions that reduce carbon and rearrange bonds to produce carbohydrate from CO2 molecules.

The first step is the addition of CO2 to a five-carbon compound (ribulose 1,5-bisphosphate). The six-carbon compound is split, giving two molecules of a three-carbon compound (3-phosphoglycerate). This key reaction is catalyzed by Rubisco, a large water soluble protein complex.

The 3-dimensional structure has been determined by X-ray analysis for Rubisco isolated from tobacco from a cyanobacterium (Synechococcus) and from a purple bacterium (Rhodospirillum rubrum. The carboxylation reaction is energetically downhill. The main energy input in the Calvin cycle is the phosphorylation by ATP and subsequent reduction by NADPH of the initial three-carbon compound forming a three-carbon sugar, triosephosphate. Some of the triosephosphate is exported from the chloroplast and provides the building block for synthesizing more complex molecules. In a process known as regeneration, the Calvin cycle uses some of the triosephosphate molecules to synthesize the energy rich ribulose 1,5-bisphosphate needed for the initial carboxylation reaction.

This reaction requires the input of energy in the form of one ATP. Overall, thirteen enzymes are required to catalyze the reactions in the Calvin cycle. The energy conversion efficiency of the Calvin cycle is approximately 90 per cent. The reactions do not involve energy transduction, but rather the rearrangement of chemical energy. Each molecule of CO_2 reduced to a sugar $[CH_2O]n$ requires 2 molecules of NADPH and 3 molecules of ATP.

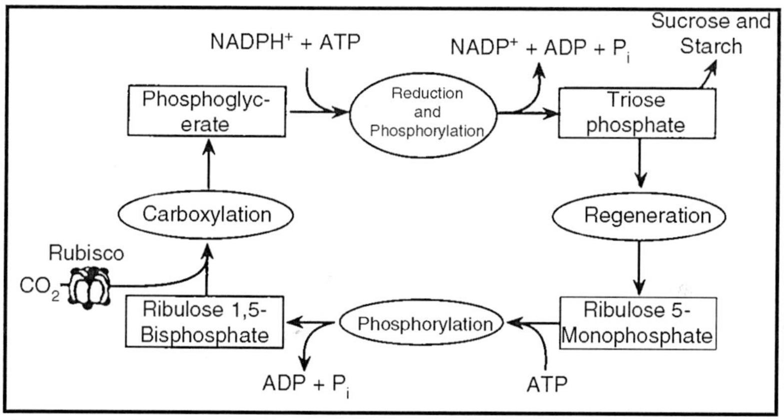

Rubisco is a bifunctional enzyme that, in addition to binding CO_2 to ribulose bisphosphate, can also bind O_2. This oxygenation reaction produces the 3-phosphoglycerate that is used in the Calvin cycle and a two-carbon compound (2-phosphoglycolate) that is not useful for the plant.

In response, a complicated set of reactions (known as photorespiration) are initiated that serve to recover reduced carbon and to remove phosphoglycolate. The Rubisco oxygenation reaction appears to serve no useful purpose for the plant. Some plants have evolved specialized structures and biochemical pathways that concentrate CO_2 near Rubisco. These pathways (C4 and CAM), serve to decrease the fraction of oxygenation reactions.

Photosynthetic Quantum Yield and Energy Conversion Efficiency

The theoretical minimum quantum requirement for photosynthesis is 8 quanta for each molecule of oxygen evolved (four quanta required by photosystem II and four by photosystem I). Measurements in algal cells and leaves under optimal conditions (e.g., low light) give quantum requirements of 8-10 photons per oxygen molecule released.

These quantum yield measurements show that the quantum yields of photosystem II and photosystem I reaction centers under optimal conditions are near 100 per cent. These values can be used to calculate the theoretical energy conversion efficiency of photosynthesis (free energy stored as carbohydrate/light energy absorbed). If 8 red quanta are absorbed (8 mol of red photons are equivalent to 1,400 kJ) for each CO2 molecule reduced (480 kJ/mol), the theoretical maximum energy efficiency for carbon reduction is 34 per cent. Under optimal conditions, plants can achieve energy conversion efficiencies within 90 per cent of the theoretical maximum. However, under normal growing conditions the actual performance of the plant is far below these theoretical values. The factors that conspire to lower the quantum yield of photosynthesis include limitations imposed by biochemical reactions in the plant and environmental conditions that limit photosynthetic performance. One of the most efficient crop plants is sugar cane, which has been shown to store up to 1 per cent of the incident visible radiation over a period of one year. However, most crops are less productive. The annual conversion efficiency of corn, wheat, rice, potatoes, and soybeans typically ranges from 0.1 per cent to 0.4 per cent.

Oxygenic Photosynthesis in Algae

Algae are photosynthetic eukaryotic organisms that, like plants, evolve O_2 and reduce CO_2. They represent a diverse group that include the dinoflagellates, the euglenoids, yellow-green algae, golden-brown algae, diatoms, red algae, brown algae, and green algae. The photosynthetic apparatus and biochemical pathways of carbon reduction of algae are similar to plants. Photosynthesis occurs in chloroplasts that contain photosystems II and I, the cytochrome bf complex, the Calvin cycle enzymes and pigment-protein complexes containing chlorophyll a, and other antenna pigments (e.g., chlorophyll b in green algae, chlorophyll c and fucaxanthol in brown algae and diatoms, and phycobilins in red algae).

Green algae are thought to be the ancestral group from which land plants evolved. Algae are abundant and widespread on the earth, living mainly in fresh and sea water. Some algae live as single celled organisms, while others form multicellular organisms some of which can grow quite large, like kelp and seaweed. Phytoplankton in the ocean is made up of algae and oxygenic photosynthetic bacteria. Most photosynthesis in the ocean is due to phytoplankton, which is an important source of food for marine life.

Oxygenic Photosynthesis in Bacteria

Cyanobacteria are photosynthetic prokaryotic organisms that evolve O2. Fossil evidence indicates that cyanobacteria existed over 3 billion years ago and it is thought that they were the first oxygen evolving organisms on earth. Cyanobacteria are presumed to have evolved in water in an atmosphere that lacked O_2. Initially, the O_2 released by cyanobacteria reacted with ferrous iron in the oceans and was not released into the atmosphere. Geological evidence indicates that the ferrous Fe was depleted around 2 billion years ago, and earth's atmosphere became aerobic. The release of O_2 into the atmosphere by cyanobacteria has had a profound affect on the evolution of life.

The photosynthetic apparatus of cyanobacteria is similar to that of chloroplasts. The main difference is in the antenna system. Cyanobacteria depend on chlorophyll a and specialized protein complexes (phycobilisomes) to gather light energy. They do not contain chlorophyll b. As in chloroplasts, the chlorophyll a is located in membrane bound proteins. The phycobilisomes are bound to the outer side of the photosynthetic membrane and act to funnel exciton energy to the photosystem II reaction centre.

They are composed of phycobiliproteins, protein subunits that contain covalently attached open ring structures known as bilins that are the light absorbing pigments. Primary photochemistry, electron transport, phosphorylation and carbon reduction occur much as they do in chloroplasts.

Cyanobacteria have a simpler genetic system than plants and algae that enable them to be easily modified genetically. Because of this cyanobacteria have been used as a model to understand photosynthesis in plants. By genetically altering photosynthetic proteins, researchers can investigate the relationship between molecular structure and mechanism.

Over the past three decades several types of oxygenic bacteria known as prochlorophytes (or oxychlorobacteria) have been discovered that have light harvesting protein complexes that contain chlorophyll a and b, but do not contain phycobilisomes. Because prochlorophytes have Chlorophyll a/b light harvesting proteins like chloroplasts, they are being investigated as models for plant photosynthesis.

PHOTOSYNTHETIC ELECTRON TRANSPORT SYSTEM

The electron transport system of photosynthetic bacteria differs from that of aerobic chemoorganotrophic bacteria. Cytochrome a and other types of cytochrome oxidase are not present in the photosynthetic electron, transport system, because photosynthesis takes place under anaerobic conditions. Hence there is no need of a cytochrome which interacts with molecular oxygen.

The electron transport system consists of an intermediate electron acceptor (I). a primary electron acceptor (X), a secondary acceptor (Y), generally believed to be ubiquinone (UQ), and b- and c-type cytochromes. The electron transport carriers are asymmetrically located in the membrane. This is necessary for setting up the hydrogen ion gradient. Immunological studies suggest that the reaction centre spans the membrane of the chromotophore. It is probably located beneath the ATPase complex. The primary acceptor X is believed to be associated with the reaction centre on the outer side of the membrane. The secondary acceptor Y, (probably UQ) takes protons from the medium. It is thus also located on the outer side. The b-type cytochrome is probably located in the interior of the membrane. The c-type cytochrome interacts with the reaction centre and is located on the inside of the membrane.

Primary Acceptor

The electron from P870 is received by an intermediate acceptor (I) and transferred to a primary acceptor (X). The exact nature of the acceptor is not known. The purified P870 reaction centre has been shown to contain non-haeme iron and ubiquinone.

This has led, to the suggestion that one of these substances, or possibly both, act as primary acceptors. X is therefore likely to be an iron-sulphur (Fe-S) protein or anifon-quinone complex. It mayor may not be ferredoxin. That the primary acceptor has a non-haeme iron component is supported by electron paramagnetic resonance (epr) data. Several iron-sulphur proteins (non-haeme iron proteins) have been purified since the isolation of ferredoxin from Clostridium by Mortengen (1962). Bacterial ferredoxins are blackish-brown in colour with absorption maxima around 390 nm.

Type b Cytochrome

There is evidence that a b-type cytochrome IS present in the photosynthetic electron transport system.

Cytochrome b has been demonstrated in Rhodopseudomons sphaeroides ($E^{\circ\prime}$ = + 50, mV) and R. capsulata ($E^{\circ\prime}$ = + 60 mV), and there are indications that it is adjacent to cytochrome c in the cyclic system. It is likely that cytochrome b may be present even in organisms where it has been previously reported to be absent.

Small amounts of cytochrome b could be masked by other substances. Chromatium vinosum and Chlorobium thiosulphatophilum lire reported to contain one molecule of cytochrome b per reaction centre. It is likely that cytochrome b has a role in cyclic photosynthetic flow in the Chromatiaceae, Chlorobiaceae as well as in the Rhodospirillaceae.

Type c Cytochrome

A number of different c type cytochromes have been found in the electron transport system of photosynthetis bacteria. In the purple non-sulphur bacterium Rhodospirillum rubrum a soluble c type cytochrome is associated with P870 of the reaction centre. This cytochrome is referred to as cytochrome c_2, and has a high mid point potential of about + 30.0 mV.

Cytochrome c_2 appears to be the electron donor to P870. In PSI of higher plants, plastocyanin (PC), a copper protein, is the electron donor to P700. Algae apparently represent an intermediate stage in evolution, since their electron donor to P700 is a c type cytochrome, although plastocyanin can also be present. Rhodospirillum also contains cytochrome cc' with two different haeme groups.

Purple sulphur bacteria like Chromatium contain cytochrome c_{552} in addition to cytochromes C_2 and cc'. Cytochrome C_{552} (MW 72,000) has two haeme groups and one FMN. The green sulphur bacterium Chlorobium has three cytochromes of the c type, but none of these has the high redox potential of cytochrome, c_2 of purple bacteria and cytochrome f of higher green plants.

PHOTOPHOSPHORYLATION

Photophosphorylation, i.e. the conversion of light energy into chemical energy, is of two types, cyclic and non-cyclic. Two photo systems, PS I and PS II are involved in the photosynthesis of green plants.

Many workers hold that only one reaction centre is involved in bacterial photosynthesis, and hence there is only one light reaction.

Electrons lost during NADH formation are believed to be replaced by electrons from the reduced substrates of the growth medium inmost bacteria. (In chloroplasts they are replaced by the splitting of H_2O)

Electron flow in bacterial photosynthesis is of the cyclic type, and is, analogous to that of PS I of higher plants, some workers; however, believe that non-cyclic photophosphorylation also occurs in bacteria.

Reaction Centre

The photosynthetic unit consists of 'antenna' or 'light harvesting' molecules for gathering light photons and a reaction, centre where energy conversion takes place. Light energy harvested by the antenna pigments IS transferred to the reaction centre.

In the chloroplasts of green plants the reaction centre chlorophylls are P700 (in PS I) and P680 (in PS II).

In green and purple photosynthetic bacteria some 40 or more bacteriochlorophyll molecules make up the photosynthetic unit. The reaction centre complex is commonly referred to as P870, although the wavelength band of maximal bleaching may vary in different species.

The reaction centre generally contains BChl a (2-5 per cent of the total) or, rarely, BChl b. In the purple non sulphur bacterium Rhodospirillum rubrum the principal light harvesting pigment is BChl a and the reaction centre molecule is P890. In Rhodopseudomonas sphaeroides the light harvesting bacteriochlorophyll absorbs maximally at 850 nm, and the reaction centre BChl is P870. In the R-26 mutant of R. sphaeroides, which lacks carotenoids, the reaction centre pigment P870 constitutes 5 per cent of the total pigment.

In green sulphur bacteria the principal light harvesting pigment is B Chl c (650. nm) or B Chl d (660. nm) in the green species and BChl e in the brown species. The reaction centre chlorophyll is B Chl a P in both. Reaction centres from Rhodopseudomonas sphaeroides, R. viridis, and Rhodospirillum rubrum contain the following components:

4 molecules of BChl a

2 molecules of bacteriopheophytin (BPh)

(BChl with 2 protons ($2H^+$) replacing Mg)

1 molecule of carotenoid (from strains that synthesize carotenoids)

1 or more quinones (ubiquinone or menaquinone)

1 atom of nonhaeme Fe.

The reaction centres of most bacterial species contain three polypeptides (MW 20,000, 24,000 and 30,000). The BChl, BPh and carotenoids are all located on the first two polypeptides, which only are essential for photochemical activity. The reaction centre of Rhodopseudomonas gelatinosa has only two polypeptides -(MW 24,000 and 34,000).

In the primary photochemical reaction light quanta absorbed by the light harvesting pigments migrate to the reaction centre BChl (P810). The energy of the absorbed photons raises chlorophyll to an excited state and an electron is expelled from the molecule. P870 thus requires a positive charge and becomes P870+.BChl (P870) → P870+ (oxidized BChl complex)

The excited P870+ transfers an electron to an intermediate electron acceptor (I) which is reduced to (I-). According to one interpretation, I consist of interacting BChl and bacteriopheophytin (BPh) molecules. It appears that one of the two BPhs of the reaction centre is a major component of I. The reduced intermediate acceptor (I-) than transfers the electron to another acceptor (X) which appears to be a quinone (X → X–). In Chromatium venosum and some other species menaquinone is the primary acceptor instead of ubiquinone.

X- releases the electron to another quinone, which is reduced. The secondary quinone rapidly transfers a pair of electrons to an external pool of quinones.

REDUCTIVE PENTOSE PATHWAY

It utilizing the ribulose dipliosphate carboxylase reaction, is also found in CO_2 fixation of higher green plants. The essential features of this pathway are as follows :

- Ribulose monophosphate (RumP) is phosphorylated to ribulose- 5-diphosphate (RudP) by the ATP-ADP reaction.
- CO_2 is used to carboxylate RudP through an unstable 6C intermediate and two molecules of 3-phosphoglyceric acid (PGA) are formed.
- PGA is reduced through a two step process requiring ATP and NADPH to 3-phosphoglyceraldehyde (PGAL). PGAL is converted to its triose phosphate isomer dihydroxyacetone phosphate (DHAP).

- PGAL condenses with DHAP to form fructose-I-6-diphosphate which then forms fructole-6-phosphate. This hexose phosphate through a series of interconversions, regnerates RumP which is re used for CO_2.

PYRUVATE SYNTHETASE PATHWAY REDUCTIVE CARBOXYLIC ACID CYCLE

In the green bacterium Chlorobium thiosulfatophilum Evans et. al (1966) described the pyruvate synthetase pathway for CO_2 fixation.CO_2 is used to form pyruvate by means of the pyruvate synthetase reaction the ultimate reductant is hydrogen sulphide.

The light dependent oxidation of HIS provides the reducing power for the reduction of ferredoxin (Fd). Acetyl CoA then accepts CO_2 and is reduced by ferredoxin to yield pyruvate,

- Formation of pyruvate by pyruvate synthetase is dependent on reduced ferredoxin.
  Acetyl-CoA+Co_2+Ferredoxin (reduced)→
  Pyruvate+CoA+Ferredoxin (oxidized)
- Conversion of pyruvate into oxaloacetate.
  Pyruvate + ATP + CO_2 → Oxaloacetate + ADP + Pi
  Oxaloacetate then enters a reversed tricarboxylic acid cycle.
- Carboxylation of succinyl CoA to yield d-Ketoglutarate (involving reduced ferredoxin).
  Succinyl CoA + co_2 + Ferredoxin (reduced)→
  α-Ketoglutarate + CoA + Ferredoxin (oxidized)
- α -Ketoglutarate is converted into citrate through oxalosuccinate. Citrate then splits into oxaloaceticacid and acetate.
  α -Ketoglutarate → Oxalossuccinate → Citrate → Oxaloacetic acid + Acetate.

The net result of each cycle is that a molecules of CO_2 are fixed (reductive fixation) and one equivalent of oxaloacetate is produced.

Three molecules of ATP are required:

- For activation of acetate,
- For carboxylation of pyruvate and
- For activation of succinate.

The reductive carboxylic acid cycle appears to be particularly suited to provide the carbon skeletons for the main products of bacterial photosynthesis, which are mainly amino acids.

PHOTOSYNTHETIC BACTERIA

The prokaryotic phototrophs include the purple bacteria (suborder Rhodospirillineae) the green bacteria (suborder Chlorobiineae), the blue, green bacteria (Cyanobacteria) and the Prochlorophyta. The purple bacteria include the families Rhodospirillaceae (purple non sulphur bacteria) and the Chromatiacae (purple sulphur bacteria).

The green bacteria include the families Chlorobiaceae (non-motile green and brown sulphur bacteria) and the Chloroflexaceae (filamentous gliding green bacteria). The Cyanobacteria and the Prochloraphyta show certain characteristics of the eukaryotic photosynthesis system.

Rhodospirillaceae (Formerly Athiorhodaceae) Purple Non Sulphur Bacteria

The purple non-sulphur bacteria do not form globules of elemental sulphur inside the cells. There are some purple bacteria which oxidize sulphide to extracellular sulphur and further to sulphate. Pfennig-(1977) has placed such species in the non-sulphur bacteria.

The principal photosynthetic pigments are bacteriochlorophylls a and b and carotenoids of groups I, 2 arid 4. There is one photocentre.The sources of hydrogen are mainly organic molecules. In some cases hydrogen sulphide serves a the source of hydrogens, but not sulphur.

The purple non-sulphur bacteria differ from the Chromatiaceae the Chlorobiaceae in that they can grow either anaerobically, in light (photosynthesis) or aerobically in the dark (oxidative respiration). This switch from photophosphorylation to oxidative phosphorylation requires the presence of both electron transport systems.

Light is absorbed by the bacteriochlorophyll molecule P800, and transferred to the reaction centre that is P890 in Rhodospirillum rubrum or P870 in Rhodopseudomonas sphaeroides. The ratio between P800 and the reaction centre molecule is 2: 1.

It bas been postulated that an electron acceptor X, and possibly a second electron acceptor Y, are reduced when P890 is oxidized. X is probably an iron-sulphur (Fe-S) protein.

Two different types of ferredoxins type (and type II, have been isolated from Rhodospirillum rubrum. Ferredoxin type II is found in cells grown heterotrophically in the dark and light. Type I is found along with type II only when cells are grown in light. Whether connection exists between X and the two types of ferredoxins is not known.

Among the quinones, ubiquinones- 10 (UQ-10) is, predominant in -the Rhodospirillaceae: The other quinones found are ubiquinone If; (Rhodopseudomonas gelatinosa) and rhodoquinone.

The quinones are reduced by the reaction centre P890. The main types of cytochromes found in the Rhodospirillaceae are the, c-type cytochromes (c550, 5, c553 and c558) and b type cytochrome. Oxidation of c553 takes place mainly at low light intensities, while the other two c type cytochromes are equally well oxidized at high light intensities. Rhodopseudomonas palustris can couple energy dependent NAD reduction in the dark to oxidize thiosulphate to sulphate. Electrons released from thiosulphate oxidation probably enter the electron transport chain at cytochrome c. The energy linked reverse flow of electrons resembles that of aerobic respiration in Thiobacillus.

Thus in, the presence of light and under anaerobic conditions, there is ATP-generating cyclic electron flow and photoreduction of NAD. Under aerobic conditions in the dark there is aerobic respiration with reversed electron flow.

Chromatiaceae (Formerly Thiorhodaceae) Purple Sulphur Bacteria

The purple sulphur bacteria form globules of elemental sulphur from sulphide inside cells. There is one photocentre. Hydrogen sulphide, sulphur, thiosulpbate or molecular hydrogen are sources of hydrogens.

Organic molecules are utilized in some cases. The principal photosynthesis pigments are bacteriochlorophylls a and b which show absorbtion bands at 375 and 590 nm in the visible region and 800, 850 and 890 in the infrared region for their chlorophylls. A large amount of carotenoids of groups 1, 3 and 4 or tetrahydrospirilloxanthin are present.

The absorption peaks of the carotenoids are between 400 and 600 nm.The primary photochemical reaction centre from the bacteriochlorophyll in Chromatium chromatophores is P890 (equivalent to P870 in photosynthetic strains). P800, is the light-harvesting cytochrome, c553, related to or identical with c555 of Chlorobium. An additional cytochrome, cc', complexes, with P870. Coenzyme Q7 is associated with the reaction centre. In addition to the cyclic flow of electrons described above, a non cyclic electron transport system has also been suggested, but the evidence is conflicting. The noncyclic system, including a P905 reaction centre and cytochrome c552, would function at high light intensities or under more reduced conditions.

It can only operate under anaerobic conditions. The P890 pathway requires more oxidizing conditions and low light intensities. The existence of a second distinct photochemical centre (P905) bas been questioned.

Chlorobiaceae (Formerly Chlorobacteriaceae) Non Motile Green and Brown Sulphur Bacteria

The Chlorobiaceae are strictly anaerobic. There is one photocentre. Hydrogen sulphide, sulphur, thiosulphate or molecular hydrogen are the sources of reducing hydrogen, and sulphate is given off as a byproduct, The Chlorobiaceae are characterized by H_2S^-, CO^-_2 and light depending metabolism, which takes place under strictly anaerobic conditions. CO_2 assimilation does not apparently take place vi the reductive pentose phosphate cycle.

Elemental sulphur is formed as an intermediate product of oxidation, and is excreted into the medium to form sulphur globules. Sulphur is then more or less completely oxidized to sulphate under conditions of limiting sulphide The Chlorobiaceae characterized by the presence of chlorobium chlorophylls BChl, c or d which form the major components. In addition small amounts of B.Chl a are present. The reaction centre pigment is BChl a Carotenoids of group 5 form accessory pigments. In 1968 Pfenning identified two brown species, Chlorobium phaeobacteroides and Chlorobium, phaeovibrioides, and since then several brown species have been isolated.

In almost all genera of the green sulphur bacteria there are green and brown coloured species that resemble each other morphologically and physiologically, but differ in their photosynthetic pigments. Separate species have been established for the green and brown counterparts of each genus.

In Chlorobium and Chloropseudomonas the primary photochemical reaction centre is P840. Electron transport chains of both cyclic and noncyclic photophosphorylation are found. Chlorobium thiosulfatophilum possesses ferredoxin (Fd) whose sole function appears to be reduction of NAD^+. Rubredoxin has been found in C. thiosulfatophilum and Chloropseudomonas ethylicum, but Dot in the purple bacteria.

It appears to link bacterial and plant ferredoxins. Three types of quinones have been isolated from C. thiosulfatophilum and Chloropseudomonas ethylicum. Chlorobium quinone is present only in sulphide adapted cells. Polar menaquinone is believed to be a precursor of chlorobium quinone. Menaquinone-7 is the third quinone.

No cytochrome b has been found. Three c-type cytochromes have been isolated from C. thiosulfatophilum: (a) c551 which acts as a thio sulphate cytochrome c reductase and may be a cytochrome Cs class enzyme, (b) c553 which appears to act as a sulphide cytochrome c reductase and (c) c555 which is similar to algal c type cytochrome (the f-type cytochromes).

Photosynthetic Pigments in Green and Brown Species of Chlorobiaceae

Colur	Light harvesting Pigment	Reaction centre pigment	Cnrotenoids
Green spp	BChl c or d	BChlap	Chlorobactene OH-chlorobactene
Brown spp	BChl c	BChlap	Isorenieratene β- sorenieratene

Cyanobacteria (Formerly Blue Green Algae) Blue Green Bacteria

The Cyanobacteria form the largest group of photosynthetic prokaryotes. They are widely distributed and show great diversity. Their photosynthetic apparatus is structurally and functionally similar to the eukaryotic chloroplast.

The light harvesting pigments of the Cyanobacteria are chlorophyll a and phycobiliproteins. These pigments are homologous with those of the chloroplasts of the Rhodophyta (red algae). The Cyanobacteria thus differ from the green and purple bacteria which contain bacteriochlorophylls as their photosynthetic pigments. They also differ from other photosynthetic bacteria in having two photosystems instead of one and in their ability to perform oxygenic photosynthesis (photosynthesis with evolution of oxygen).

The source of hydrogen is water. Respiration is aerobic. The Cyanobacteria are gram negative. Many are motile. Motility is of the gliding type as in motile green bacteria.

The photosynthetic apparatus is located in a series of thylakoids or membranous sacs which contain chlorophyll a and carotenoids, the 'photochemical reaction' centres and the photosynthetic electron transfer chain.

In the 50 nm space between the thylakoids are disc shaped objects, '40 nm' in diameter, called phycobilisomes. They contain phycobiliproteins (PBPs), the major light harvesting pigment. In Synecchococcus and Pseudanabaena there are 2-5 parallel thylakoids in the cortical region of the bacterial cell. In most Cyanobacteria the thylakoids are convoluted, and occupy a much larger area of the cell.

The thylakoid membranes are functionally distinct from the cell membrane: the hitter never has phycobilisome attached to it. In the unicellular cyanobacterium. Gloeobacter violacens there are no thylakoids. The phycobiliproteins are contained in an 80 nm wide electron dense cortical layer attached to the inner surface of the cell membrane.

Components of the Photosynthetic electron transport system (e.g. ferredoxin, plastocyanin and cytochrome f) have been located in the thylakoid membrane. The only chlorophyll found in the Cyanobacteria is chlorophyll a.

This is an important homology with chloroplasts. There are two photosystems I and II. Photosystem I almost exclusively receives the light energy harvested by chlorophyll, and only cyclic photophosphorylation takes place, β-carotene is always present, and is usually accompanied by either or both oxycarotenoids, zeanthin and echinone.

In addition most Cyanobacteria synthesize group specific pigments, notably carotenoid glycosides. Phycobiliproteins are the major light-harvesting pigments of the Cyanobacteria. (These are also found in the red algae). Light energy absorbed by the PBPs is transfered mainly to photosystem II.

Each pap is composed of two different polypeptide chains α and β, each carrying at least one bilin chromatophore attached io it covalently. The α and β monomers form the native protein oligomers, (αβ) n.

Cyanobacteria contain at least, three different types 6f PBPs phycocyanin (PC) (A max 620 nm), allophycocyanin (AP) (λ max 650 nm) and allophycocyanin B (APB) (A max 670 nm). Many Cyanobacteria contain a fourth PBP, usually phycoerythrin (PE) (λ max 565 nm).Some strains of Anabaena contain phycoerythrocyanin (PEC) (A max 565 nm).

The chloroplasts of the red algae contain PC, AP, APB and PE, but not PEC. In Cyanobacteria there are four major membrane lipid components monoglactosyl diglyceride (MGDG), digalactosyl diglyceride (DGDG), sulphoquinovosyl diglyceride (SQDG) and phosphatidyl glycerol MGDG, DGDG and SQDG are found in chloroplast membranes of eukaryote cells.

None of these glycolipids occur in purple bacteria. Green bacteria contain only MGDG, which is localized in the chlorobium vesicle, the light harvesting organelle.

Prochlorophyta

The prochlophyta is a newly established group. Like the Cyanobacteria the group has oxygenic photosynthesis, with two linked photosystems. The electron donor is water, and oxygen is the end product of oxidation. The photosynthetic pigments are chlorophylls a and b and carotenoids.

Evolution of Photosynthesis - On the basis of fossil evidence there are indications that biochemically advanced microbial life, existed on the earth more than 3, billion years ago. The precursors of prokaryotic cells must have existed even earlier and have been called progenotes or protocells. The evolution of progenotes into prokaryotes is believed to have occurred 3 to 4 billion years ago. Broda (1975) has presented the following sequence for the appearance of metabolic activities, in prokaryotes during the course of evolution:

Anaerobic fermentation

Primitive bacterial photosynthesis

Complex plant type photosynthesis, as in Cyanobacteria

Aerobic respiration

The earliest prokaryote cells were presumably fermenting bacteria They lived in an environment in which organic compounds were freely available, and in which the atmosphere was reducing. The bacteria were anaerobic scavengers which broke down compounds with a high free energy content.

The organic compounds were thus used as sources of energy, which was utilized by the organisms. The molecules with the lower freeenergy content were released as waste products. This process of fermentation is still used as a source of energy by such an-aerobic bacteria as Clostridium.

The next step was the development of photosynthesis in which solar energy was trapped to synthesize molecules with a high free energy content. Photosynthesis is essentially a process in which electro magnetic energy is converted to chemical energy, which is easily convertible to ATP.

The energy stored in the high free energy content molecules can be used in later activities according to Dickerson (1980), sulphate respiration (e.g. Desulfovibrio) evolved in response to sulphate-releasing photosynthesis (Chlorobiaceae green sulphur bacteria; Chloroflexaceae green filamentous bacteria;Chromtiaceae purple sulphur bacteria and Rhodosplrillaceae purple nonsulphur bacteria) early on the primitive earth.

Oxygen respiration (e.g. Pseudomonas. Paracocus and mitochondrial respiration) evolved later in response to oxygen releasing photosynthesis (Cyanobacteria blue green bacteria and chloroplasts).

Chlorobiaceae

The green sulphur bacteria were the earliest photosynthetic bacteria. They utilized hydrogens from H_2S to photoreduce NAD^+ directly to NADH $^+$ H^+. releasing sulphate as a byproduct. The bacteria apparently did not have the Calvin Benson cycle for reducing CO_2 they utilized the reductive tricarboxylic cycle, which resembles the mitochondrial Krebs cycle run in reverse. In Chlorobium, cytochrome c as an electron carrier is strictly photosynthetic.

8

Respiration The Hormones

PLANT-HORMONES

The plant hormones was set up initially at institute of the BBSRC. Long Ashton will be closing in March 2003 so the Plant-Hormones site has been moved to a new commercial service. The general information on plant hormones and links to the plant-hormones has been set up. This list has been set up as a freely accessible electronic communication forum for use by plant-hormone scientists.

The list is intended to promote communication between professionals in the plant-hormone field. Discussion topics should include gibberellins, auxins, cytokinins,abscisic acid, etc. All aspects of research may be discussed; synthesis extraction, purification, analysis, enzymology, molecular biology, hormone action etc. Requests for help on particular aspects may also be posted to the list as well as requests for standards, spectra, identification of spectra etc. The list may also be used for announcement of meetings, calls for papers and posters, visits of scientists to a particular area.

ABSCISIC ACID

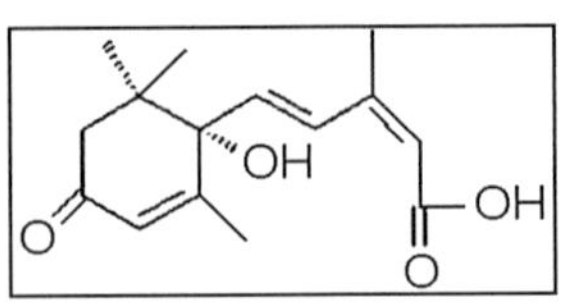

Nature of Abscisic Acid

Abscisic acid is a single compound unlike the auxins, gibberellins, and cytokinins. It was called"abscisin II" originally because it was thought to play a major role in abscission of fruits. At about the same time another group was calling it"dormin" because they thought it had a major role in bud dormancy. The name abscisic acid (ABA) was coined by a compromise between the two groups.

Though ABA generally is thought to play mostly inhibitory roles, it has many promoting functions as well.

History of Abscisic Acid

In 1963, abscisic acid was first identified and characterized by Frederick Addicott and his associates. They were studying compounds responsible for the abscission of fruits (cotton). Two compounds were isolated and called abscisin I and abscisin II. Abscisin II is presently called abscisic acid (ABA). Two other groups at about the same time discovered the same compound. One group headed by Philip Wareing was studying bud dormancy in woody plants. The other group led by Van Steveninck was studying abscission of flowers and fruits from lupine. Plant physiologists agreed to call the compound abscisic acid.

Biosynthesis and Metabolism

ABA is a naturally occurring compound in plants. It is a sesquiterpenoid (15-carbon) which is partially produced via the mevalonic pathway in chloroplasts and other plastids. Because it is sythesized partially in the chloroplasts, it makes sense that biosynthesis primarily occurs in the leaves. The production of ABA is accentuated by stresses such as water loss and freezing temperatures. It is believed that biosynthesis occurs indirectly through the production of carotenoids. Carotenoids are pigments produced by the chloroplast which have 40 carbons. Breakdown of these carotenoids occurs by the following mechanism:

Violaxanthin is a carotenoid which has forty carbons. It is isomerized and then split via an isomerase reaction followed by an oxidation reaction. One molecule of xanthonin is produced from one molecule of violaxanthonin and it is uncertain what happens to the remaining biproduct. The one molecule of xanthonin produced is unstable and spontaneously changed to ABA aldehyde. Further oxidation results in ABA.

Activation of the molecule can occur by two methods. In the first method, an ABA-glucose ester can form by attachment of glucose to ABA. In the second method, oxidation of ABA can occur to form phaseic acid and dihyhdrophaseic acid. The transport of ABA can occur in both xylem and phloem tissues. It can also be translocated through paranchyma cells. The movement of abscisic acid in plants does not exhibit polarity like auxins. ABA is capable of moving both up and down the stem.

Functions of Abscisic Acid

The following are some of the phyysiological responses known to be associated with abscisic acid:

- Stimulates the closure of stomata (water stress brings about an increase in ABA synthesis).
- Inhibits shoot growth but will not have as much affect on roots or may even promote growth of roots.
- Induces seeds to synthesize storage proteins.

- Inhibits the affect of gibberellins on stimulating de novo synthesis of a-amylase.
- Has some effect on induction and maintanance of dormancy.

Induces gene transcription especially for proteinase inhibitors in response to wounding which may explain an apparent role in pathogen Defence.

CYTOKININS

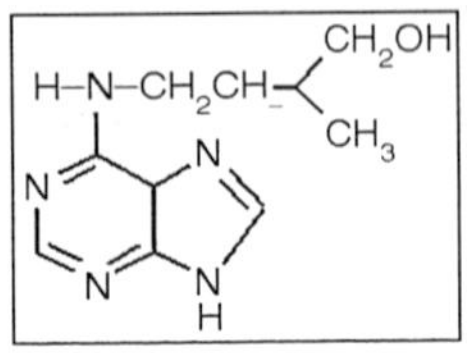

NATURE OF CYTOKININS

Cytokinins are compounds with a structure resembling adenine which promote cell division and have other similar functions to kinetin. Kinetin was the first cytokinin discovered and so named because of the compounds ability to promote cytokinesis (cell division). Though it is a natural compound, It is not made in plants, and is therefore usually considered a"synthetic" cytokinin (meaning that the hormone is synthesized somewhere other than in a plant). The most common form of naturally occurring cytokinin in plants today is called zeatin which was isolated from corn.

Cytokinins have been found in almost all higher plants as well as mosses, fungi, bacteria, and also in tRNA of many prokaryotes and eukaryotes. Today there are more than 200 natural and synthetic cytokinins combined. Cytokinin concentrations are highest in meristematic regions and areas of continuous growth potential such as roots, young leaves, developing fruits, and seeds.

HISTORY OF CYTOKININS

Gottlieb Haberlandt discovered that a compound found in phloem had the ability to stimulate cell division. Johannes van Overbeek discovered that the milky endosperm from coconut also had this ability. He also showed that various other plant species had compounds which stimulated cell division. In 1954, Jablonski and Skoog extended the work of Haberlandt showing that vascular tissues contained compounds which promote cell division.

The first cytokinin was isolated from herring sperm by Miller and his associates. This compound was named kinetin because of its ability to promote cytokinesis. Hall and deRopp reported that kinetin could be formed from DNA degradation products. The first naturally occurring cytokinin was isolated from corn by Miller.

It was later called zeatin. Almost simultaneous with Miller Letham published a report on zeatin as a factor inducing cell division and later described its chemical properties. It is Miller and Letham that are credited with the simultaneous discovery of zeatin. Since that time, many more naturally occurring cytokinins have been isolated and the compound is ubiquitous to all plant species in one form or another.

BIOSYNTHESIS AND METABOLISM OF CYTOKININS

Cytokinin is generally found in higher concentrations in meristematic regions and growing tissues. They are believed to be synthesized in the roots and translocated via the xylem to shoots. Cytokinin biosynthesis happens through the biochemical modification of adenine. The process by which they are synthesized is as follows:

A product of the mevalonate pathway called isopentyl pyrophosphate is isomerized. This isomer can then react with adenosine monophosphate with the aid of an enzyme called isopentenyl AMP synthase. The result is isopentenyl adenosine-5'-phosphate (isopentenyl AMP). This product can then be converted to isopentenyl adenosine by removal of the phosphate by a phosphatase and further converted to isopentenyl adenine by removal of the ribose group. Isopentenyl adenine can be converted to the three major forms of naturally occurring cytokinins. Other pathways or slight alterations of this one probably lead to the other forms. Degradation of cytokinins occurs largely due to the enzyme cytokinin oxidase. This enzyme removes the side chain and releases adenine. Derivitives can also be made but the pathways are more complex and poorly understood.

CYTOKININ FUNCTIONS

A list of some of the known physiological effects caused by cytokinins are listed below. The response will vary depending on the type of cytokinin and plant species.

- Stimulates cell division.
- Stimulates morphogenesis (shoot initiation/bud formation) in tissue culture.

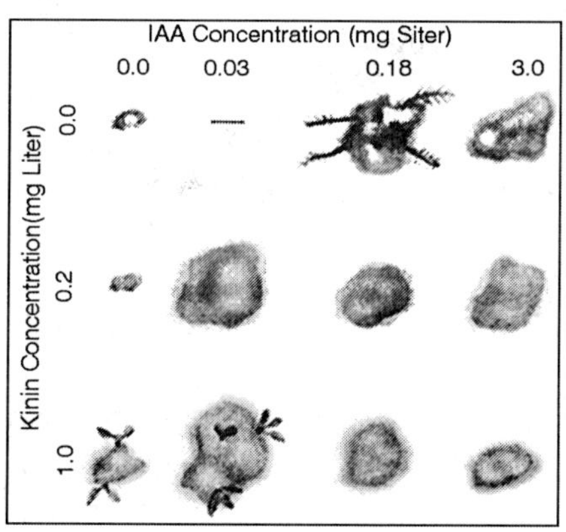

- Stimulates the growth of lateral buds-release of apical dominance.
- Stimulates leaf expansion resulting from cell enlargement.
- May enhance stomatal opening in some species.
- Promotes the conversion of etioplasts into chloroplasts via stimulation of chlorophyll synthesis.

The illustration above shows the effect of cytokinin and auxin concentration on tissue culture experiments.

ETHYLENE

NATURE OF ETHYLENE

Ethylene, unlike the rest of the plant hormone compounds is a gaseous hormone. Like abscisic acid, it is the only member of its class. Of all the known plant growth substance, ethylene has the simplest structure. It is produced in all higher plants and is usually associated with fruit ripening and the tripple response.

$$H_2C{=}CH_2$$

HISTORY OF PLANTS

Ethylene has been used in practice since the ancient Egyptians, who would gas figs in order to stimulate ripening. The ancient Chinese would burn incense in closed rooms to enhance the ripening of pears. It was in 1864, that leaks of gas from street lights showed stunting of growth, twisting of plants, and abnormal thickening of stems (the triple response).

In 1901, a russian scientist named Dimitry Neljubow showed that the active component was ethylene. Doubt discovered that ethylene stimulated abscission in 1917. It wasn't until 1934 that Gane reported that plants synthesize ethylene.

In 1935, Crocker proposed that ethylene was the plant hormone responsible for fruit ripening as well as inhibition of vegetative tissues. Ethylene is now known to have many other functions as well.

Biosynthesis and Metabolism

Ethylene is produced in all higher plants and is produced from methionine in essentially all tissues. Production of ethylene varies with the type of tissue, the plant species, and also the stage of development. The mechanism by which ethylene is produced from methionine is a 3 step process. ATP is an essential component in the synthesis of ethylene from methionine. ATP and water are added to methionine resulting in loss of the three phosphates and S-adenosyl methionine. 1-amino-cyclopropane-1-carboxylic acid synthase facilitates the production of ACC from SAM.

Oxygen is then needed in order ro oxidize ACC and produce ethylene. This reaction is catalyzed by an oxidative enzyme called ethylene forming enzyme. The control of ethylene production has received considerable study.

Study of ethylene has focused around the synthesis promoting effects of auxin, wounding, and drought as well as aspects of fruit-ripening. ACC synthase is the rate limiting step for ethylene production and it is this enzyme that is manipulated in biotechnology to delay fruit ripening in the"flavour saver" tomatoes.

Functions of Ethylene

Ethylene is known to affect the following plant processes:

- Stimulates the release of dormancy.
- Stimulates shoot and root growth and differentiation (triple response)
- May have a role in adventitious root formation.
- Stimulates leaf and fruit abscission.
- Stimulates Bromiliad flower induction.
- Induction of femaleness in dioecious flowers.
- Stimulates flower opening.
- Stimulates flower and leaf senescence.

AUXINS

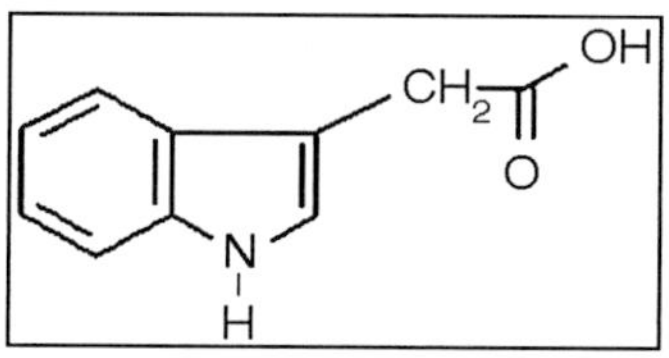

Nature of Auxins

The term auxin is derived from the Greek word auxein which means to grow. Compounds are generally considered auxins if they can be characterized by their ability to induce cell elongation in stems and otherwise resemble indoleacetic acid (the first auxin isolated) in physiological activity. Auxins usually affect other processes in addition to cell elongation of stem cells but this characteristic is considered critical of all auxins and thus"helps" define the hormone.

History of Auxins and Pioneering Experiments

Auxins were the first plant hormones discovered. Charles Darwin was among the first scientists to dabble in plant hormone research. In his book"The Power of Movement in Plants" presented in 1880, he first describes the effects of light on movement of canary grass coleoptiles.

The coleoptile is a specialized leaf originating from the first node which sheaths the epicotyl in the plants seedling stage protecting it until it emerges from the ground.

When unidirectional light shines on the coleoptile, it bends in the direction of the light. If the tip of the coleoptile was covered with aluminum foil, no bending would occur towards the unidirectional light. However if the tip of the coleoptile was left uncovered but the portion just below the tip was covered, exposure to unidirectional light resulted in curvature towards the light. Darwin's experiment suggested that the tip of the coleoptile was the tissue responsible for perceiving the light and producing some signal which was transported to the lower part of the coleoptile where the physiological response of bending occurred. He then cut off the tip of the coleoptile and exposed the rest of the coleoptile to unidirectional light to see if curving occurred. Curvature did not occur confirming the results of his first experiment.

It was in 1885 that Salkowski discovered indole-3-acetic acid (IAA) in fermentation media. The isolation of the same product from plant tissues would not be found in plant tissues for almost 50 years. IAA is the major auxin involved in many of the physiological processes in plants. In 1907, Fitting studied the effect of making incisions on either the light or dark side of the plant. His results were aimed at understanding if translocation of the signal occurred on a particular side of the plant but his results were inconclusive because the signal was capable of crossing or going around the incision.

In 1913, Boysen-Jensen modified Fritting's experiment by inserting pieces of mica to block the transport of the signal and showed that transport of auxin towards the base occurs on the dark side of the plant as opposed to the side exposed to the unidirectional light. In 1918, Paal confirmed Boysen-Jensen's results by cutting off coleoptile tips in the dark, exposing only the tips to the light, replacing the coleoptile tips on the plant but off Centreed to one side or the other.

Results showed that whichever side was exposed to the coleoptile, curvature occurred towards the other side. Soding was the next scientist to extend auxin research by extending on Paal's idea. He showed that if tips were cut off there was a reduction in growth but if they were cut off and then replaced growth continued to occur. In 1926, a graduate student from Holland by the name of Fritz Went published a report describing how he isolated a plant growth substance by placing agar blocks under coleoptile tips for a period of time then removing them and placing them on decapitated Avena stems. After placement of the agar, the stems resumed growth

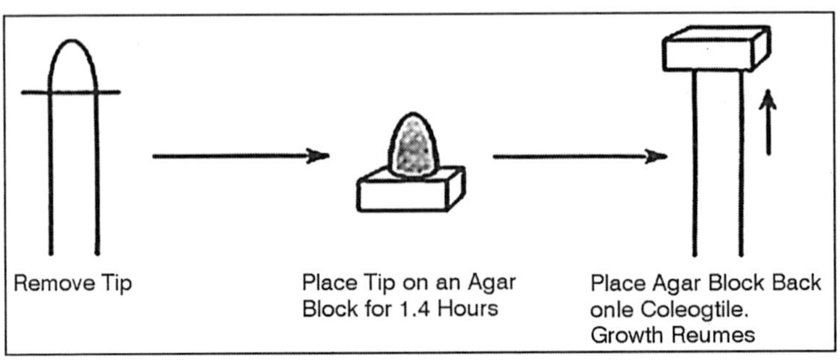

In 1928, Went developed a method of quantifying this plant growth substance. His results suggested that the curvatures of stems were proportional to the amount of growth substance in the agar. This test was called the avena curvature test.(see below) Much of our current knowledge of auxin was obtained from its applications. Went's work had a great influence in stimulating plant growth substance research. He is often credited with dubbing the term auxin but it was actually Kogl and Haagen-Smit who purified the compound auxentriolic acid (auxin A) from human urine in 1931.

Later Kogl isolated other compounds from urine which were similar in structure and function to auxin A, one of which was indole-3 acetic acid (IAA) initially discovered by Salkowski in 1985. A committee of plant physiologists was set up to characterize the group auxins. The term comes from the Greek auxein meaning"to grow." Compounds are generally considered auxins if they are synthesized by the plant and are substances which share similar activity to IAA.

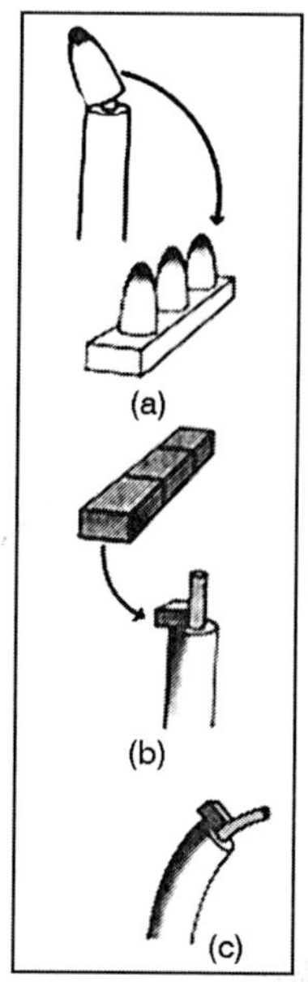

BIOSYNTHESIS AND METABOLISM OF AUXIN

IAA is chemically similar to the amino acid tryptophan which is generally accepted to be the molecule from which IAA is derived. Three mechanisms have been suggested to explain this conversion: Tryptophan is converted to indolepyruvic acid through a transamination reaction. Indolepyruvic acid is then converted to indoleacetaldehyde by a decarboxylation reaction. The final step involves oxidation of indoleace-taldehyde resulting in indoleacetic acid. Tryptophan undergoes decarboxylation resulting in tryptamine. Tryptamine is then oxidized and deaminated to produce indoleacetaldehyde. This molecule is further oxidized to produce indoleacetic acid. As recently as 1991, this 3rd mechanism has evolved. IAA can be produced via a tryptophan-independent mechanism. This mechanism is poorly understood, but has been proven using trp(-) mutants. Other experiments have shown that, in some plants, this mechanism is actually the preferred mechanism of IAA biosynthesis.

The enzymes responsible for the biosynthesis of IAA are most active in young tissues such as shoot apical meristems and growing leaves and fruits. The same tissues are the locations where the highest concentrations of IAA are found. One way plants can control the amount of IAA present in tissues at a particular time is by controlling the biosynthesis of the hormone. Another control mechanism involves the production of conjugates which are, in simple terms, molecules which resemble the hormone but are inactive. The formation of conjugates may be a mechanism of storing and transporting the active hormone. Conjugates can be formed from IAA via hydrolase enzymes. Conjugates can be rapidly activated by environmental stimuli signaling a quick hormonal response. Degradation of auxin is the final method of controlling auxin levels.

This process also has two proposed mechanisms outlined below: The oxidation of IAA by oxygen resulting in the loss of the carboxyl group and 3-methyleneoxindole as the major breakdown product. IAA oxidase is the enzyme which catalyzes this activity. Conjugates of IAA and synthetic auxins such as 2,4-D can not be destroyed by this activity. C-2 of the heterocyclic ring may be oxidized resulting in oxindole-3-acetic acid. C-3 may be oxidized in addition to C-2 resulting in dioxindole-3-acetic acid. The mechanisms by which biosynthesis and degradation of auxin molecules occur are important to future agricultural applications. Information regarding auxin metabolism will most likely lead to genetic and chemical manipulation of endogenous hormone levels resulting in desirable growth and differentiation of important crop species. Ultimately, the possibility exists to regulate plant growth without the use of hazardous herbicides and fertilizers.

Functions of Auxin

The following are some of the responses that auxin is known to cause:

- Stimulates cell elongation
- Stimulates cell division in the cambium and, in combination with cytokinins in tissue culture
- Stimulates differentiation of phloem and xylem.
- Stimulates root initiation on stem cuttings and lateral root development in tissue culture.
- Mediates the tropistic response of bending in response to gravity and light.
- The auxin supply from the apical bud suppresses growth of lateral buds.
- Delays leaf senescence.

- Can inhibit or promote (via ethylene stimulation) leaf and fruit abscission.
- Can induce fruit setting and growth in some plants.
- Involved in assimilate movement towards auxin. possibly by an effect on phloem transport.
- Delays fruit ripening.
- Promotes flowering in Bromeliads.
- Stimulates growth of flower parts.
- Promotes (via ethylene production) femaleness in dioecious flowers.
- Stimulates the production of ethylene at high concentrations.

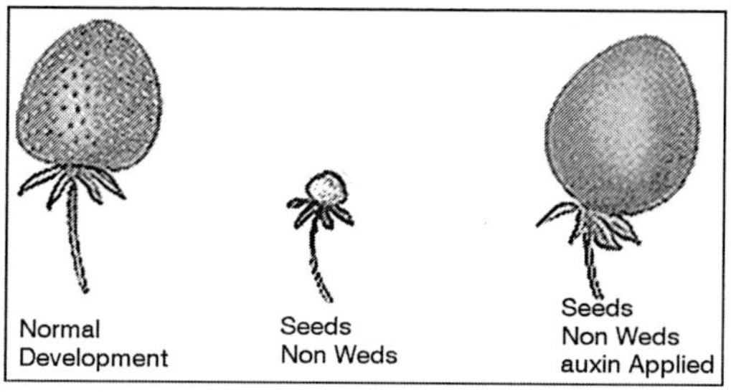

Above describes the effect of auxin on strawberry development. The achenes produce auxin. When removed the strawberry does not develop.

GIBBERELLINS

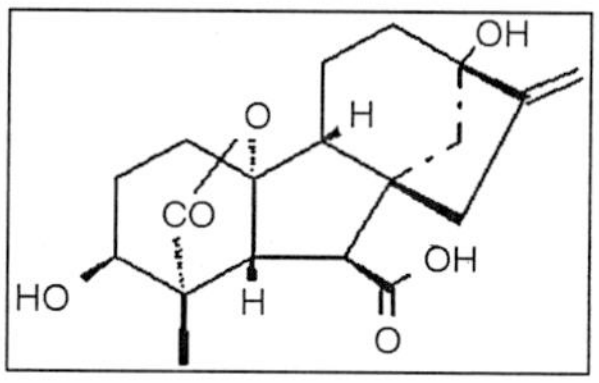

NATURE OF GIBBERELLINS

Unlike the classification of auxins which are classified on the basis of function, gibberellins are classified on the basis of structure as well as function. All gibberellins are derived from the ent-gibberellane skeleton. The structure of this skeleton derivative along with the structure of a few of the active gibberellins are shown above. The gibberellins are named GA1....GAn in order of discovery. Gibberellic acid, which was the first gibberellin to be structurally characterised, is GA3. There are currently 136 GAs identified from plants, fungi and bacteria GA's are widespread and so far ubiquitous in both flowering (angiosperms) and non-flowering (gymnosperms) plants as well as ferns.

GIBBERELLIN BIOSYNTHESIS AND METABOLISM

Gibberellins are diterpenes synthesized from acetyl CoA via the mevalonic acid pathway. They all have either 19 or 20 carbon units grouped into either four or five ring systems. The fifth ring is a lactone ring as shown in the structures above attached to ring A. Gibberellins are believed to be synthesized in young tissues of the shoot and also the developing seed. It is uncertain whether young root tissues also produce gibberellins.

There is also some evidence that leaves may be the source of some biosynthesis. The pathway by which gibberellins are formed is outlined below and illustrated in figure. 3 acetyl CoA molecules are oxidized by 2 NADPH molecules to produce 3 CoA molecules as a side product and mevalonic acid. Mevalonic acid is then Phosphorylated by ATP and decarboxylated to form isopentyl pyrophosphate. 4 of these molecules form geranylgeranyl pyrophosphate which serves as the donor for all GA carbon atoms.

This compound is then converted to copalylpy-rophosphate which has 2 ring systems Copalylpyrophosphate is then converted to kaurene which has 4 ring systems. Subsequent oxidations reveal kaurenol (alcohol form), kaurenal (aldehyde form), and kaurenoic acid respectively.

Kaurenoic acid is converted to the aldehyde form of GA12 by decarboxylation. GA12 is the 1st true gibberellane ring system with 20 carbons. From the aldehyde form of GA12 arise both 20 and 19 carbon gibberellins but there are many mechanisms by which these other compounds arise.

Certain commercial chemicals which are used to stunt growth do so in part because they block the synthesis of gibberellins. Some of these chemicals are Phosphon D, Amo-1618, Cycocel (CCC), ancymidol, and paclobutrazol. During active growth, the plant will metabolize most gibberellins by hydroxylation to inactive conjugates quickly with the exception of GA3. GA3 is degraded much slower which helps to explain why the symptoms initially associated with the hormone in the disease bakanae are present. Inactive conjugates might be stored or translocated via the phloem and xylem before their release (activation) at the proper time and in the proper tissue.

FUNCTIONS OF GIBBERELLINS

Active gibberellins show many physiological effects, each depending on the type of gibberellin present as well as the species of plant.

Some of the physiological processes stimulated by gibberellins are outlined below:

- Stimulate stem elongation by stimulating cell division and elongation.
- Stimulates bolting/flowering in response to long days.
- Breaks seed dormancy in some plants which require stratification or light to induce germination.
- Stimulates enzyme production (a-amylase) in germinating cereal grains for mobilization of seed reserves.
- Induces maleness in dioecious flowers (sex expression).
- Can cause parthenocarpic (seedless) fruit development.

GROWTH AND DEVELOPMENT

Growth is an increase in size of basic parts of plants (due to production of extra cells), whereas development is the production of new structures in the progress towards maturity - flowers, fruits etc.

PLANT HORMONES VS ANIMAL HORMONES

In that green plants make growth in response to stimuli from their surroundings, they show sensitivity, *i.e.* one of the seven characteristic processes of living organisms. However, green plants respond to fewer stimuli than animals, and they make a slower and more restricted response.

The 3 stimuli to which green plants responds are:

1. Light.
2. Gravity.
3. Water/touch (peasticks etc)NOT CO_2.

If a plant responds by growing in a particular direction, it is called a tropism. Positive tropisms are towards the stimulus, negative tropisms are away from the stimulus.

REGULATION OF DEVELOPMENT

Stimulation of Roots in Cuttings

When cuttings are taken, the shoot which is selected is removed from the plant and disconnected from a supply of water and minerals from the roots. It is therefore important to encourage the development of roots, to aid the establishment of the"cutting" as an independent plant. "Hormone rooting powders" contain the substance indoleacetic acid IAA which has been identified as the same chemical as"auxin" which causes curvature in roots and shoots. Usually the newly exposed surface of the cutting is dipped into a powder containing the hormone, mixed with inert powder, and then it is placed into a pot containing moist compost or soil, which is gently firmed around the cutting and placed in a cool moist environment. There are a variety of other chemically similar compounds which may be used for this purpose.

HORMONE WEEDKILLERS

Although there are a wide variety of chemicals which can be toxic to plants or which damage their exposed parts, such as leaves and shoots, these are of limited use due to problems of toxicity to humans and other animals, and danger due to their oxidising nature. These substances may not affect parts of plants which are underground, and there is a likelihood that new growth will sprout again from these parts. There are also many situations in which more specificity is required: killing some plants but not others, and leaving soil fit for subsequent growth of plants which are required.

Hormone weedkillers have a variety of advantages over more traditional substances:

- They should be more toxic to plants than animals.
- They should be carried to underground plants and kill then too.
- Some weedkillers have a greater effect on"broad-leaved weeds" than on grasses and also cereal crops (monocots). For this reason they are called selective weedkillers.

These may be used to control weeds amongst the grass on lawns, or to reduce competition from weeds in cereal crops. Examples include compounds with complicated chemical names, popularly abbreviated to 2.4-D and 2,4,5-T. Other compounds such as Dalapon are more effective against monocots than dicots, so they could be used in the opposite situation. In fact these weedkilllers cause sensitive plants to grow too fast, damaging their cells so that their conducting tissue does not develop properly, or they use up their food reserves and die.

Weed

A plant growing in the wrong place - *i.e.* not planted by a gardener or grower. In other words, a plant which is able to compete and grow well in the prevailing conditions - probably in its own natural habitat!

Consequences of Using Selective Weedkillers

It must be admitted that the widespread use of selective weedkillers in European and Western agriculture has improved the efficiency of crop production (productivity of arable farms), in response to policies promoted by government ministries. This has been quite successful. resulting in the excessive production of cereal and other crops - so called"grain mountains".

However this apparent economic improvement has been at the expense of biodiversity in the countryside, as many"weed" species are food sources and act as shelter for many species of wildlife. Many plants which used to flower alongside crops and weeds, adding to the visual impact of the countryside, have been adversely affected by herbicides sprayed onto fields. Similarly, the survival of animal species, such as birds and butterflies, which add immeasurably to the public enjoyment of the wider environment, are consequently threatened more by the lack of food and protection than by the killing power of the chemicals involved.

Recently, there have been a series of developments aimed at reversing this trend, including"set-aside" where farmers are paid compensation in order to leave sections of their farms fallow in order to revert to"natural" conditions which should support normal fauna and flora. Regulations about maintenance of hedges and borders around fields have also been relaxed in order to increase the areas which can act as habitats for wildlife.

Development of Fruits

The ripening of fruits is known to be affected by a number of plant hormones. Commercially, growers sometimes use chemical sprays to prevent fruit from falling from trees. The substances used are auxins such as IAA or substances such as 2,4-D which is also used as a herbicide. Depending on weather conditions, this normally happens with a proportion of the fruit crop, leading to uncertainty as to the likely productivity of the crop. The consistency of cropping and ease of picking can thus be increased. Normally, a flower which is pollinated develops into a fruit, containing seeds derived from the fertilised egg cell. However, in some circumstances, fruit develops without pollination taking place. In seedless varieties of fruits this parthenocarpic fruit development occurs when other parts of the fruit produce auxins which would otherwise come from the seeds.

Other chemical substances acting as growth regulators can be used to induce flowering and alter the number and shape of fruit forming. The gas ethane (ethylene) has also been shown to have a hormone-like activity on fruit and is used to control the ripening of stored fruit such as bananas and apples which are best picked and shipped before maturity.

9

Biological Fixation of Nitrogen The Mitochondria

INTRODUCTION

Mitochondria have been known for many years. They were first discovered and described by Altmann in 1886 and were put more or less definitely on the cytological map by Benda in 1903. They can be easily demonstrated with suitable dyes. Altmann's aniline fuchsine-picric acid technique shows them up very well. Regaud's iron-hematoxylin method is also very good. The mitochondria can even be seen in the living cell by staining them intravitally with Janus green and, of course, they show up extremely well with the phase-contrast and interference microscopes. There is not much difficulty, then, in establishing their existence and nature. Their dimensions vary, of course, but they range in most cells from about 0.5 to 2ì or longer.

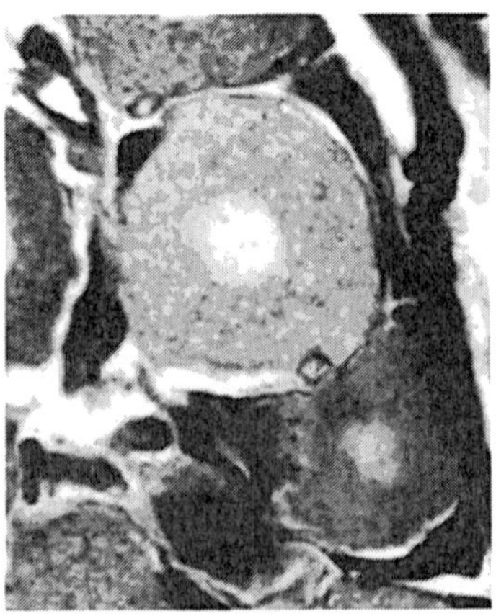

The filamentous mitochondria, for example, which are present in connective tissue cells may be much longer than this. Extremely small and even submicroscopic mitochondria may also exist. Extremely smallsized mitochondria are, in fact, known and Rhodin has described structures ranging from 0.1 to 0.5ì which are presumably mitochondrial in nature. Green has pointed out that the various enzymes and proteins which mitochondria contain can be accommodated in a very much smaller particle than any known mitochondria and still perform all the functions required of them.

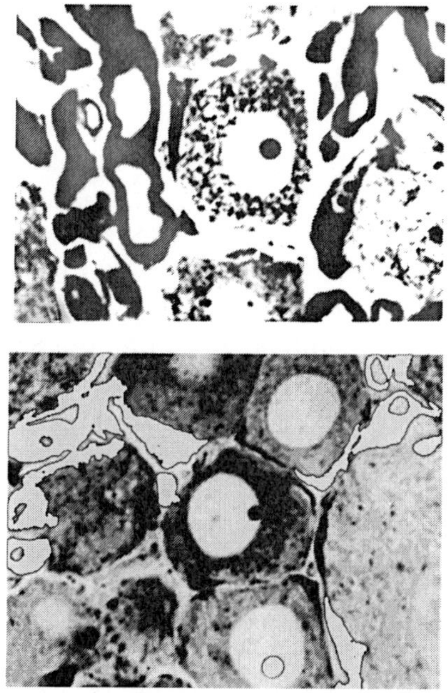

Fig. Relation between Nucleolus and Mitochondria in Spinal Ganglion Neurons.

Some of these much smaller mitochondria may not have all the details of organization of larger mitochondria but there is no reason why they should not contain all the enzymes and other compounds which are important for mitochondrial structure and function.

It is of interest to note that, in living tissue culture cells, the mitochondria undergo continuous movement. This has been noted by quite a number of workers and the movements have been described frequently. These movements may be fitful individual movements by the mitochondrion itself, i.e., bending, wriggling and writhing, or may consist of movement of the whole mitochondrion through the cytoplasm of the cell.

The cause of the bending and twisting movements of the mitochondria is not exactly known. At one time it was thought to result from the contracting of the polypeptide chains which are part of the membrane of the mitochondria. It may be due to the fact that the mitochondria were engaged in active ion pumping or being twisted by cytoplasmic movements. In the second type of movement in which there is a transport of the whole mitochondrion from one part of the cell to another, the movement is possibly related to streaming movements of the cytoplasm or to electrical forces.

The mitochondria have been described as making journeys from the cell membrane to the nuclear membrane and back again almost as if they were discharging either an electric charge or even perhaps some compound on the membrane that they touch. There are also records that, when the mitochondria come in contact with the nuclear membrane the nucleolus sometimes moves across the cell and comes in contact with the nuclear membrane at the same point at which the mitochondrion is touching it and this suggests that there may be some exchange of compounds or substances during this period. This has been demonstrated in spinal ganglion cells by Dr. Tewari in the author's laboratory. Filamentous mitochondria have frequently been observed to break up into batonettes and these have been observed to break up into granules. The reverse process has also been found to occur, granules have been seen to join up into batonettes and these into longer filamentous mitochondria. What the significance of this breaking up is we do not know, but a little later we will refer to this process again in the light of what will be said about the metabolic activities of mitochondria.

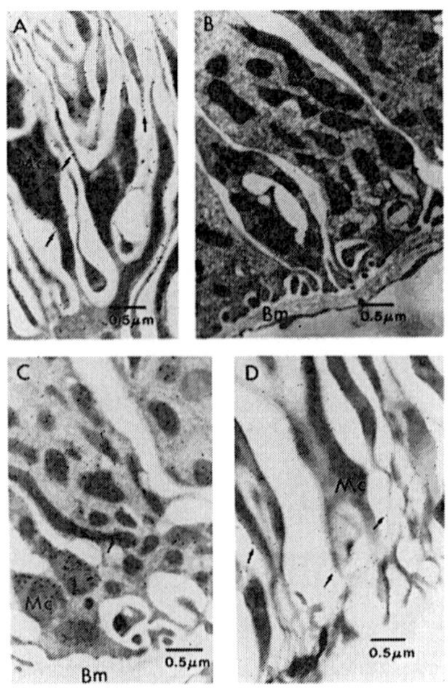

Fig. Elongated Mitochondria in Kidney

When mitochondria were studied in ultrathin sections of tissue, it was found by Palade in 1953, Sjöstrand, 1953 and by Rhodin in 1954, that they had an interesting internal structure. These workers found first of all that there was a single membrane round the outside, that inside this membrane was another membrane so that the two made a double membrane and, after a little controversy, it was agreed that the inner of these two membranes was extended to form a number of bars or plates which projected into the interior of the mitochondria, in some cases touching or almost touching the other side. These plates or bars or tubes had double structure again, since they were a reflection of the inner membrane of the mitochondria.

They were described by Palade as the "cristae mitochondriales". Changes in these cristae appear to be related to function, for instance Palade has drawn attention to the fact that the amount of cytochrome in the mitochondrial fraction of a homogenate is directly related to the number of cristae present in the mitochondria. When metabolic activity is high, for example, in rapidly contracting skeletal and heart muscle, the mitochondria have many densely packed cristae. In smooth muscle, however, where the activity is greatly reduced, the cristae present in the mitochondria are relatively sparse. There are claims that, e.g., in mouse tumor cells and in paramecium, cristae may be everted into the cytoplasm. Presumably in this case the outer membrane is folded into cristae.

It is of interest that the double membranes which surround the mitochondria show the same 80 A unit structure which is characteristic of the cell membrane and Robertson has suggested that it is possible that the mitochondria may have originated by an invagination of the membrane which was nipped off to form the mitochondrion.

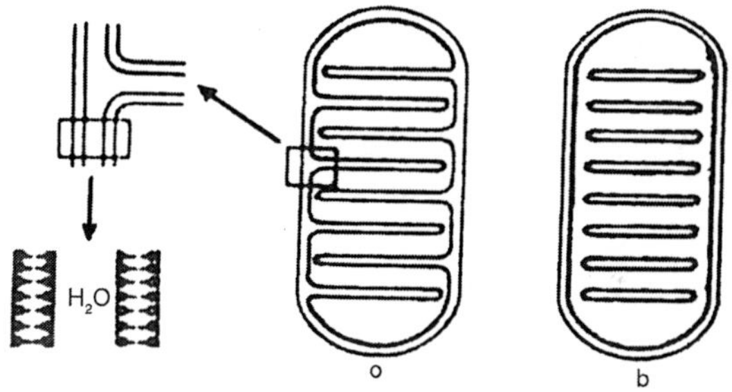

Fig. Structure of Mitochondria

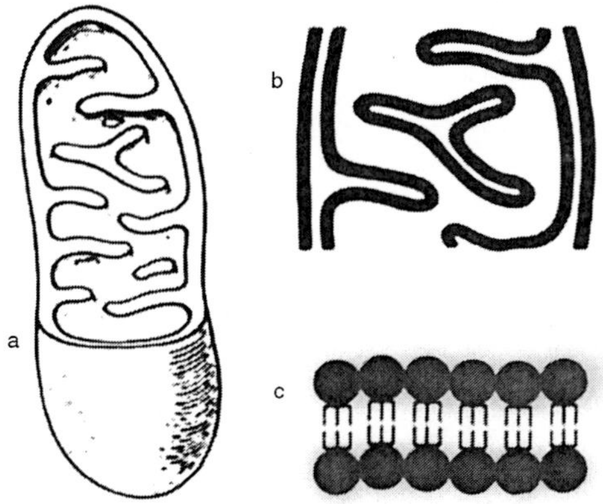

Fig. Three Dimensional View of Structure of Mitochondria which Appears as a Bag filled with Fluid.

This is an interesting suggestion but it is not readily acceptable to most cytologists and is, in fact, considered unlikely by many of them. The membranes of the mitochondria appear to consist of two protein layers separated by a double layer of lipids. This is again characteristic of cell membrane structure. Bensley in 1934 was the first one to isolate mitochondria by differential centrifugation and he was able to demonstrate that mitochondria contained a considerable amount of protein, lipid and fat. Bensley's chemical composition of mitochondria is

Proteins and unknowns	65%
Glycerides	29%
Lecithin and cephalin	4%
Cholesterol	2%

Subsequent studies have shown up to 30 per cent of phospholipid (lecithin and cephalin). Mitochondria undergo a very considerable change under various conditions. Bensley claimed that mitochondria (as demonstrated by standard staining reactions) were greatly reduced in liver cells of starved animals, but Sheridan in our laboratory and Fawcett have demonstrated by electron-microscope techniques a great increase of mitochondria in starvation. Sheridan has further demonstrated a still further increase of mitochondria in the liver cells of guinea pigs suffering from scurvy. Mitochondria are also affected in number and form by narcotics, enzyme poisons, old age and so on. Sometimes mitochondria store unusual compounds and sometimes they store compounds which are usual but to an abnormal degree. The double membrane enclosing the mitochondria is of a semipermeable nature and as a result of this mitochondria have been described as osmometers.

There is no doubt that molecule of various sizes pass with different degrees of readiness through the mitochondrial membrane. It has been shown that large molecules pass through fairly easily while some small molecules pass through either not at all or only with very great difficulty. It has even been said by Emmelot and Bos that this permeability, particularly of liver mitochondrial membranes, is affected by thyroxine. Also there are claims by some authors that mitochondrial membranes have pores. Mitochondria may accumulate water and swell. They may do this in starvation and in various pathological conditions. When this occurs they produce the condition which the pathologists describe as cloudy swelling. Cloudy swelling has been known to occur sometimes in neo-plasms and sometimes after poisoning with various toxins.

The mechanics of cloudy swelling are that the mitochondria become very large and lose their cristae, the matrix stains less strongly and sometimes the outer membranes of the mitochondria may disappear and the mitochondria may fuse and so form structures known as chondriospheres. Under some pathological conditions mitochondria may become disrupted, undergo lysis and portions may be ejected from the cell. Mitochondria are capable of abnormal storage. Accumulation of ferritin by mitochondria was demonstrated by Kuff and Dalton in 1957 and they may also accumulate iron pigments or bile pigments. Brachet has described ferritin particles in young erythroblasts where he says they penetrate the mitochondria and then break into smaller iron-containing particles. When the mitochondria eventually burst, the granules become scattered throughout the cytoplasm and take part in the formation of hemoglobin. Silver granules and keratinous materials have been also seen inside mitochondria. They have been found to contain melanin and Graffi in a series of papers from 1939 to 1941 claims that they also contain or store carcinogenetic hydrocarbons. Sometimes this storage affects the function of the mitochondria and sometimes the accumulation of material is due to malfunction of the organelle. They may also store neutral fats or lipids. In plants mitochondria have been recorded as storing starch.

As a result of storage mitochondria undergo physical changes. The membrane may become single and the cristae decrease in size and may be lost altogether. If the stored substance is eliminated, the mitochondria resume their normal appearance. There is a vast literature on the role of mitochondria in the production of almost any kind of product—protein, yolk, fat, glycogen and so on—but this is not the place to review this work.

There have been many scornful remarks in the literature about the supposed synthetic activity of mitochondria, but the study of the enzymic equipment of mitochondria shows that they have the equipment to effect a wide range of synthetic activities. It has been mentioned that the isolation of mitochondria by the differential centrifugation of homogenized cells was first carried out in 1935 by Bensley and by Hoerr. It was not until 10 or 15 years later that this technique became used extensively and this time by biochemists who set to work to find out the metabolic composition of mitochondria, i.e., the nature and content of enzymes concerned with metabolic processes.

THE CHEMICAL NATURE OF MITOCHONDRIA

In Bensley's work it was suspected that mitochondria contained a good deal of lipid material. One finding which suggested this was that mitochondria were stained by a method which is similar to that used for staining *Mycobacterium tuberculosis* and *Mycobacterium leprae*. These bacteria are stained by treating them for some minutes with a hot phenolic solution of basic fuchsine and once stained in this way they resist the de-staining effects of acid alcohol. This is the origin of their designation as acid-fast bacteria and their staining idiosyncrasies are believed to be due to a waxy or lipoidal coat. Mitochondria stained by hot acid fuchsine resist the decolouring action of picric acid for longer periods than most other cell constituents and this suggests that the mitochondria might also have a membrane which contains material of a lipoidal nature or at any rate which is of similar composition to that of the bacteria mentioned. Mitochondria are difficult to demonstrate with conventional methods of staining if the fixative has contained acetic acid, alcohol, ether, chloroform, acetone, or other fat solvents. Also mitochondria have been found to stain with osmic acid and, incidentally, to resist de-staining by extraction with turpentine—a property of certain types of lipoprotein complexes. Baker has found that his "acid-haematin" test for lipids applied to a variety of tissues, nearly always gave a positive, reaction with the mitochondria. Thus there had accumulated for some years direct and indirect evidence that mitochondria contained appreciable amounts of lipid. Bensley's experiments, which, incidentally, were carried out some years before Baker's staining studies, showed that mitochondria which had been produced by homogenization and differential centrifugation contained appreciable amounts of lipid (30 per cent).

METABOLIC SUBSTANCES FOUND IN MITOCHONDRIA

Perhaps we should give a brief summary of what was suspected about the metabolic significance of these organelles prior to the revolutionary studies which were made with isolated mitochondria beginning during the middle of the 1940's. One of the characteristic staining reactions of these organelles is that they give a green or green-blue stain with Janus green B which is diethylsafranine azodimethylaniline and it has been claimed by Cowdry that this reaction is primarily due to the diethylsafranine monocarboxylic acid component since this compound gives a very good and specific reaction with these structures. It was observed many years ago by T. B. Roberston that if one drop of a saturated solution of safranine was added to a solution of trypsin, a coloured precipitate was formed and subsequently it was demonstrated that this coloured precipitate had proteolytic activity.

Then Marston demonstrated that other azo dyestuffs would react in a similar way, particularly neutral red which is a dimethyldiaminotoluazine hydrochloride. Marston suggested that these results indicated that the reaction of mitochondria with Janus green might signify that they contained proteolytic enzymes. Now, the concept of the staining reaction with Janus green has been pretty well proved by Lazarow and Cooperstein to indicate that mitochondria play an important part in cellular oxidations and that the production of a pink colour from the Janus green by the mitochondria is due to the DPN specific dehydrogenases. The fact that Janus green B does not stain the mitochondria permanently green but that the green colour gradually becomes reduced to the pink and then to the colourless form has been known for many years. Joyet-Lavergne demonstrated more than twenty years ago that mitochondria contain an oxidase system which oxidizes cobaltous to cobaltic salts and that these latter stain the mitochondria green. It was noted by Gatenby that in the snail, *Lymnaea*, the mitochondria are coloured yellow in the natural state presumably by a carotenoid pigment; also extracted lipomitochondria frequently have a yellow appearance which too is probably due to carotene. It is of interest in this connection that it has been demonstrated by the present author and by

Joyet-Lavergne that mitochondria give a blue reaction with antimony trichloride in chloroform solution, an acknowledged reaction for vitamin.

A. Carotene is also a provitamin A, so these histochemical results indicate the presence of vitamin A in mitochondria. Criticism of Joyet-Lavergne's results was made by Gomori because the former author had used alcohol as a fixative. The present author has, however, always applied antimony trichloride in chloroform solution direct to fresh unfixed tissues and this suggests that this result is a true one, in any case biochemical tests have now confirmed that mitochondria (prepared by homogenization and differential centrifugation of the cells) contain 27-32 per cent of their weight of lipids and that 100 mg. of this lipid contains approximately 249-910 U.S.P. units of vitamin A. JoyetLavergne has suggested that mitochondria contain a redox system in which vitamin A plays a part.

A number of authors, Leblond, the present author and Giroud and his co-workers have demonstrated that mitochondria of some organs react with acetic acid-silver nitrate solution (which has been demonstrated as being a specific reagent for vitamin C) to give a positive reaction and this indicates that vitamin C may be present in them. One has to accept the intracellular localization of vitamin C with a certain amount of discretion in view of the very destructive effect of this reagent on the cell cytoplasm. Electron-microscope studies in the present author's laboratory have demonstrated that this reagent has a most drastic effect on the ultrastructure of the cell and we cannot be sure that the localization of vitamin C in the mitochondria is, in fact, a real thing.

It is of interest, however, that Chayen has found that the mitochondria of plant cells give this reaction very intensely and very specifically. It may be that mitochondria in plant cells and some animal cells do, in fact, contain vitamin C, but we need further studies before this can be confirmed. Other vitamins appear to be present in mitochondria, members of the vitamin B complex have been found to be present, in some cases, in the form of coenzymes. The actual vitamins of the B complex recorded are vitamin B1 (thiamine), riboflavin, nicotinic acid (niacin), pantothenic acid and pyridoxine.

However, although these vitamins are present in mitochondria, they are not present in any greater concentration than in the other parts of the cell so that they are not exclusively contained in these organelles. These studies were made on cell homogenates and, of course, it is possible that the presence of the vitamins in the other fractions of the homogenate may be due to the fact that they have been leached out of the mitochondria by the saline solution which is used in the homogenization process in these particular experiments and further work would have to be done before we could be certain about this.

The present author, Joyet-Lavergne and Giroud have demonstrated that mitochondria contain glutithione or protein-bound SH and this is further evidence that mitochondria can play an important part in oxidation-reduction mechanisms in the cell. It is of interest that mitochondria have been found in large quantities in the phloem cells of plants which are concerned with transport and may be concerned with this process. This suggestion is made in view of Conway's views that redox systems can play an important part in ion pumping. Another substance which was demonstrated histochemically in mitochondria by the present author, using the Schultz reaction, was cholesterol. This was particularly evident in the mitochondria of cells of the adrenal cortex. Similar but less intense reaction was shown by the mitochondria of the liver and it has been demonstrated that isolated liver mitochondria contain about 2 per cent cholesterol and it is possible they contain more in the adrenal cortex.

Bensley has recorded the presence of a red pigment in the mitochondria and also in the submicroscopic particles of the liver cell. He believed that this pigment is derived from the oxidation of unsaturated fats and possibly the phospholipids of the mitochondria and this led him to suggest that in the liver cell the mitochondria in particular are, possibly, the seat of highly active oxidative processes which involve the metabolism of fats. The possible relation of mitochondria to oxidation-reduction processes was indicated by the publications of Ludford. He demonstrated that, if methylene blue was added to tissue cultures, the mitochondria of the cells stained a brilliant blue colour but this could be inhibited by potassium cyanide. If the cells were exposed to a bright light, the blue colour was rapidly bleached.

It is of interest that the mitochondria, although they have been demonstrated to contain a good deal of fat and lipid, do not give a positive reaction with Sudan III. They contain protein (quite a high proportion of it), but it is of interest that the earlier workers, using Millon's reagent, produced a negative reaction for protein. Subsequently Bensley and Gersh using a Millon's reagent of a different formula were able to obtain positive results from the mitochondria of many tissues and they found that they were particularly well stained in frozen, dried sections and particularly in those of *Amblystoma* liver. The same authors demonstrated that, in undenatured, frozen, dried sections of *Amblystoma* liver, the mitochondria were destroyed if the sections were exposed to artificial gastric juice and artificial pancreatic juice.

Despite all this chemical information most of which was in existence by the beginning of the 1940's, there was no certainty as to the function of mitochondria and it was not until 1946 and 1947 that the late George Hogeboom and his colleagues at Bethesda completely revolutionized our ideas of their function by demonstrating that the major proportion of the succinic dehydrogenase and an appreciable proportion of the cytochrome oxidase activity of the liver cell were present in the mitochondria. This at once suggested that these organelles were the major sites of aerobic respiration in the cell. It is of interest that as long ago as 1915, Dr. Kingsbury had suggested that mitochondria were concerned with the respiration of the cell. His reasons for this were largely due to his observations that anesthetics such as ether and chloroform, which depressed cellular respiration and the respiration of the animal in general, also broke up mitochondria in the cell.

The work of Hogeboom and his colleagues was carried out on homogenates of liver which had been produced by grinding up liver with saline. This gave poor preservation of the form of the mitochondria and subsequently 0.8 *M* sucrose was used—this preserved the nature and form of the mitochondria very well. In the beginning there was some doubt as to whether the material being assayed was in fact mitochondria or not, but when sucrose was used this doubt gradually disappeared. Eventually electron-microscope studies of the structure of the granules isolated by homogenization and differential centrifugation demonstrated beyond doubt that they contained the same structures as the mitochondria of normal cells.

Fig. Localization of Respiratory Chain Enzymes (black spheres) in the Mitochondrial Membrane

Another possible source of error, however, began to haunt the biochemists and was frequently verbalized by cytologists and this was that the mitochondria did not really contain the oxidative enzymes, mentioned in the foregoing, but that they were being absorbed or adsorbed by them from the homogenate. This possibility was negated at least in part by adding more enzyme to the homogenate and demonstrating that it could be recovered almost 100 per cent from the supernatant and so was not taken up by the mitochondria. Thus the localization of the enzymes in the original cell was probably in the mitochondria.

Subsequently, Kennedy and Lehninger demonstrated that in addition to containing cytochrome oxidase and succinic dehydrogenase, mitochondria also catalyze the condensation of pyruvic acid with oxaloacetic acid, the formation of succinic acid from á-ketoglutaric acid and the formation of malate from succinate. These three reactions represent three steps of fundamental importance in the Krebs tricarboxylic acid cycle of which we will have more details shortly. Mitochondria also contain coenzyme I of which nicotinamide is an important constituent and cytochrome reductase which is a flavoprotein. These two enzymes are links between the Krebs tricarboxylic acid cycle and the cytochrome system and it, therefore, appears that mitochondria probably contain the whole enzymic equipment necessary for aerobic respiration of the cell. In fact, it was originally demonstrated that a centrifugate of cells which included only nuclei and mitochondria were capable of carrying through the whole of the oxidation of glycogen to CO_2 and water; subsequently it was demonstrated that isolated mitochondria on their own could do this. Transaminase activity was subsequently demonstrated in mitochondria although in only the same concentration as in the rest of the cytoplasm.

Transaminase is concerned with protein synthesis and it is of interest that pyruvic, oxaloacetic and á-ketoglutaric acids are the corresponding á-keto acids of the amino acids, alanine, aspartic and glutamic and need only to be transaminated to produce them. Since the former compounds are themselves formed as intermediates in the course of the Krebs cycle, this indicates a mechanism whereby mitochondria might synthesize fresh protein material and so increase in size themselves or synthesize protein for other parts of the cell. This possibility is further extended by the fact that RNA is known to be present in varying amounts in mitochondria, although the exact amount is uncertain and there is a possibility of contamination of mitochondria with RNA from the rest of the homogenate when making such estimates. However, there is no doubt that mitochondria do have the equipment for synthesising protein although to what extent they do this *in vivo* is not known.

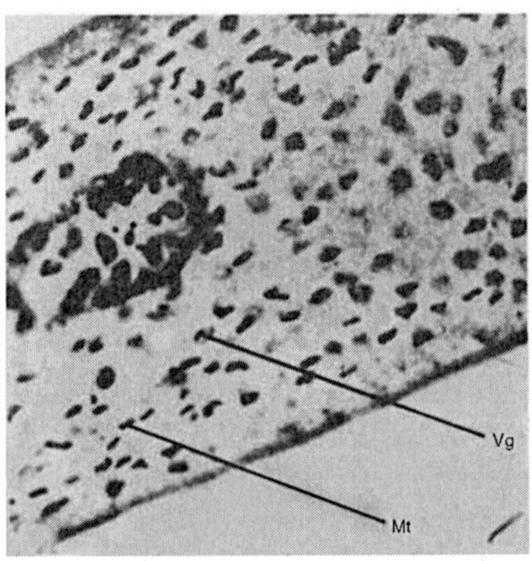

Fig. Mitochondria in the Protozoan Opalina.

All the fatty acid oxidase activity in the cell is also in the mitochondria and so is 80 per cent of the octanoxidase activity. Since these discoveries, an enormous amount of work has been done on the mitochondrial enzymes and this work was summarized by Hogeboom (unfortunately now deceased), Kuff and Schneider from the *National Institutes of Health* and has been published in Volume 6 of *The International Review of Cytology*, 1957. It may be of interest to record here other enzymes and compounds which they listed as being present in mitochondria.

It is of interest too, that a considerable amount of ribonuclease and deoxyribonuclease are present in mitochondria—this seems unusual since only a very small amount of DNA and a relatively small amount of RNA are present in the mitochondria compared with other parts of the cell. However, it is possible that the bodies which contain RNAase and DNAase are not true mitochondria but are lysosomes; we will discuss these shortly. In the meantime, it is of interest to note that it was eventually demonstrated by fragmenting mitochondria and separating the membranes from the contents that cytochrome oxidase and succinic dehydrogenase activity were located primarily in the membranes.

Subsequently it has been shown that the ability of the mitochondria to carry out the Krebs cycle and to combine this with oxidative phosphorylation is dependent on the presence of intact double membranes in the mitochondria. Now, it has been demonstrated and mentioned earlier that in the structure of mitochondria there is an outer membrane and an inner membrane to the mitochondrion, so that it is surrounded by a double membrane. Also, the inner membrane is folded inward to give elongated cristae and these cristae, because they consist of a fold of the inner membrane, are themselves double membranes. There appears to be little difference biochemically between the double membrane structures of the cristae and the double membrane structure of the outer wall of the mitochondria.

An important factor in the process of oxidative phosphorylation is the transport of electrons. Dr. David E. Green has described the process of electron transport as carried out fundamentally with the aid of an electron transport particle—this is a submitochondria particle.

The latter can be isolated in a form so that it can no longer carry out the full citric acid cycle, but it can still carry out the process of oxidative phosphorylation which it performs by the oxidation of succinate and DPNH (reduced diphosphopyridine nucleotide, a codehydrogenaste) with freshly prepared by homogenization and differential centrifugation and are then exposed to alternate freezing and thawing, their form undergoes considerable change. First of all, the external membrane ruptures and some of the cristae break off and eventually the mitochondrion is completely disintegrated; however, fragments of double membrane structure are left which, according to D. E. Green, may be derived either from the cristae or from the outer envelope of the original mitochondrion. These double membrane fragments represent collections of Dr. Green's electron transport particles.

If other methods are used for preparing these particles and the particles so produced do not show a double membrane structure, then oxidation of succinate and DPNH coupled with phosphorylation does not occur. Apparently the double membrane structure is essential for this process to take place. In the structures which do not have the double membrane structure, the capacity for electron transport is still present but oxidation phosphorylation cannot take place.

About 35 per cent of the dry weight of the electron transport particle of Green is composed of lipid and the lipid is concentrated in packets of lipoprotein. These are interspersed between two molecules of cytochrome or between a flavoprotein and cytochrome. In other words, all the oxidation reduction members of the election transfer chain are connected to one another by lipoproteins. These lipoproteins, according to Green, act not only as structural devices, holding various enzymes together, but they also conatin a compound which can shuttle electrons back and forth so that then can pass from cytochrome through to flavoprotein and reverse. This compound, which Green and his colleagues have described, is a new coenzyme called coenzyme Q.

It is a completely water insoluble benzoqtuinone derivative, which is connected in the lipoproteins of the electron transport particles and is related to vitamin K. Coenzyme Q can undergo oxidation and reduction in a similar fashion to cytochromes; for instance, the quinone part of the molecule can be reduced to become a hydroquinone and the hydroquinone can be reoxidized. It is pretty certain that this coenzyme plays an important part in the electron transfer chain of the particle. The significance of the electron transport in these mitochondrial membranes may seem obscure but, as Green points out, one should think of this problem of electron transport in terms of its coupling—the fact that the process can be coupled to the esterification of inorganic phosphate. In this process a monophosphoric ester of the hydroquinone or, what Green described as a semiquinone form of coenzyme Q would be formed.

This ester would then react with adenosine diphosphate by a process of transphosphorylation and ATP would be forme and simultaneously the semiquinone or hydroquinone would be oxidized to the quinone form in coenzyme Q by the ferric form of cytochrome c. The production of ATP is, in fact, the reason for the existence of such a system. This explanation of the way in which the oxidative processes of the mitochondria actually are used to produce ATP by oxidative phosphorylation is an extremely fundamental discovery and Green and his co-workers are to be complimented very much on the extraordinary patience and care which they have put in over a large number of years to work out this very complex problem in mitochondrial physiology.

In view of the studies by Green and his colleagues which have just been described, it is of interest to consider some of the studies of Martias. This author has for some time put forward the view that there may exist in mitochondria two routes for the transport of hydrogen ions or electrons, respectively, between the pyridine nucleotides (DPN and TPN) and cytochrome c. He thought that only one of these would be linked with enzymes which brought about phosphorylation. The route which is best known leads from DPN to cytochrome c reductase and then to cytochrome c. Martias believes that this particular route is not one which is concerned with the formation of highenergy phosphates.

On the other hand, he proposed a second alternative route in which vitamin K (phylloquinone) is an important factor. Until recently this concept of an electron transport system involving vitamin K in mitochondria has not proved to be particularly popular, Martias and Strouff have finally isolated a phylloquinone reductase that is a flavoprotein having a number of similarities to cytochrome c reductase. There is, however, a difference between them. Phylloquinone reductase and cytochrome c reductase can both react in the reduced state with phylloquinone, but with cytochrome c only cytochrome c reductase will have any effect. Vitamin K reductase has an extremely high activity even in relatively low concentrations.

If we consider that there are two routes in mitochondria which can be used for electron transport, then the problem to be solved is what decides whether these electrons and hydrogen atoms are directed into either of these particular pathways. Under normal circumstances they seem to be directed exclusively along the pathway which is coupled with oxidative phosphorylation and which leads via vitamin K reductase to vitamin K, cytochrome b and so through to cytochrome c.

This pathway via vitamin K seems to be sensitive to any alteration of the internal structure of the mitochondria, possibly because some of its constituents, particularly the vitamin K, are associated with the double membranes of the cristae. If their combination is disturbed, the electrons take the emergency route which is via cytochrome c reductase to cytochrome c. It is of interest in connection with this theory that dicoumarol is capable of decoupling oxidative phosphorylation from respiration and, since dicoumarol is known to be a vitamin K antagonist, its action is possibly explicable by the fact that it blocks the pathway between vitamin K and its reductase.

It is also of interest that thyroxine is a hormone which causes uncoupling of respiration from oxidative phosphorylation. Since it is known that, if thyroxine is added to mitochondria, it causes a swelling and change in the internal structure, this would be perhaps further evidence that the internal structure of the mitochondria is the one which is particularly concerned with the flow of electrons along the route leading to oxidative phosphorylation. It may well be that hormones which affect respiration do so by affecting the structural nature of the mitochondria, apart from their effects on the endoplasmic reticulum which have already been mentioned. The precise relation between Green's coenzyme

Q and Martias' vitamin K phylloquinone reductase is not absolutely clear, but it is very likely they are identical or closely related systems.

LYSOSOMES

De Duve in his studies on ultracentrifugation of cytoplasmic particles has demonstrated that some particles appear to contain acid phosphatase, cathepsin and ß-glucuronidase and ribonuclease, DNAase and cholinesterase. It is of interest that these compounds appear to be present in separate particles—they require higher centrifugal forces for sedimentation than do the cytochrome-oxidase bearing mitochondria. De Duve has pointed out that it is of special interest that such hydrolytic enzymes are in special particulate bodies differing from others of the cytoplasm. He is not clear as to how this should be interpreted but has pointed out that, if hydrolytic enzymes are free to act within the living cell, as, for instance, they can do in homogenates, they would interfere with the efficiency of the synthesis and may even affect the structural integrity of the cell; he then suggests that segregation of hydrolytic enzymes in this way is one method by means of which this activity is either kept in check or localized in specific parts of the cell.

De Duve once described these bodies as "suicide bags" and suggested that on the death of the cell these enzymes are released completely into the cytoplasm and play a part in the process of autolysis. The present author has suggested that these enzymes may be less effectively contained in the lysosomes in senescence and that this leakage may be partly responsible for the cellular process of aging. It is of interest that lysosomes have been demonstrated *in situ* with the electron microscope and stained to demonstrate their acid phosphatase activity. No typical mitochondrial structure can be seen in these bodies and they appear to be of a different nature from mitochondria.

THE METABOLISM OF CARBOHYDRATES

The role of mitochondria in oxidative phosphorylation has already been mentioned and their role in the metabolism of carbohydrates through the presence in their substance of enzymes concerned with the Krebs cycle and the cytochrome system has been indicated. We should now consider the problem of carbohydrate metabolism and the role that mitochondria and other parts of the cytoplasm play in it.

Carbohydrate metabolism is extremely important for cell synthesis and is the main source of energy for cell activities. Before attempting to localize the various activities of carbohydrate metabolism in the actual parts of the cell, we should consider briefly what the metabolism of carbohydrates involves. There are two types of metabolism, anaerobic and aerobic. The anaerobic route is demonstrated very well by muscle and most of the information on this type of metabolism of carbohydrates has been obtained by studies of this tissue.

The result of anaerobic metabolism is the production of lactic acid and the liberation of a good deal of CO_2. However, although we think in terms of anaerobic metabolism for muscle, we have to realise that muscle itself has a first class blood supply which appears to be increased by various physiological mechanisms when muscle is forced to do work and that muscle in the process of contraction uses a rather surprisingly large amount of oxygen.

Lactic acid has been shown to accumulate in muscle extracts and in isolated muscles kept under anaerobic conditions. If we consider the accumulation of lactic acid in an animal in vivo, we find that after moderate work the accumulation of lactic acid goes up slightly but remains at a pretty steady level. However, if strenuous work is done, then the amount of lactic acid goes up extremely steeply and slowly comes back to normal. The reason for this is that under normal circumstances muscle can obtain oxygen fast enough to reoxidize the lactic acid as rapidly as it is formed and only a small amount of lactic acid accumulates. However, it is possible for muscle to do more work than it can supply oxygen for and it can do this by oxidizing carbohydrates anaerobically and so accumulating lactic acid. Eventually this lactic acid has to be converted with the aid of oxygen but this can take place over a longer period. Some of the lactic acid is converted into glycogen in the liver and the rest is oxidized.

The fact that it is possible to accumulate lactic acid and slowly oxidize this after the work is finished provides a mechanism by means of which an "oxygen debt" can be produced. Under aerobic circumstances it is not lactic acid which is formed in the metabolism of carbohydrates but pyruvate. However, this does not accumulate and it is oxidized almost as rapidly as it is formed—as we shall see in a minute there is a very complicated system for oxidizing this pyruvate. The only time when pyruvic acid does accumulate in the tissues is in the absence of vitamin B_1 or thiamine, a fact which was demonstrated years ago in Oxford by R. A. Peters. Under anaerobic conditions pyruvic acid undergoes the process known as oxidative decarboxylation with the aid of the cocarboxylase (thiamine pyrophosphate). It yields acetate, carbon dioxide and lactate. It has been stressed that this reaction is fundamentally of an oxidative nature and leads to the term oxidative decarboxylation which is in important process both in carbohydrate and protein metabolism.

We have mentioned the production of pyruvate but have not yet considered the process by means of which this compound is produced—this process is known as glycolysis and it represents the first stage in the metabolism of carbohydrates. Glycolysis is, in effect, the reverse of photosynthesis. In photosynthesis the energy of sunlight is used to combine CO_2 and water into carbohydrates; in the process of glycolysis, the glucose which is formed from carbohydrates such as glycogen and other polysaccharides is converted into CO_2 and water with the liberation of energy. We can express this as $C_6H_{12}O_6 + 6O_2 = H_2O + 6CO_2$ + energy. This oxidation of glucose to give CO_2, water and energy is not a single step but involves a very large number of steps of considerable complexity. During the various steps, small packets of energy are released at a rate at which the cell can use them, whereas if there was a sudden explosive release of energy by the oxidation of glucose the cell would probably not be able to use this relatively large amount of energy in a coordinated way and a good deal of it would probably be wasted. The first stages of glycolysis involve the phosphorylation of glucose and this is done with the aid of ATP as follows (the enzyme concerned in this process is placed above the arrow):

$$\text{glucose} + \text{ATP} \xrightarrow{\text{hexokinase}} \text{glucose-6-phosphate} + \text{ADP}$$

Hexokinase is not just one enzyme, there are in fact a number of hexokinases that catalyze phosphorylation of hexoses. The phosphorylation of glucose results in the transferring of a high-energy phosphate from the ATP to glucose and so to form a phosphate ester which is poor in energy—this type of reaction is called an exergonic reaction and is essentially irreversible. It is of interest that one of the properties of glucose-6-phosphate which differs from glucose is the fact that the phosphate ester has difficulty in penetrating cell membranes whereas glucose itself, apparently, crosses without any difficulty and it has been suggested that this hexokinase reaction is one way in which glucose can be locked in a cell. It is also an essential prerequisite for the resynthesis of glycogen. Many things can happen to glucose apart from phosphorylation and ultimate conversion into CO_2 and water via the glycolytic and Krebs cycle system. Amongst these are its dehydrogenation by a glucose dehydrogenase to form gluconic acid. This is done by a diphosphopyridine nucleotide (DPN) linked (codehydrogenase) reaction. In mammals, it is believed that this is not a usual pathway for glucose to follow. If glucose can be locked into position in a cell by being converted into a phosphate, it is obvious that there must be in existence a mechanism which can release it again since it needs, for instance, to be fed from the liver periodically into the blood stream to keep the blood glucose level at a relatively constant figure. This is carried out by a specific enzyme, glucose-6-phosphatase,

$$\text{glucose-6-phosphate} + H_2O \xrightarrow{\text{G-6-Pase}} \text{glucose} + PO_4.$$

This glucose-6-phosphatase is probably present in all tissues which release glucose from cells, but it does not appear to occur in skeletal muscle.

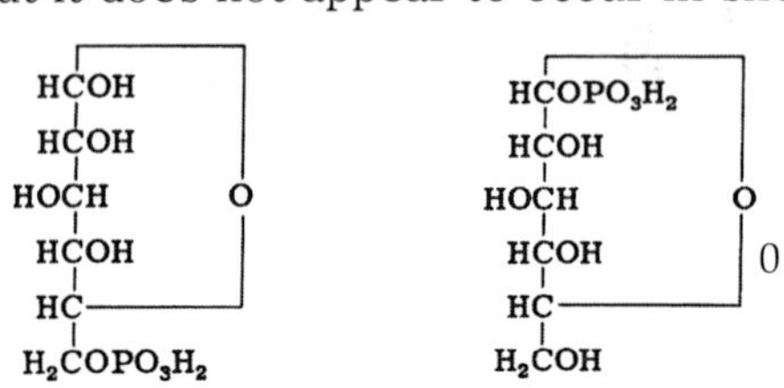

Fig. Glucose-6-phosphate α-Glucose-1-phosphate

Glucose-6-phosphate may be converted into glucose-1phosphate, in other words the phosphate is simply shifted around from the 6-position to the 1-position on the glucose molecule.

This change of position of the phosphate group is catalyzed by an enzyme known as phosphoglucomutase and the reaction changing the phosphate group from one part of the molecule to the other is an easily reversible reaction. The formation of glucose-1-phosphate from glucose-6-phosphate is a stage in the synthesis of glycogen, for instance, if glucose is accumulating in a cell it is converted to glucose-6-phosphate and then to glucose-1-phosphate.

The molecule of glucose-1-phosphate then polymerizes into glycogen. The reverse process can occur—glucose-6-phosphate can be obtained from glycogen by first producing glucose-1phosphate and then by producing glucose-6-phosphate from the glucose-1-phosphate. We have, however, diverted a little from the direct line of our story of the glycolytic cycle.

After the formation of glucose-6-phosphate from glucose, the next stage in the glycolytic cycle is the conversion of glucose-6phosphate into fructose-6-phosphate. This reaction is readily reversible and is catalyzed by an enzyme called phosphohexose isomerase.

Fructose-6-phosphate may also be formed directly from fructose and it is known that an enzyme, fructokinase, which is present in brain and muscle can produce this effect. The next step is the further phosphorylation of fructose-6-phosphate; another phosphate group is added in the 1-position to give fructose 1-6-diphosphate. The enzyme responsible for this is phosphofructokinase and the reaction is carried out with the aid of ATP:

$$\text{fructose-6-phosphate} + \text{ATP} \xrightarrow{\text{PFkinase}} \text{fructose-1-6-diP} + \text{ADP}$$

This reaction is also exergonic: Fructose-1-6-diphosphate is also known as hexose diphosphate and in the next stage this is broken down by what is described as the "aldolase" reaction into 3 phosphorylated compounds—triosephosphates. The first of these is ketose triosephosphate, the second is dihydroxyacetone phosphate, the third is phosphogly ceraldehyde. The triosephosphate can be converted into phosphogly ceric acid; dihydroxyacetone phosphate can be converted into phosphoglyceraldehyde or alternatively it can be reduced to á-glycerophosphate with the aid of DPNH and á-glycerophosphate dehydrogenase. This reaction is potentially important for the synthesis of lipids since from the á-glycerophosphate, phosphatidic acid can be synthesized and phosphatidic acid can be the starting point for the synthesis of lecithin, cephalin and also of fats. Phosphoglyceric acid with the aid of the enzyme enolase becomes converted into phosphoenol-pyruvic acid.

Pyruvic acid penetrates cell membranes very well and can thus leave the cell and in theory can be distributed in any cell in the body. Conversely all the steps which we have mentioned can be reversed and pyruvate can be converted back into glucose-6-phosphate. When pyruvic acid is reduced it gives lactic acid.

Lactic acid itself can be converted back to pyruvic acid. Liver cells are capable of reversing the whole glycolytic series of reactions and can produce glucose and glycogen again from lactic acid. Muscle can reverse lactic acid to glucose-6-phosphate and glucose-1-phosphate and glycogen but cannot produce non-phosphorylated glucose.

In these first stages of carbohydrate metabolism, a certain amount of energy is liberated but this is not much more than about one-tenth of the total amount of energy which is produced by the production of CO_2 and water from glucose. The glycolytic part of the cycle is the least energy producing part. In addition to its conversion into lactic acid or its oxidation, pyruvic acid can be converted to alanine, an amino acid, by the process of transamination. Thus here one can see a link between carbohydrate and protein metabolism.

If there is ample oxygen, the pyruvic acid produced by this first stage of carbohydrate metabolism can be oxidized. This is a complex process in which in the first stage the pyruvic acid is converted by the process of oxidative decarboxylation into acetyl coenzyme A and CO_2. Thiamine pyrophosphate (cocarboxylase) is an essential enzyme for this process.

Acetyl coenzyme A is a 2-carbon substance and it condenses with oxaloacetic acid which is a 4-carbon dicarboxylic acid to yield a 6-carbon tricarboxylic acid, namely, citric acid. The enzyme catalyzing this is called the "condensing enzyme." Thus starts a series of changes which ultimately lead to the formation of CO_2 and water. The production of citric acid is followed by the loss and recapture of water and it becomes converted to cis-aconitic acid (with the aid of aconitase, glutathione and ferrous iron) which on further hydration is converted into isocitric acid. This compound then loses hydrogen and thereby becomes oxidized to oxalosuccinic acid (this reaction is catalyzed by isocitric dehydrogenase) and decarboxylation turns it into á-ketoglutaric acid (the enzyme responsible is oxalosuccinic decarboxylase and oxidized manganese). á-Ketoglutaric acid is decarboxylated and then oxidized by the loss of two hydrogen atoms to succinic acid (the enzyme concerned is á-ketoglutaric dehydrogenase).

Succinic acid is also oxidized by the loss of H_2 (with the help of succinic dehydrogenase) to fumaric acid. The latter by the addition of the elements of water becomes converted by fumarase into malic acid and the malic acid by dehydrogenation (enzyme malic dehydrogenase) is converted in oxaloacetic acid and there we are back at the beginning of the cycle again. The oxaloacetic acid is ready to combine with another molecule of acetyl coenzyme A to produce citric acid once more. During this process 3 molecules of CO_2 are given off and 5 molecules of H_2.

It can be seen that quite a complex series of reactions take place in what has been called the "tricarboxylic" or "citric acid cycle," or the "Krebs cycle." All three terms are applicable. The final combination of the hydrogen atoms liberated by the Krebs cycle with oxygen is brought about by the cytochrome system. It is of interest that cytochrome is a protein which contains a form of heme, an iron containing pigment which is also present in hemoglobin. It is probably more widely distributed than any other type of heme protein since it occurs in the cells of all organisms that use oxygen, irrespective of whether they are animals or plants or whether in the case of animals they are vertebrates or invertebrates, or protozoa.

In the cytochrome system a variety of compounds and enzymes are involved. Cytochrome oxidase is a widely distributed enzyme since its occurrence is comparable in distribution to cytochrome. It has not yet been obtained pure and it has been found to be bound to the insoluble material when cells are homogenized and spun down. This is not surprising when we appreciate that cytochrome oxidase is associated with the mitochondria. There is also another compound involved called flavoadenine dinucleotide (FAD) for short) which is associated with various proteins to form a variety of different enzymes. A flavine containing enzyme was first isolated and called a yellow ferment as long ago as 1932 by Warburg and Christian.

Flavoproteins are usually associated with metals of various sorts and molybdenum, copper and iron are three which have been found to be necessary for their adequate functioning. It is not quite clear where the metal atoms are situated on the molecule but they appear to be essential for the action of the enzymes with which they are associated. There are two types of riboflavin containing enzymes, one type is an electron acceptor from reduced DPN or TPN and it can transfer these either to oxygen or to the cytochromes, whereas the other type of flavoprotein accepts electrons directly from metabolites. Among the important flavoproteins is cytochronie reductase which comes in two types—one for reduced DPN (DPNH) and the other for reduced TPN (TPNH). The oxidation of these two compounds (DPNH and TPNH) in the cell can thus be carried out by the cytochrome system. Now we are in a position to describe the next stages which take place in the oxidation of glucose. The 5 pairs of hydrogen atoms which are passed down to the cytochrome system from the Krebs cycle are combined with the cytochrome and result in its reduction. The enzyme concerned is cytochrome reductase, the flavoprotein already mentioned. Cytochrome is then oxidized with the aid of cytochrome oxidase, which removes the hydrogen atoms. They become combined with atmospheric oxygen to produce water and thus the long journey of oxidation of carbohydrates is done, 5 molecules of water and 3 molecules of CO_2 being produced from each molecule of pyruvate.

An analysis of oxygen uptakes of various tissues in the body gives an indication of the degree to which their cells are metabolizing. As a matter of interest, it might be noted that of the tissues examined, retina and kidney had the highest metabolism and liver was next, then the rate decreased progressively from adrenal, lung, bone marrow, diaphragm, heart, lymph nodes, skeletal muscle, skin to eye lens which had the lowest level of oxygen consumption. The important point to remember is that the complex of reactions described above is localized to a great extent in the mitochondria.

One of the important functions of this oxidative cycle just described is its relationship to phosphorylation. We have described earlier the work of Green who showed that oxidative phosphorylation could take place without the whole Krebs cycle occurring, but in intact mitochondria the whole cycle normally goes through and phosphorylation is an important by-product of these reactions. This process of phosphorylation results in the production of high-energy phosphate esters, such as ATP.

Since ATP is one of the principal energy containing compounds of the body, it is very important in the whole energy cycle of the cell. It is, for example, the main source of energy in muscular contraction and for many of the synthetic processes of other cells. This briefly is the story of carbohydrate metabolism in the cell. What we want to try to do now is to demonstrate where this complex of activity is situated in the living cell. It appears that most of the processes of respiration and glycolysis actually take place in the cytoplasm and mitochondria.

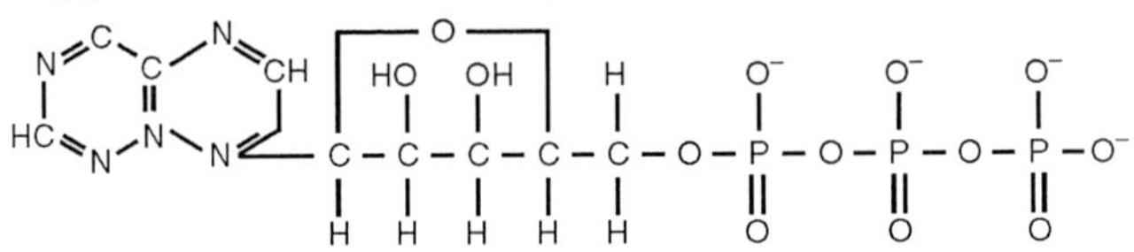

Fig. Adenosine Triphosphate (ATP)

Some years ago, the present author and R. J. Allen demonstrated that zymohexase which is really a complex of two enzymes and is concerned with the splitting of hexose diphosphate (the aldolase reaction) was localized in the cytoplasm of the cell and in the case of muscle fibres in between the fibrils in the sarcoplasmic material rather than in any of the formed elements. It is noted too that the results of differential centrifugation of cell homogenates have demonstrated that most of the enzymes responsible for the glycolytic cycle have been found to be present either in the supernatant or in "microsomes" and those concerned with the cytochrome system and Krebs cycle ire localized specifically in the mitochondria.

Now, since the microsomes mostly represent fragments of the endoplasmic reticulum, we can assume that most of the enzymes found in them can also be found in the membranes of the endoplasmic reticulum. In addition to the aldolase complex which has been demonstrated to be present in the supernatant, it has been found that phosphorylase and phosphoglucomutase and glucose 6-phosphate dehydrogenase have also been found in the supernatant and glycolysis has in fact been found to take place in this fluid—a fact which confirms that all the glycolytic enzymes necessary for this process must be present there. However, glucose-6-phosphatase has been found not to be localized at all in the supernatant but exclusively in the microsomes (fragmented endoplasmic reticulum). There is some evidence too that hexokinase is present in the microsomes. On the other hand, DPNH and DPN and TPN-linked cytochrome c reductases which are found in high concentration in mitochondria have also been found to be present in the microsomes. It might be possible to suggest from these facts a tentative scheme which would help us to understand to some extent the relationship between the reactions occurring in the cytoplasm itself and those in the mitochondria.

Let us assume that glucose passes across the cell membrane, perhaps it is taken in by pinocytosis as has been suggested by a number of authors or, possibly, if and when the membranes of the reticulum are continuous with the outside of the cell, it may simply pass up the cisternae between the membranes of the reticulum and enter into the cell through the membranes of this reticulum. Let us assume that the glucose has passed into the cell by the process of pinocytosis. Then, with hexokinase present in the cytoplasm, it can be converted into glucose-6-phosphate and subsequently run through the cycle to pyruvate.

At this point (another system which we will discuss in a minute) associated with the mitochondria comes into play. If the glucose has to pass across the cell membrane or across the membranes of the endoplasmic reticulum, it should also be able to do this without too much difficulty. If any glucose-6-phospate should occur in the cisternae it will probably be able to pass freely into the cytoplasm because of the presence of glucose-6-phosphatase in the membrane. It may be that there is hexokinase in the fluid within the cisternae and that, perhaps, all the glucose there (if any) is first phosphorylated and then released in a timed fashion into the cytoplasm through the action of glucose-6-phosphatase in the membranes.

Perhaps this is a way of controlling the feeding of the carbohydrate fuel into the furnace. It will be remembered that glucose-6-phosphate passes cell membranes with difficulty unless there is the appropriate hydrolytic enzyme on the membrane. The glucose, once into the cytoplasm by whatever route, can be synthesized into glycogen or it can be brought to pyruvate by the enzymes of the glycolytic cycle which are known to occur in the cytoplasm. Once pyruvate is prepared, it is changed into acetyl coenzyme A and is ready for the next stage. For this it has to come into contact with the mitochondrial membrane. Whether it has to pass through the mitochondrial membrane and come in contact with the cristae we do not know, but the existing evidence suggests that the envelope and cristae of mitochondria have fundamentally the same enzymatic structure. Furthermore, there are pores and even large openings in some mitochondrial envelopes and there is thus little difficulty in metabolites passing in or out.

Now, it has been mentioned that a great proportion of the enzymes concerned with the Krebs tricarboxylic acid cycle are localized in the mitochondrial membranes and since acetyl coenzyme A appears to be the most commonly used fuel of this system, it is at this point that the mitochondria largely take over the final oxidative stages. Thus, as far as we can tell at the moment either in the cytoplasm of the cell itself or in the endoplasmic reticulum, glycolysis occurs with the production of either acetyl coenzyme A or pyruvic acid. We do not know at what stage the latter are fed to the mitochondria.

They must go largely to the mitochondria because these organelles contain the greatest proportion of the Krebs cycle enzymes—it is these latter which complete the oxidative breakdown of carbohydrate that was set into motion by hexokinase. The rate of penetration of these fuels into, or their absorption onto the mitochondria will be conditioned both by the nature of the mitochondria and the total area of mitochondria available. If under normal conditions we have a large number of mitochondria occupying the cell, we can assume, other things being equal, that oxidation should be proceeding at a fast rate.

However, where one gets a large increase of number of mitochondria as, for example, in starvation this may be compensatory hypertrophy and not indicative of increase in oxidative activity. We do not know whether the nature of the mitochondrial surface varies from time to time but we know that the surface area varies. Mitochondria, for instance, fragment from the filamentous and rodlike condition to the granular state under a variety of circumstances. These include mechanical damage to the cells, rough handling, influence of bacteria and other toxins, as a result of anesthesia and anoxia and so on. It can be shown mathematically that there is a greater surface area available the more fragmented the mitochondria become. Because of this either the pyruvate and/ or acetyl coenzyme

A can feed more rapidly into or become attached in larger amounts to the mitochondria when they are in a fragmented condition simply because there is a greater surface area. Thus the mitochondria in virtue of their ability to fragment and reform into the filamentous condition can act as a throttle controlling the rate of aerobic metabolism in the cell. They may also have an additional means of doing this, a sort of fine control. Since it is fairly certain that enzyme molecules are aligned along the cristae, these structures represent another surface where reactions can take place and reduction in the size and number of cristae would also have a throttling effect on the metabolism or synthesis due to mitochondria.

This can be seen in operation in the case of a mitochondrion which has accumulated a good deal of fat or other product of chemical reactions. In such a case, the cristae are reduced or absent altogether as if metabolic processes are brought to a virtual standstill because of the accumulation of reaction products. This brings us to another process of control, a chemical method known as "feedback" which will be discussed shortly. Siekeiwitz believes that both glycogen synthesis and breakdown take place at the surface between the cytoplasm and the ergastoplasmic membranes in association with the enzymes present at those surfaces. He points out, however, that only glucose-6-phosphatase and DPNH and TPNH cytochrome c reductases (these latter are believed to act as coenzymes for glycolysis in the early stages of glucose oxidation) have been found to be associated with the microsome fraction, but he believes that the site of effective action of the other enzymes might be at the interface between the membranes and the matrix of the cytoplasm. He suggests that hexokinase might be activated at the membrane surface of the endoplasmic reticulum. Other cofactors such as glucose-1-6-diphosphate and adenosine monophosphate possibly also bind their appropriate enzymes to the E.R. membranes.

Siekewitz has also discussed certain biochemical aspects of the control of glucose metabolism. First of all he points out that, if the concentration of glucose in the lumen of the endoplasmic reticulum is in equilibrium with the glucose concentration in the blood (this assumes at least temporary continuity between the lumen or cisternae of the E.R. and the exterior of the cell), there exists then a mechanism whereby the glucose level in the blood would definitely affect intracellular glucose equilibrium.

Thus a reduced production of glucose from the diet would lead to reduction of glucose in the blood and this would cause a reduction of the amount of glucose in the fluid within the endoplasmic reticulum. The latter result would lead to increased phosphatase activity which would cause an increase in the breakdown of glycogen (e.g., in the liver cells) and a production of glucose which would pass out from the endoplasmic reticulum into the blood stream.

Siekewitz points out that hexokinase, phosphoglu comutase and phosphorylase might have their activity enhanced if they were attached to the endoplasmic reticulum membranes. In the case of phosphorylase, active phosphorylase B has to undergo a conversion to active phosphorylase A before it can have any catalytic effect. An enzyme phosphorylase B-kinase carries out this conversion by phosphorylating the enzyme in the presence of ATP. Siekewitz suggests that this kinase may be part of the membrane of the endoplasmic reticulum and that the activating process for phosphatase in the cell might consist of moving it out of the cytoplasm onto the site of the E.R. membranes.

Siekewitz points out that it is possible that hormones control this type of movement of enzymes within the cell and its internal membranes. At this point one might remember the work quoted earlier by Brandes and the present author in which it was demonstrated that shifts of acid phosphatase activity took place from the Golgi apparatus to the nucleus in ventral-lobe prostate cells following castration and that acid phosphatase activity was restored to the Golgi apparatus following implantation of the male sex hormone. Also of interest are the further studies of Brandes in which he has demonstrated that in castrated animals there is a rearrangement of the membranes of the endoplasmic reticulum. These studies demonstrate a definite morphologico biochemical effect on the part of the male sex hormone. Siekewitz suggests that the hormones concerned with carbohydrate metabolism do not act directly on the enzyme as such but bring it and the substrate and various cofactors together at a suitable surface and then complex them together there. Other enzymes may be localized on the endoplasmic reticulum e.g. 5 nucleotidase. The nucleus, as will be described later, appears to be enclosed in a fold of the double membrane of the endoplasmic reticulum and not to have a membrane of its own. Since the E.R. spaces might be connected directly to the exterior of the cell, it is possible that glucose coming from outside the cell could be converted into glucose-6-phosphate in the E.R. lumen and thus be prevented from passing through the E.R. membrane.

Thus it could pass along all the ramifications of the E.R. canals and come in direct contact with the nuclear fold of the reticulum. If this membrane contains glucose-6-phosphatase, it could penetrate through into the nucleus without having to pass through the cytoplasm at all. Histochemical preparations show that the glucose-6-phosphatase reaction in a particular histological section is not always positive for all nuclei in the section and it is of interest that Siekewitz has pointed out that enzymes may be present or activated at the E.R. membrane only when the glucose concentration reaches a critical level.

It is of interest that histochemical studies of the cells of many organs demonstrate the fact that many dephospho rylating enzymes as well as glucose-6-phosphatase show an association with nuclear membranes—perhaps here lies the mechanism whereby even low levels of glucose-6-phosphate could penetrate readily through into the interior of the nucleus. The endoplasmic reticulum could provide a pathway straight to the nucleus which would prevent glucose from getting in contact with or passing through the cytoplasm where it could be attacked by glycolytic enzymes. This may be the mechanism by which a supply of glucose to the nucleus is ensured. Glucose could also be supplied to the nucleus from the cytoplasm by the process of glycolysis.

In this case, glucose being formed from glycogen as glucose6-phosphate would be dephosphorylated and pass through the E.R. membrane and into the lumen where possibly it would be rephosphorylated to prevent its passing back again and would thus move along the lumen to the nucleus. Hence the nucleus could get its glucose directly from the cytoplasm via the endoplasmic reticulum or directly from the outside (the latter, however, only if a direct connection really exists).

In many cells the mitochondria can also be seen to be completely surrounded by endoplasmic reticulum. It is possible that the mitochondria themselves obtain their glycolytic fuel directly as a result of glucose passing through the E.R. membranes undergoing glycolysis there and the glycolytic products feeding directly to the mitochondria.

The control of the rate of different types of metabolism, particularly respiration, in the cell can depend on two main factors. First, a structural factor which brings into apposition the appropriate reactants and which varies with the extent of the surfaces available for these processes to take place and, second, it may also depend on some chemical feedback where an excessive production of one kind of compound inhibits its continued production or slows down further synthesis. Alternatively, the metabolism in any one particular direction may be affected by the absence of a limiting amount of some specific substance or compound in the reaction chain.

Sir Hans Krebs in a recent article entitled *"Rate Limiting Factors in Cell Respiration"* discussed the control of energy utilization and pointed out that, in unicellular organisms, energy can be obtained directly from oxidation if air is present but if air is not present then anaerobic fermentation takes its place and energy is obtained by this source. In this instance it is the supply of air or oxygen which regulates which of these mechanisms comes into use and oxygen is, in fact, the "rate limiting factor." In higher animals the ability to undergo fermentation or an aerobic oxidation is still present and can be particularly well demonstrated in muscle. Krebs pointed out that the chemical systems in the cell which are concerned with the function of regulation are all fairly simple reactions but that there is an elaborate interlocking of these reactions. By this he means that the individual reactants may take part not only in more than one reaction but in very many different processes.

One of the difficulties in sorting out such a complex of activity is that not all the component reactions of this elaborate interlocking series are known and we are dealing with a heterogeneous system in which there are many varied membranes and different spatial arrangements of the various reactants. He also points out that regulation is probably a matter of reaction velocities, some of which will be accelerated and some slowed down and the question that has to be decided is the degree to which any of these are ratelimiting. Krebs illustrates this point by considering the amount of oxygen used by 4-ml. sheep heart homogenate, which contained about 10 per cent of tissue.

Thus one can demonstrate that the addition of glycogen adds nothing to the oxygen uptake so the amount of glycogen present is not a limiting factor in this system. The same applies if glucose is added instead of glycogen. On the other hand, when acetate, pyruvate, or other intermediates of the tricarboxylic acid cycle are added there is an appreciable increase in the rate of oxygen uptake. The fact that the oxygen uptake in this system can be increased if suitable substrates are added to the mixture demonstrates that the electron transport system from DPNH to oxygen is not being used to its full capacity, thus it cannot be the factor which is limiting the uptake of oxygen. Certain special substrates which are known to reduce either DPN or flavoprotein seem to be able to increase oxygen consumption.

Thus the limiting factor appears to be that, the mechanism for the transport of hydrogen from DPN or flavoprotein is not being used to its full capacity if these special substrates are not present. The limiting step therefore is really the first stage in the electron transport system. If it is found that a particular substrate increases the rate of respiration this is due to the fact that the substrate reacts more readily or easily with DPN or flavoprotein than any endogenous substrate already present.

This is the reason why pyruvate or á-ketoglutarate or succinate are responsible for the stimulation of respiratory rate in the experiment quoted. However, even if we accept this we are still faced with the identification of the factor that decides the rate of reaction between substrate and DPN or flavoprotein.

Studies with dinitrophenol which decouples oxidative phosphorylation from respiration is of interest and helps to throw light on this problem. Perhaps we should first say a word or two about this action of dinitrophenol. The uncoupling of oxidative phosphorylation from respiration has been compared to putting a car into neutral gear and still leaving the engine running. If the respiratory activities are regarded as the engine and the phosphorylation as the process of making the car go, then dinitrophenol uncouples the engine from the transmission of the car, the engine continues to turn but the car does not move; in the cell the respiration goes on quite happily but no ATP is formed. Normally ATP is formed from ADP and inorganic phosphate and thus the factor which limits the rate of oxygen consumption and oxidation of pyruvate in the normal system such as we have described is not really the amount of enzymes present but actually the level of either ADP or inorganic phosphate.

In the experiments carried out by Krebs, inorganic phosphate was present in a fairly good concentration and further quantities added to the system did not stimulate respiration. Therefore it is almost certain that the limiting factor must be the amount of ADP which is available. These experiments demonstrate the type of chemical control which a single compound can exert on a whole chain of reactions. One should also remember that another mechanism, a structural one that controls the rate of respiration, is the rate at which glucose can enter the cell and this can be hormonally controlled, although the method of action of the hormone is not exactly known.

It is of interest that one of the factors which probably affects the rate at which glucose can enter the cell (if, in fact, the endoplasmic reticulum is continuous with the outside of the cell) is the degree of complexity of the endoplasmic reticulum. If this structure develops many ramifications, as presumably it seems able to do in certain cells such as spermatocytes, as demonstrated by Fawcett, then the surface area available for glucose to enter into the cytoplasm of the cell and so be metabolized is enormously increased or, conversely, it may be decreased by a reduction in complexity of the reticulum. To return to the subject of respiration and ATP formation, the reaction for this process can be given as follows:

$$C_6H_{12}O_6 + 6\ O_2 + 38\ ADP + 38\ H_3PO_4 \rightarrow 6\ CO_2 + 44\ H_2O + 38\ ATP$$

This is the general reaction and is a summary of all the complex intermediatary reactions which in the end simply produce carbon dioxide, water and ATP. Since, as Slater and Houlsman have pointed out, cells contain only relatively small amounts of ADP, as soon as it is all converted into ATP the process of respiration will stop—ADP is thus the limiting factor. However, when the cell is stimulated to do work there is a breakdown of ATP according to the formula given by Slater and Houlsman:

$$38\ ATP + 38\ H_2O \rightarrow 38\ ADP + 38\ H_3PO_4 \rightarrow \text{work}$$

Respiration can go on so long as there is some left to be resynthesized into ATP. The addition of further ATP will, of course, keep respiration going.

It is of interest that, if mitochondria that have been separated from the cell by differential centrifugation are permitted to stand for some hours at room temperature, the phenomenon of uncoupling (which can also be brought about by dinitrophenol) of the oxidative phosphorolytic system from respiration takes place. This type of mitochondrial preparation is described as "aging" mitochondria and it is tempting to speculate whether in senescing tissues there may not be a progressive uncoupling of respiration from oxidative phosphorylation or a progressive hydrolysis of ADP so that less and less of this becomes available for synthesis of ATP. It has been possible to isolate from mitochondria which have been aged in this way, a heme compound which will actually produce this uncoupling reaction. It has been described and given the name "mitochrome" and is fundamentally a pigment.

However, there appears to be a lipid component in this "mitochrome" heme-protein preparation which is the actual factor responsible for uncoupling and the heme protein of the mitochrome is not, in fact, the uncoupling factor at all. Mitochrome is very similar in structure and form to cytochrome and is probably derived from it. Certain unsaturated fatty acids such as oleic acid are also found to be active as uncoupling agents and the lipid isolated from the mitochrome particle also appears to contain an unsaturated fatty acid. The fact that an uncoupling agent can be produced *in vitro* this way is of considerable importance since it seems possible that the formation of such a substance in mitochondria might take place in vivo and may itself function as a controlling agent for respiration and oxidation.

FAT METABOLISM

The general view of metabolism of fatty acids at present is that they are broken down beginning at the end of the chain where the carboxyl group is and then the chain is progressively degraded as two carbon pieces are removed by a process of oxidation.

This type of oxidation of fatty acids is known as ß-oxidation and is so called because the fatty acid is attacked oxidatively at the ß-carbon atom in the first instance. Among the products of this oxidation is the formation of acetyl coenzyme A and acyl coenzyme A. The latter can be subjected to further oxidation with the production of more acetyl coenzyme A (CoA).

Although very little has been said about the mechanism of degradation of all these fatty acids, it is of interest that the enzymes which catalyze these reactions are all located in or on the mitochondria. The acetyl CoA can enter the Krebs cycle by condensing with oxaloacetic acid and the final oxidation thus follows the same path as the carbohydrates.

The acetyl CoA formed from pyruvic acid (i.e., from carbohydrate breakdown) can be used to synthesize fats and likewise so can acetyl CoA produced as a result of protein metabolism. We see, therefore the reason why the Krebs tricarboxylic acid cycle has been spoken of as the meeting place of protein, fat and carbohydrate metabolism.

Where precisely in the mitochondria the enzymes responsible for ß-oxidation of fatty acids are centered is not known for certain. It is very likely that they are more associated with the outer membrane of the mitochondrion than with the cristae so that the problem of penetration of the fatty acid through the membrane of the mitochondrion does not become so important. On the other hand, there is some evidence that pores occur in the mitochondrial membrane and if this is the case it is possible for the long-chain fatty acids to pass into the interior of the mitochondria and to become subject to ß-oxidation at this site by enzymes located in the cristae.

Hoberman has shown that in the mitochondria, deuterium labeled DPNH, which has been reduced during the oxidation of fat in the organelles, is not available for reactions that take place in the outside cytoplasm. This suggests that the enzymes concerned with the oxidation are localized within the cristae. It is of interest that recent electron micrographs of the adrenal cortex have shown large areas of the surface where the mitochondrial membrane is incomplete and the interior of the organelle is thus open to the penetration of the largest molecules and even particles of fat. Novikoff's scheme for the differential distribution of biochemical activities in liver cells.

Fat is metabolized largely in the way already described, but the synthesis of fat is also an important part of the activity of the cell. Fat cells have an important mechanical function to perform and fat itself is a valuable reserve store of energy. Fats, fatty acids and phosphorylated fats (phospholipids such as lecithin) are also important structural units of cell, mitochondrial and other membranes and their synthesis becomes an important cellular activity. Fatty acids are made up of long chains of carbon atoms, usually an even number, with a COOH group stuck at one end. Fat is formed from a fatty acid molecule by a combination between the latter and a molecule of alcohol such as glycerol.

The link takes place through the COOH group which reacts with the OH of the alcohol to eliminate a molecule of water. Fatty acids may be short or long or intermediate chained, a typical short-chain fatty acid is acetic acid which has only 2 carbon atoms and a typical long-chain fatty acid is palmitic acid which has 16 carbon atoms.

In the synthesis of these long-chain fatty acids, the starting point appears to be acetic acid. This combines with CoA to form acetyl CoA. This condenses with CO_2 to form malonyl CoA (the coenzyme A ester of malonic acid). This latter compound then condenses with another molecule of acetyl CoA to form an intermediate substance which becomes reduced to form a 4-carbon fatty acid which condenses with a molecule of malonyl CoA to give a 6-carbon fatty acid and so the carbon chain is built up. Then, as described, combination of the appropriate fatty acid with an alcohol such as glycerol gives an ester known as a fat.

Phospholipins such as lecithin, which play a very important part in the structure of the cell membranes, also have to be synthesized by the cell. These compounds are not only esters of a fatty acid but are simultaneously esters of phosphoric acid. In this synthetic activity ATP plays an important part. It starts the ball rolling by phosphorylating one of the OH groups of glycerol. To the other two OH groups, CoA esters of palmitic acid are attached.

At this point the phosphoric acid group that was put on in the first step drops off. The molecule then reacts with a compound known as cytidine diphosophocholine.

This compound, which is really a coenzyme, drops the diphosphocholine part of its molecule nicely into the spot on the OH group which had been vacated by the first phosphoric group and so the lecithin is formed. Other coenzymes are concerned with the production of the other steps in the synthesis but these will have to be studied in more specialized works of biochemistry. Dr. D. E. Green has pointed out that one of the surprising things about the synthesis of fatty acids is that the synthesis stops at 16-carbons. It is rare to get 12- and 14-carbon chain fatty acids and 18- or more carbon chains scarcely ever form. What tells the cell to break off the synthesis at that point is certainly an intriguing problem.

Although one would expect that mitochondria would be the principal fatty acid synthesizers of the cell, the belief at the moment is that they are not the site of synthesis and that it takes place at the surface of the endoplasmic reticulum. However, in the production of fats from the fatty acids and especially in the case of lecithin synthesis where ATP is required, the mitochondria make a contribution to the synthesis because of their ability to produce the latter material. Mitochondria do play a direct role in fat metabolism but, this role as mentioned earlier in this chapter, appears to be in fat degradation rather than in fat synthesis.

ISOLATION OF MITOCHONDRIA

Although it is possible to measure respiration by intact tissues it is often useful to be able to isolate uncontaminated, intact and functional mitochondria. Here, we will consider some of the principles for isolating cell organelles. This discussion is not meant to be exhaustive nor will it be a cookbook with recipes to be followed step-by-step.

Choice of Material

If the aim of the experiment is to investigate a specific physiological problem, then the plant material is often well-defined. However, in many cases there is some flexibility in the choice of material and it is therefore wise to follow certain rules of thumb:

- Choose a tissue that is easy to obtain in relatively large amounts (at least 100–200 g fresh weight), such as tubers or hypocotyls.
- Choose a tissue for which a good protocol is available. It can take quite a lot of time and effort to optimize an isolation protocol.

Note that the isolation of mitochondria from green tissues presents special problems because of the large amount of chloroplasts. Roots are a notoriously difficult tissue for isolation of mitochondria.

Disruption of the Tissue

The aim is to rupture as many of the cells in the tissue as possible in order to release their cytoplasm with the mitochondria. Whereas many mammalian tissues are easy to homogenize, plant tissues are much tougher mainly because of the cell walls. There are a number of standard homogenizers, including mixers with rotating knives or razor blades, juice extractors, ball mills, and mortar and pestle; the most suitable method for the specific plant material should be chosen. This may involve some initial optimization experiments.

Homogenization Medium

The medium for homogenizing rat liver only contains an osmoticum and a buffer. That will not work with plant tissues because of the disruption of the large central vacuole (absent in mammalian cells), which releases its fairly acidic content often containing phenols and other potentially harmful substances. Let us consider the components of the optimal homogenization medium for isolating intact and functional organelles from plant material, one by one.

Osmoticum: This is required to prevent the mitochondria from losing their matrix content because of swelling and bursting. For the isolation of plant mitochondria, there is a tradition of using sucrose or mannitol at 0.2–0.5 M; whereas 0.15 M KCl is standard for the isolation of mammalian mitochondria. In neither case does this mimic precisely the conditions in the cytosol, but it works.

Buffer: The pH in the cytosol of plant cells is around 7.5, and the homogenate should be buffered at pH 7–8. A buffer should therefore be included and a plethora of different buffers are available. Choose an inexpensive one with a pKa of 7–8 (e.g., 4-morpholinopropane sulfonic acid = MOPS) and be sure to use enough to counterbalance the acidic vacuolar content. The tissue type, the tissue/medium ratio, and the buffer concentration determine whether the acidic vacuolar content is properly neutralized. Obviously, special tissues like lemons or CAM plants harvested late at night or in the early morning, require special treatment!

Metal ion chelators: EDTA is usually included to bind mainly Ca^{2+} and Mg^{2+}, which may otherwise activate proteolytic enzymes.

Antioxidants: Antioxidants (reductants) such as ascorbic acid, dithiothreitol, or cysteine are used mainly to protect sulfhydryl groups on proteins against oxidation.

Bovine serum albumin (BSA): This protein from the blood of cows is usually included to protect the mitochondria: BSA binds fatty acids (its function in the blood) released by the action of phospholipases on phospholipids. BSA may also bind phenolics that are released from the vacuole. Both of these groups of compounds can otherwise be troublesome later. In addition, BSA may prevent proteases from degrading mitochondrial proteins by serving as an alternative substrate present at high concentrations.

Polyvinylpyrrolidone (PVP): A polymer that is often included, especially when using leaf material, to adsorb phenolic compounds.

Filtration

To remove larger tissue fragments, cell walls, and often starch grains, the homogenate is always filtered, usually through a nylon net with a mesh size of 30–100 μm.

Purification of Mitochondria

The first step of the purification is usually differential centrifugation. This means that the filtered homogenate is centrifuged first at low speed for a short time (e.g., 3000 g for 5–10 min.) and the pellet (containing cell wall fragments, starch grains, nuclei, intact plastids) is discarded. The supernatant is then centrifuged faster and for a longer time (e.g., 10,000 g for 10–20 min.) and the pellet (termed *crude mitochondria*) is collected. It contains intact and damaged mitochondria as well as peroxisomes and plastid membranes. The supernatant is discarded. As soon as the mitochondria have been pelleted and thus removed from potentially damaging soluble substances in the homogenate, the medium can be greatly simplified. Usually the so-called wash medium contains only osmoticum and weak buffer (no need for a strong buffer since there is no vacuole content to neutralize). However, it may be wise to include EDTA to bind divalent cations as well as BSA to adsorb any fatty acids released by the action of phospholipases. The resuspended crude mitochondria are often diluted to 50–100 ml with wash medium and subjected to another round of slow and fast centrifugation to remove more contaminants. The aim of the next step is to separate the intact mitochondria from the contaminants. This usually involves a density gradient centrifugation with Percoll either alone or together with other dense materials. Percoll (colloidal silica coated with PVP) is inert and osmotically inactive. It has the further advantage that a Percoll gradient is self-generating because the Percoll microspheres pellet very slowly during the centrifugation. The crude mitochondria are layered on top of the wash medium containing the appropriate concentration of Percoll (usually 25–30 per cent). If the Percoll concentration and the speed and duration of the centrifugation have been chosen correctly, the intact mitochondria will form a distinct band, well away from the contaminants. This band can be collected, washed free of Percoll by repeated dilution and centrifugation, and either used immediately, or frozen with 5 per cent (v/v) dimethylsulfoxide (to preserve integrity) in liquid nitrogen and stored at -80°C until use. This preparation can be stable for months, retaining the ability to form a proton electrochemical gradient over the inner membrane.

The yield is 1–50 mg mitochondrial protein per 100 g fresh weight, depending on the tissue. The purification of mitochondria from green leaves presents special problems and a good protocol is described in Day et al.. There is, at present, no good protocol for the isolation of both pure and functional mitochondria from Arabidopsis leaves, but a protocol for the purification of mitochondria from Arabidopsis cell cultures is given in Kruft et al.. Finally, Boutry et al. demonstrate how purified mitochondria can be isolated from small amounts of plant tissue (down to 0.5 g fresh weight).

Subfractionation of Mitochondria

The mitochondrion contains four compartments, each with a unique complement of proteins: the outer membrane (containing around 7 per cent of the protein), the intermembrane space (perhaps 1–2 per cent), the inner membrane (50–60 per cent) and the matrix (30–40 per cent). It is sometimes desirable to be able to isolate these, for instance when trying to localize a given function within the mitochondrion and is usually done in the above figure.

Since the surface area of the inner membrane (IMM) greatly exceeds that of the outer membrane (OMM), the latter breaks, releasing the contents of the intermembrane space (IMS) together with some OMM. Addition of osmoticum causes the mitoplasts—mitochondria minus (some) OMM—to contract, making them less susceptible to mechanical damage. The mitoplasts can be pelleted and subfractionated into IMM vesicles (submitochondrial particles or SMP) and matrix by sonication followed by differential centrifugation. When sonication is done in a high-salt medium, the SMP are inside-out—that is, they have the opposite orientation to that of the IMM in intact mitchondria.

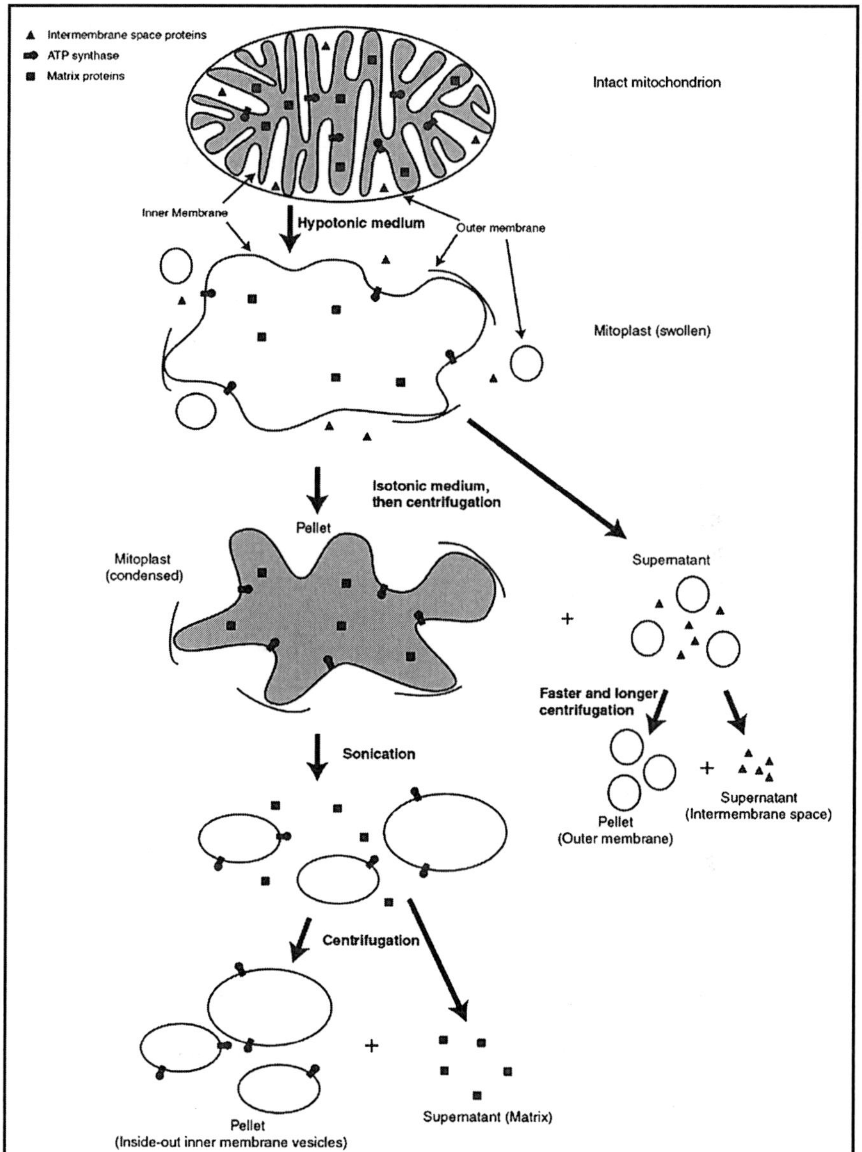

Fig. Subfractionation of mitochondrion compartments. The hypotonic medium causes the mitochondria to swell because the matrix takes up water.

The SMP are contaminated with some OMM. The OMM are isolated from the mitoplast supernatant by centrifugation. The yield of OMM is small, but relatively uncontaminated by IMM. Intermembrane proteins are found in the OMM supernatant. Symbols: Small squares—matrix protein (e.g., malate dehydrogenase), triangles—intermembrane space protein (no known marker for plant mitochondria), square-and-circle—F_oF_1-ATPase (ATP synthase) in the inner membrane with the F_1 facing the matrix.

- The outer membrane is ruptured, either by osmotic swelling or by selective solubilization with digitonin, a special detergent. In both cases, it is critical to avoid rupturing the inner membrane because the release of matrix from even a few per cent of the mitochondria will cause very large contamination of the intermembrane space, which is small in volume.
- Pellet the mitochondria. The supernatant contains intermembrane space plus outer membrane from at least a fraction of the mitochondria (usually only 10–20 per cent). The outer membrane can be pelleted from this supernatant and the intermembrane space proteins are left in solution.
- The pelleted mitochondria—often called mitoplasts, which are mitochondria without outer membrane—are then disrupted, (e.g., by sonication), and the inner membranes pelleted. That leaves the matrix proteins in the supernatant. The inner membrane usually reseals after rupture and forms sealed vesicles, (so-called submitochondrial particles) the sidedness of which can vary depending on the disruption method and conditions. Note that the inner membrane fraction usually contains some outer membrane contamination.

Methods for Establishing Purity, Intactness, and Functionality of the Fractions

It is often important to determine the degree of purity of the mitochondria and their subfractions The intactness of the mitochondria is usually established with enzyme assays that take advantage of the fact that an intact membrane is not permeable to certain molecules. For instance, the outer mitochondrial membrane is impermeable to cytochrome *c*, a 12.5 kDa protein that is too large to pass through the outer membrane pores, and NADH cannot pass through an intact inner membrane. Functionality is usually tested using an oxygen electrode as illustrated in textbook, Figure. If the mitochondria show high rates of oxygen consumption, high respiratory control ratios (>3), and ADP:O ratios close to the theoretical values they are in very good shape. Note that, for mitochondria with an appreciable alternative oxidase activity (measured in the presence of the cytochrome oxidase inhibitor, KCN), the respiratory control ratio and ADP:O ratio should be measured in the presence of an alternative oxidase inhibitor such as salicylhydroxamic acid.

Table. ADP:O ratios in isolated plant mitochondria

Substrate	ADP:O ratio
Malate	2.4– 2.7
Succinate	1.6–1.8
NADH (external)	1.6–1.8
Ascorbate	0.8–0.9

Mitochondrial Activities Can Be Determined *In Situ*

It is relatively straightforward to measure the activity of enzymes in the soluble matrix fraction isolated from mitochondria. However, this fraction is very diluted compared to the matrix inside intact mitochondria (300–500 mg protein/ml!) and protein–protein interactions may have been lost.

When measuring the activity of enzymes in fractions containing membranes—intact mitochondria and submitochondrial particles—a detergent can often be used to remove the permeability barrier that the inner membrane constitutes. However, it is also possible to use the ionophore alamethicin to permeabilize the inner membrane to molecules <1000 Da, which makes it possible to measure the activity of matrix and inner membrane enzymes in their native proteinaceous environment.

Using alamethicin, it is even possible to measure the activity of cytosolic and mitochondrial enzymes inside intact cells since the ionophore also permeabilizes the plasma membrane.

THE Q-CYCLE EXPLAINS HOW PROTONS CAN BE PUMPED ACROSS THE INNER MITOCHONDRIAL MEMBRANE

The electron transport chain (ETC) in the mitochondrial inner membrane of most eukaryotes consists of four complexes, complex I to complex IV. Complexes I, III and IV pump protons over the inner mitochondrial membrane when transporting electrons. For each electron pair (corresponding to one NADH or one oxygen atom), four protons are pumped by complex I, four by complex III, and two by complex IV. The detailed three-dimensional structures have been determined for the mammalian complexes III and IV, but not for complex I. The structural information has improved the understanding of the electron paths of complexes III and IV, but the proton-pumping function of the latter is still not fully understood. For complex I, neither electron nor proton paths are known in any detail.

Ubiquinone (Q), also called coenzyme Q, is a mobile electron and proton carrier in the inner mitochondrial membrane, structurally and functionally similar to plastoquinone in the thylakoid membrane of plastids. The ability of Q to carry both electrons and protons—and that it can accept and donate them step-wise—is utilized in electron transport and proton pumping.

To be reduced, it can receive two electrons from complex I or succinate dehydrogenase (but also, for example, the type II internal NADH dehydrogenase) and take up two protons from the matrix, forming QH_2. However, at the oxidizing side (complex III), there is actually both oxidation and reduction taking place. A QH_2 will be oxidized to Q at the outer Q-site, releasing two electrons. One of these electrons goes to cytochrome *c*, whereas the other is recycled to reduce an oxidized Q in the inner Q-site—one step to becoming a semiquinone. The next time a QH_2 is oxidized at the outer Q-site, the semiquinone in the inner site will receive a second electron, becoming a QH_2. So, from two QH_2, we have made one Q and one QH_2, which again can be oxidized. In each cycle (i.e., for each electron passing to cytochrome *c*), two protons are taken up from the matrix and two protons released to the intermembrane space. So, the Q-cycle mediates the pumping of an extra proton per electron transported, increasing energy conservation.

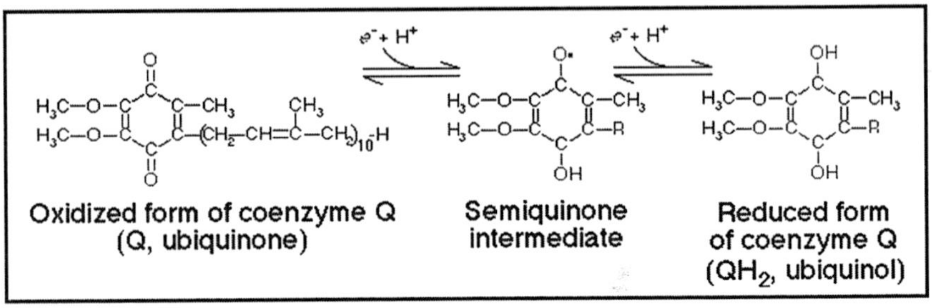

Fig. The structure of Q and its reduced form, ubiquinol (QH_2).

A Q can take up two electrons and two protons and become QH_2. Alternatively, the reduction can go stepwise, with one electron plus one proton, at a time. Then the intermediate form, ubisemiquinone, will exist as a relatively short-lived intermediate. Oxidation of QH_2 is the reversal of the process. The Q molecule has a long hydrocarbon tail that anchors it within the membrane, depicted as R in the semiquinone and QH_2 structures.

Complex III and Q together carry out proton pumping. In this picture we have also included Q reduction via the succinate dehydrogenase (complex II) so that we can see a more complete process. The process contains two steps.

The reactants of each step are depicted in blue, the products are denoted in red, and intermediates are black. The reactant Q molecules of the first step, one reduced, and two oxidized, are the same as the products of step two, i.e. we return to the same situation. In each step, one electron is transported from succinate dehydrogenase to cytochrome *c*, but two protons from the matrix are transported to the intermembrane space. The electron path in succinate dehydrogenase involves an FAD and several iron-sulfur centers (FeS) bound to the protein.

In complex III, one electron from QH_2 goes to the mobile carrier cytochrome *c* via the Rieske FeS and cytochrome c_1 (bound to the complex). The other electron goes to a second Q-binding site via b-type cytochromes (cytochrome b_L and cytochrome b_H), where it is donated to a Q (step 1) or a semiquinone (step 2).

The subscripts L and H stand for "low potential" and "high potential", respectively. The higher the potential, the more readily the molecule accepts electrons. If we look at the whole reaction from succinate to cytochrome *c*, we see that for each succinate, four protons are transported from the matrix to the intermembrane space.

However, at the same time, the two electrons are also transported out. So, this corresponds to the net pumping out of two positively charged species. Since, energetically, the mitochondrial membrane potential (which determines the energetic cost of transporting a positive charge out) is much larger than the pH gradient (which determines the cost for electroneutral export of a proton), the complex III proton pumping is less energy-demanding than, for example, complex I, which pumps four protons in their completeness.

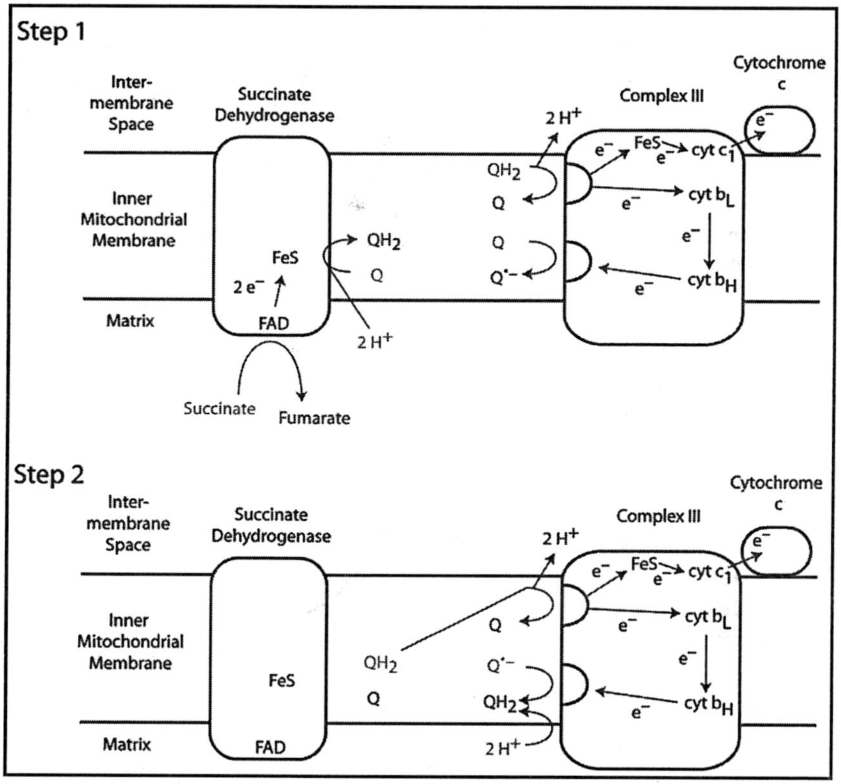

Fig. Reduction and Oxidation of Ubiquinone (Q) and the Q-cycle.

MULTIPLE ENERGY-CONSERVATION BYPASSES IN OXIDATIVE PHOSPHORYLATION OF PLANT MITOCHONDRIA

The electron transport chain (ETC) in the mitochondrial inner membrane of most eukaryotes consists of four large protein complexes, complex I to complex IV. The electron transport activity of complexes I, III and IV (the latter two, often jointly called the cytochrome pathway) is coupled to extrusion of protons across the inner mitochondria.

The electrochemical proton gradient that is formed and maintained in the process is used by the ATP synthase (also called complex V) to make ATP .The plant ETC, however, is highly branched.

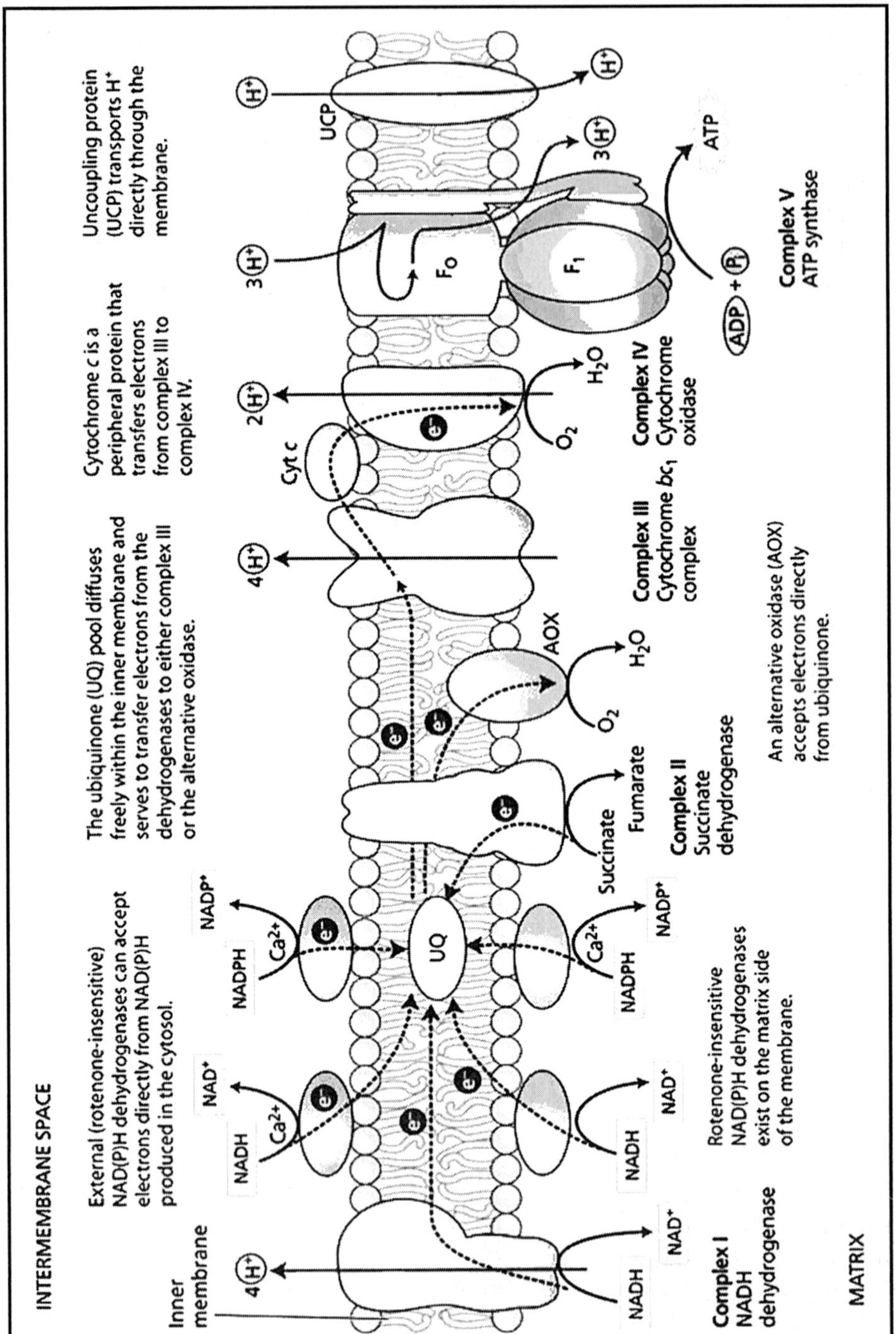

Fig. Organization of the Electron Transport chain and ATP Synthesis in the Inner Membrane of Plant Mitochondria.

As a consequence, the plant oxidative phosphorylation system has several alternative pathways of electron transport via type II NAD(P)H dehydrogenases and the alternative oxidase (AOX), and additionally, an alternative proton transporter called the uncoupling protein (UCP). These proteins mediate bypasses around the proton-translocating multi-protein complexes (I–V): type II NAD(P)H dehydrogenases bypass complex I, AOX bypasses complexes III and IV, and UCP bypasses the ATP synthase. The type II NAD(P)H dehydrogenases and the AOX transport electrons without pumping protons, whereas the UCP allows proton flow from the intermembrane space to the matrix without ATP synthesis.

The consequence of this is that flux through the energy conservation bypasses will neither contribute to respiratory ATP production, nor will they be controlled by cellular adenylate status. This Topic will describe the special features of these energy bypass proteins and discuss the consequences of their existence to the plant.

In addition to the four standard protein complexes found in nearly all other mitochondria, the electron transport chain of plant mitochondria contains five additional enzymes. None of these additional enzymes pumps protons. Additionally, the uncoupling protein directly bypasses the ATP synthase by allowing passive proton influx. This multiplicity of bypasses in plants, where animals have only the uncoupling protein, gives a greater flexibility to plant energy coupling. Specific inhibitors—rotenone for complex I, antimycin for complex III, cyanide for complex IV, and salicylhyroxamic acid (SHAM) for the alternative oxidase—are important tools used to investigate mitochonondrial electron transport. Some also have commercial uses, such as rotenone—which is used as an insecticide and to remove unwanted fish from lakes. Because plants have the alternative pathways, they can survive exposure to inhibitors of the respiratory complexes.

NAD(P)H Dehydrogenases in Plant Mitochondria

Except for yeast, the ETC of all eukaryotes—including plants—contains a proton-pumping NADH dehydrogenase, complex I, which is inhibited specifically by rotenone. Additionally, plant, fungal and protist mitochondria contain different setups of rotenone-insensitive enzymes capable of oxidizing NADH and/or NAD(P)H and passing the electrons on to ubiquinone. These NAD(P)H dehydrogenases are classified as type II, whereas complex I belongs to type I. The inner mitochondrial membrane in plants contains such NAD(P)H dehydrogenases, facing either the matrix or the intermembrane space, thus oxidizing NAD(P)H either from the matrix or the cytoplasm.

The outer membrane of all mitochondria contains NADH-ferricyanide and NADH-cytochrome *b* reductase activities that are insensitive to rotenone and to the complex III inhibitor, antimycin A. However, the outer membrane activities are not linked to the inner membrane ETC and will not be treated further here.

There are Three NAD(P)H Dehydrogenases on the Inner Matrix Surface of the Inner Membrane

Much of the information about the number of matrix-facing NAD(P)H dehydrogenases has come from experiments using inside-out inner membrane vesicles. Such vesicles have the matrix surface facing the medium and the external NAD(P)H dehydrogenases hidden inside the vesicles and unable to interfere with the assays. Recently, this method has been complemented with a technique where one uses a peptide called alamethicin to make a channel for the substrate through the inner membrane. Experiments with these methods show the presence of three NAD(P)H dehydrogenase activities on the matrix side of the inner membrane:

- Complex I pumps protons, is Ca^{2+}-independent and is inhibited by rotenone and diphenylene iodonium (DPI). Complex I has a high affinity for NADH [low Km(NADH)].
- ND_{in}(NADH) does not pump protons, is Ca^{2+}-independent and DPI insensitive. The enzyme has a tenfold lower affinity for NADH than complex I [i.e., a tenfold higher Km(NADH)].
- ND_{in}(NADPH) does not pump protons, is Ca^{2+}-dependent and very sensitive to DPI.

Because of the contrasting affinities of complex I and ND_{in}(NADH) for NADH, only complex I is engaged at low matrix concentrations of NADH. However, when the NADH concentration in the matrix increases (e.g., if the ETC is restricted by lack of ADP in the cell), ND_{in}(NADH) can become engaged.

When electrons are transported through the type II NAD(P)H dehydrogenases, only the two proton pumping sites of the cytochrome pathway are involved in the flow of electrons from matrix NADH to oxygen. As a consequence, the theoretical ADP/O yield is 1.5 unless the AOX or the UCP are active, in which case it will be even lower.

The External NAD(P)H Dehydrogenases are Ca^{2+}-Dependent

Plant mitochondria can oxidize cytosolic NADH and NAD(P)H directly via external dehydrogenases. The NADH and NAD(P)H dehydrogenase activities are catalyzed by separate enzymes. None of the external dehydrogenases pumps protons, which means that, at most, two sites of proton-pumping are involved when the electrons move from the dehydrogenases to oxygen. So, as for ND_{in}(NADH), the theoretical maximum ADP/O yield is 1.5.

The oxidation of exogenous NADH or NAD(P)H by most plant mitochondria requires Ca^{2+}, and as a result, is inhibited by EDTA and EGTA, chelators of di- and trivalent cations (Coleman and Palmer 1971; MΨller et al. 1981).

In vitro, the Ca^{2+}concentration needed for half-maximal activity of NADH oxidation is 0.1–0.5 μM, depending on ionic environment and respiratory state This is in the same range as the 0.1–0.2 μM free Ca^{2+} thought to be present in the cytosol of a resting plant cell. Therefore, the external NAD(P)H oxidation may become activated by the increased Ca^{2+}-concentration associated with cellular excitation due to external conditions.

Mammalian mitochondria lack external NAD(P)H dehydrogenases. Instead, they have a Ca^{2+}-dependent glycerol-3-phosphate dehydrogenase on the outer surface of the inner membrane, which donates electrons directly to the ubiquinone pool. The dihydroxyacetone phosphate formed is converted back into glycerol-3-phosphate by a cytosolic NAD-dependent glycerol-3-phosphate dehydrogenase. In this way, the two low-molecular-weight compounds, glycerol-3-phosphate and dihydroxyacetone phosphate, shuttle redox equivalents from the cytosol, across the outer membrane and to the inner membrane to yield the same amount of ATP as is produced by the plant mechanism.

Yeast have both external NADH dehydrogenases and a glycerol-3-phosphate shuttle, and since a mitochondrial glycerol-3-phosphate dehydrogenase has been found in plant mitochondria, plants may also have both systems. Another way to oxidize cytosolic NAD(P)H that is found in mammalian mitochondria is to use the malate/aspartate shuttle to tranfer reductant (a collective term for high-energy electrons carried by, for example, NAD(P)H, ferredoxin or an organic acid redox couple like malate/oxaloacetate) into the matrix as NADH that can be oxidized by the internal NAD(P)H dehydrogenases.

However, it is not clear whether this shuttle can work for import of cytosolic reductant into plant mitochondria. In the opposite orientation, NADH is shuttled from the matrix to the cytosol via the malate/oxaloacetate shuttle.

In this way, the external NADH dehydrogenase can also work as a bypass around complex I. Support for this bypass has been seen in the form of an upregulation of external NADH dehydrogenase activity in mutants where complex I or an internal type II NAD(P)H dehydrogenase has been inactivated.

The Molecular Identities of the Rotenone-Insensitive NAD(P)H Dehydrogenases

The type II NADH dehydrogenases from *E. coli* and yeast are small, single polypeptide proteins of around 50 kDa. They oxidize only NADH and contain FAD as an electron-accepting prosthetic group. Attempts to purify the type II NAD(P)H dehydrogenases have been unsuccessful in identifying the plant proteins. However, the characterization of potato homologues to the type II NADH dehydrogenases in *E. coli* and yeasts have provided candidate plant proteins to try and assign to the different reported activities. One protein, NDA1, is internal and the other, NDB1, external of the inner mitochondrial membrane. The latter was shown to be an external calcium-dependent NAD(P)H dehydrogenase. Like a related fungal NAD(P)H dehydrogenase, NDB1 contains a motif for Ca-binding, indicating that in the activation of plant and fungal NAD(P)H dehydrogenase by Ca^{2+}, the ion binds directly to the enzyme.

In Arabidopsis, a weed for which the nuclear genome sequence is known, small families of both NDA-type and NDB-type proteins are encoded by nuclear genes. One of the NDA-type proteins has been shown to be an internal NAD(P)H dehydrogenase. Additionally, a third type of type II NAD(P)H dehydrogenase was found in Arabidopsis, the NDC1. This protein is most closely related to cyanobacterial enzymes, suggesting that the gene entered the eukaryotic domain with the plastid, later to be transferred to the nucleus.

The Alternative Oxidase

The presence of the AOX is one of the features that sets plant (and fungal) mitochondria apart from mammalian mitochondria. This enzyme has therefore received a lot of attention from plant scientists and we now have a fairly good understanding of the regulation of its structure and activity.

The AOX is a quinol-oxygen oxidoreductase and it does not pump protons. The oxygenase transfers electrons from ubiquinol (the reduced form of ubiquinone) to oxygen and generates water as the end-product of the reaction, so four electrons must be transferred to oxygen. The precise mechanism is not known, but the protein has been shown to contain a catalytical di-iron centre. The functional form of the enzyme is a dimer with the two polypeptides either covalently or non-covalently bound to each other. The predicted structure places the protein embedded into the inner leaflet of the inner mitochondrial membrane, with the active site and the regulatory cysteines exposed to the matrix.

When electrons from NADH oxidation flow through the AOX, at least two of the three sites of proton translocation (complexes III and IV) are bypassed, so energy conservation in the form of ATP is much smaller when the AOX is active. The activity of the AOX must therefore be carefully regulated. There are often several AOX isoforms in each plant species—for example, five genes encoding AOX proteins are present in Arabidopsis. The relative amounts of three isoforms of AOX vary with development, and cellular redox status, as well as external factors. The AOX is present in the inner membrane as a dimer in which the two subunits can be covalently linked through a disulfide bridge between cysteine residues.

The reduced non-covalently linked dimer is the active form, and this form appears to be the predominant one *in vivo*. It was recently shown that a mitochondrial thioredoxin system regulates AOX.

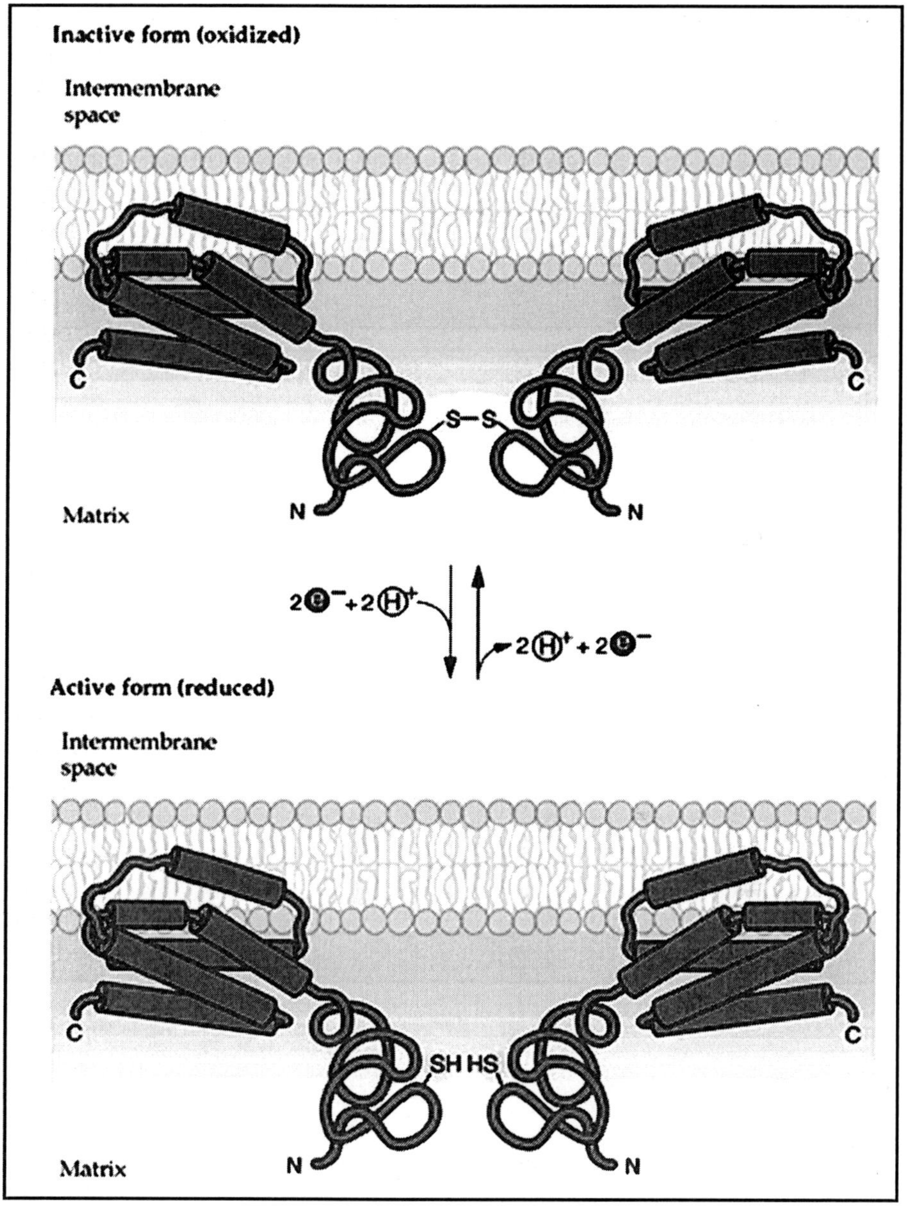

Fig. The Structure and Regulatory Features of the Alternative Oxidase in Plant Mitochondria. The C and N indicate the C- and N-terminal ends of the protein, Respectively.

Thioredoxins regulate enzymes by reducing disulfide bridges, and it is likely that the regulation involves the redox-active cysteines. The presence of thioredoxin regulation connects the AOX activation level to the mitochondrial NAD(P)H pool and the oxidation of citric acid cycle metabolites by NADP-utilizing enzymes.

The activity of AOX is also stimulated by 2-oxo organic acids—in particular, pyruvate, which activates by interacting with the same cysteine that is involved in dimerization. This activation may be a feed-forward mechanism that upregulates the capacity of the ETC when the glycolytic end-product is available. Both the cytochrome pathway and the AOX pathway use ubiquinol as the substrate. In the absence of pyruvate, the cytochrome pathway is saturated at 40 per cent Q_{red}/Q_{tot} (40 per cent of the UQ is reduced), at which point the AOX pathway is not engaged at all. The AOX pathway starts accepting electrons at 50 per cent reduction and is not fully saturated at 80 per cent.

In the presence of pyruvate, the AOX becomes engaged at a much lower Q_{red}/Q_{tot} (20 per cent), whereas the cytochrome pathway is unaffected. It is puzzling, however, that the normal *in vivo* concentration of pyruvate appears to be higher than the concentration of pyruvate required to give maximal stimulation of AOX activity *in vitro*. It thus appears that there is little regulation of AOX activity via pyruvate under physiological conditions.

Q_r is ubiquinol (reduced ubiquinone) and Q_t is total ubiquinone (reduced + oxidized). Myxothiazol is an inhibitor of complex III and, in its presence, only the AOX is active.

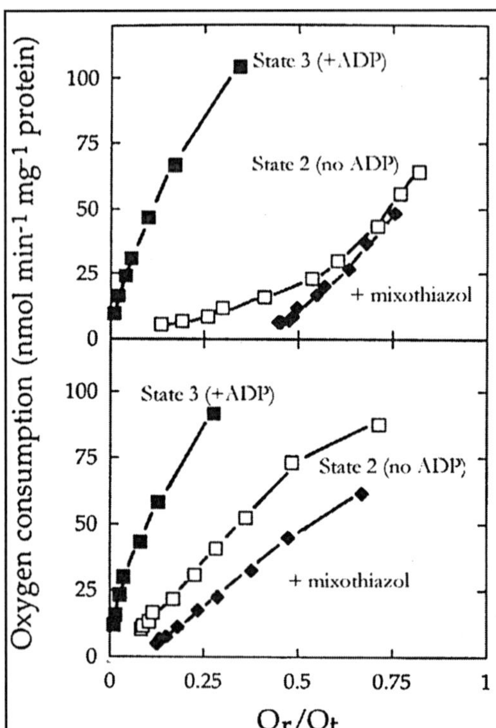

Fig. Dependence of respiratory rate on quinone redox state in purified soybean cotyledon mitochondria oxidizing succinate in the absence (*top* graph) or presence (*bottom* graph) of pyruvate.

AOX Activity Can Be Measured *In Vivo*

For many years, benzhydroxamic acid derivatives, [e.g., salicylhydroxamic acid (SHAM)], were used to quantify the capacity and activity of the AOX pathway in both isolated mitochondria and intact tissues. AOX capacity was determined from the rate of respiration in the presence of KCN (which blocks the cytochrome pathway), and corrected for residual respiration in the presence of both KCN and SHAM. AOX *in vivo* activity was determined from the decrease in the rate of respiration when SHAM was added in the absence of KCN. The assumption was that the cytochrome pathway was always used first and that the AOX would be engaged only when the cytochrome pathway was saturated. However, as we have seen above, this assumption is not correct. The consequence of this is that *in vivo* activity of AOX cannot be measured using inhibitors alone.

A better method for measuring AOX *in vivo* activity is now available. It makes use of the fact that cytochrome *c* oxidase and AOX discriminates differently between ^{18}O relative to ^{16}O.

The so-called oxygen discrimination of the sample (tissue or mitochondria) can be measured by mass spectrometry: (1) with KCN to resolve the oxygen discrimination by AOX; (2) with SHAM to resolve the oxygen discrimination by cytochrome *c* oxidase; and (3) without inhibitors to resolve the discrimination value resulting from the activity of both enzymes.

From these values, the percentage contribution by AOX and cytochrome *c* oxidase to the total uninhibited rate of respiration can be calculated. An example of the use of the oxygen discrimination technique with an intact tissue is given below. A remaining problem in the estimation of AOX activity *in vivo* is that measurements can only be made in darkness (to avoid interference from photosynthetic oxygen exchange).

Plant Mitochondria also Contain an Uncoupling Protein

Mammalian mitochondria contain neither type II NAD(P)H dehydrogenases nor AOX. However, they have a UCP that increases the proton permeability of the inner mitochondrial membrane and, in that way, dissipates the proton gradient.

This is another mechanism for reducing the ATP production and increasing heat production, for example, for thermoregulation. Surprisingly, plant mitochondria also contain a UCP. We do not know for certain why plant mitochondria should require two different mechanisms for achieving the same end result.

The UCP is related to anion carriers. Its *in vitro* transport activity is dependent on free fatty acids and is inhibited by nucleotides like ATP and GTP but stimulated by superoxide. The latter result indicates that the UCP may protect against oxidative stress.

The Physiological Impact of Energy Bypass Proteins

The respiratory rates of some thermogenic flowers are the highest among plants and, in fact, exceed even those of warm-blooded animals. The external NADH dehydrogenase, AOX and UCP appear particularly active in thermogenic flowers where they are primarily responsible for heat production. At the same time it is clear that the respiratory rate of most plant tissues is insufficient to generate an appreciable increase in tissue temperature.

AOX is up-regulated by cold stress, although this is clearly not to heat the tissue. However, during cold stress, reactive oxygen species (ROS) are formed, and also other stress treatments that lead to elevated ROS will induce AOX. ROS (e.g., superoxide and hydrogen peroxide) are formed as by-products of electron transport under aerobic conditions.

ROS can cause damage to proteins, lipids, and DNA and the cell must therefore limit their formation. AOX helps prevent overreduction of the ETC and thus lowers ROS production. Supporting this, Maxwell et al. (1999) demonstrated the importance of the AOX for limiting ROS formation in living plant cells.

Overexpression of AOX lowered the steady state level of ROS, whereas suppression increased it. This result also points to the mitochondrion as an important site of ROS generation in plant cells. With the recent discovery of ROS stimulation of plant UCP and the increased ROS resistance in plants overexpressing UCP, the picture has become more complex.

By decreasing the resistance for electron transport through the proton-pumping complexes, UCP activity may, like AOX, lead to decreased overreduction and ROS formation. Another factor speaking in favour of UCP having a role in ROS protection is that the AOX is inhibited by an ROS product. For further information on mitochondrial involvement in ROS metabolism.

Both type II NAD(P)H dehydrogenases and AOX may act as reductant overflow enzymes by reoxidizing surplus NADH to oxygen. This can prevent inhibition of the citric acid cycle under conditions in which the ADP concentration is low, and thus the proton-pumping complexes are restricted by a high membrane potential.

In this way, the citric acid cycle can supply carbon skeletons for biosynthesis (e.g., 2-oxoglutarate for nitrogen assimilation), even if the ATP level in the cell is high. The concept of overflow is also relevant for photorespiratory conditions, under which there is a massive flow of glycine that has to be oxidized by the mitochondria.

As much as half of the NADH produced in this process is probably exported, but the other half has to be oxidized promptly. There are several reasons to believe that a reductant overflow via energy bypass proteins is involved. The ND_{in}(NADH) and AOX capacity in leaf mitochondria is higher than in mitochondria from heterotrophic tissue, and both proteins are down-regulated by dark treatment. Consistent with these observations, AOX activity *in vivo* is up-regulated by light, AOX capacity is induced by greening of etiolated tissue, and expression of the *nda*1 gene [encoding an internal NAD(P)H dehydrogenase] in potato and Arabidopsis is light-dependent.

Mutant tobacco plants, lacking the proton-pumping complex I can still survive, apparently aided by the action of ND_{in}(NADH), which oxidizes matrix NADH with a lower ATP yield. These plants, however, show decreased photosynthetic rates, indicating that the ND_{in}(NADH) cannot fully compensate for the missing complex I.

Also, external NAD(P)H dehydrogenases have been suggested to support photosynthesis. These enzymes may reoxidise NAD(P)H that is shuttled out of the chloroplast in order to balance the NAD(P)H to ATP ratio in the stroma. In summary, there are many indications that mitochondria—especially the energy bypass proteins—have roles in supporting photosynthesis, for example by oxidizing excess NAD(P)H.

The AOX pathway is involved in the acidification/deacidification cycle of CAM plants. CAM plants fix CO_2 in the form of malate during the night. This is done by cytosolic PEP carboxylase, and the malate produced is stored in the vacuole. The CO_2 fixed is released during the daytime by malic enzymes, located in cytosol and mitochondria. Robinson et al. (1992) used the oxygen discrimination technique to measure the contribution of the AOX pathway to whole-leaf respiration in the CAM-plant, *Kalanchoë daigremontiana.* They found that the respiratory rate of the leaves increased by 35 per cent in the early hours of the day. This increase was due to increased AOX activity. A more detailed time course showed that the AOX activity increased two- to threefold during the early day at the same time as the malate was metabolized. As soon as the malate had disappeared by mid-day, the AOX activity returned to the low level observed during the night. A reasonable conclusion is that a substantial amount of the malate from the vacuole is oxidized via malic enzyme in the mitochondria. This generates NADH in the matrix, which is oxidized to a large extent via the AOX. The released CO_2 can then be refixed by the Calvin cycle in the chloroplast.

Table. Respiration, oxygen isotope discrimination and partitioning of respiratory pathways in kalanchoe leaf discs during acidification adn deacidification (in the first 6th on the light period).

Parameter	Acidification	Deacicification
Measured		
Unhibited respiration		
Rate (μmol O^2 uptake $m^{-2}s^{-1}$)	1.81	2.56
Discrimination (o/oo)	22.38	25.00
Calculated*		
Cytochrome pathway flux	1.33	1.40

(µmol O_2 uptake $m^{-2}s^{-1}$)		
Alternative pathway flux (µmol O_2 uptake $m^{-2}s^{-1}$)	0.52	1.16

Respiration, oxygen isotope discrimination and partitioning of respiratory pathways in *Kalanchoë* leaf discs during acidification and deacidification (in the first 6 h of the light period)

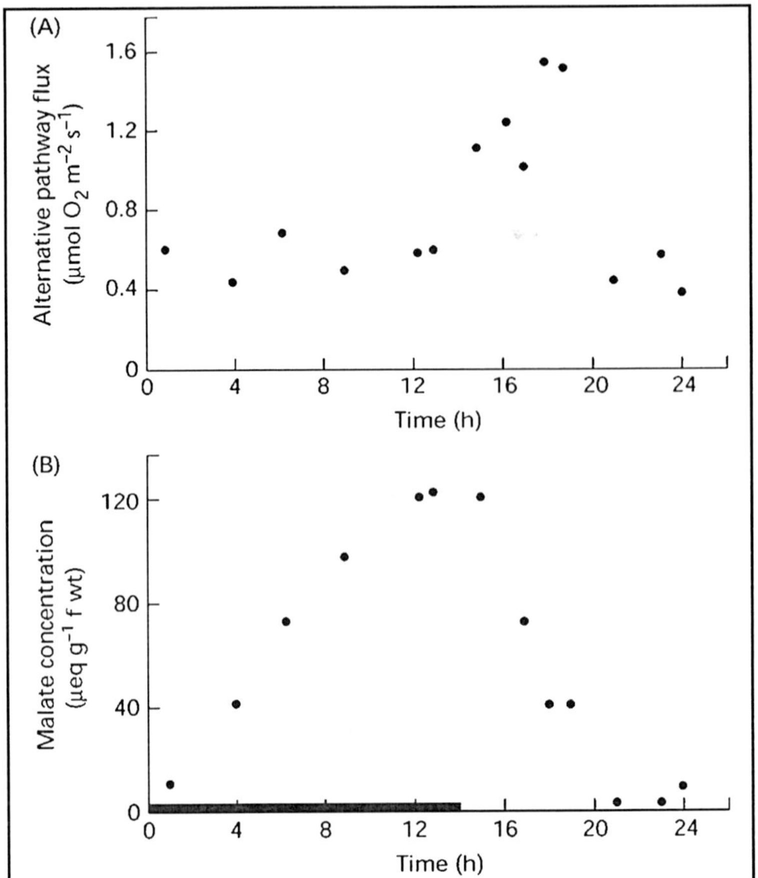

Fig. The Participation of the AOX in Malate Oxidation in *Kalanchoë Diagremontiana* during a night-day cycle.

(A) Changes in flux through the AOX pathway in leaf discs. (B) Changes in malate concentration of leaf discs. The solid bar at the bottom of the figure represents the dark period.

Bibliography

M. Amin: *Advanced Bioenergetics and Biodynamics*, Capital Publishing Company, 2011.

Suhasini Besra: *Growth and Bioenergetics of Anabas Testudineu (Bloch) (An Air-Breathing Climbing Perch of South-East Asia)*, Narendra, 1997.

S. Sundara Rajan: *Introduction to Bioenergetics*, Anmol, 2003.

U K Mishra: *Physiology of Animal Growth and Bioenergetics*, Satish Serial Pub House, 2008.

Ashok Kumar Sharma: *Biochemistry of Biomolecules*, Random Pub, 2011.

Smriti Ganguly: *Biochemistry of Biomolecules*, Pearl Books, 2007.

Shubhrata R. Mishra: *Biomolecules*, Discovery Pub, 2003.

VK Ahluwalia: *Biomolecules*, Manakin Press, 2015.

T. Devasena: *Biomolecules*, MJP Publishers, 2010.

Rajendra Kaushik and Asha Sharma: *Biomolecules : Structure and Functions*, Oxford Book Company, 2009.

A.K. Goel: *Chemistry of Biomolecules*, Pearl Books, 2010.

Julia Laskin and Chava Lifshitz: *Principles of Mass Spectrometry Applied to Biomolecules*, Wiley India, 2013.

Bhaskar Bala: *Energy Security and Economic Development in India*, Teri Book Store, 2012.

Bhaskara Rao Digumarti and Harshitha Digumarthi: *Energy Security*, Discovery, 2001.

Doraiswarmy Rashmi: *Energy Security: India Central Asia and the Neighbourhood*, Manak Publications, 2013.

Ghosh G K.: *Solar Energy: The Infinite Source*, APH, Publication, 2008.

Malakar S.N.: *India's Energy Security and the Gulf*, Academic Excellence, 2006.

Mallik Amitav: *Renewable Energy Technologies: Special Focus on Distributed Power Generation Potential for Applications to Rural Sectors in India*, Academic Foundation, 2008.

Muni S.D.: *India's Energy Security: Prospects for Cooperation with Extended Neighbourhood*, Rupa, 2002.

Parvez Razia: *Energy Security and Conservation in India*, New Century Publications, 2013.

Peimani Hooman: *The Challenge of Energy Security in the 21st Century*, Bookwell Publications, 2013.

Shinde Kunal: *Get Rid of Diabetes with the Help of Solar Energy*, Indus Publication, 2005.

Vandana S.: *Alternative Energy*, APH, Publication, 2002.

Vinodan C.: *Energy Security Choices for India*, New Century Publications, 2015.

V.N. Meena Devi, G.S. Vijayalakshmi and P. Nagendra Prasad: *A Handbook of Bio Energy Crops*, Pointer Pub, 2008.

Sudhir T. Devare: *A New Energy Frontier: The Bay of Bengal Region*, Manohar, 2010.

B. Muzumdar: *A Text Book of Energy Technology : Both Conventional & Renewable Source of Energy*, APH, 1999.

A.B. Saxena: *A Textbook of Energy Environment Ecology and Society (RGPV)*, New Age International, 2011.

Anubha Maheshwari and Geeta Parmar: *A Textbook of Energy, Ecology, Environment and Society*: , Anmol, 2002.

P. C Maithani and Deepak Gupta: *Achieving Universal Energy Access in India : Challenges and the Way Forward*, Sage, 2015.

Ruchita Beri and Uttam Kumar Sinha: *Africa and Energy Security : Global Issues, Local Responses*, Academic Foundation, 2009.

S. Vandana: *Alternative Energy*, APH, 2002.

S.R. Manjunatha: *An Introduction to Environmental Energy Resources*, Cyber Tech Publications, 2010.

S.N. Virnave: *Atomic Energy and Indias Nuclear Capability*, Anmol, 2001.

Christine M. Grimm: *Aura Healing Handbook : Learn to Read and Interpret the Aura Perceive Energy Fields in Color and Utilize Them for Holistic Healing : Walter Lubeck*, Sri Satguru, 2000.

Beena Sharma: *Awakening of Serpent Energy Through Tantra*, Shubhi, 2009.

S.V. Govindan: *Ayurvedic Massage for Health and Healing Ayurvedic and Spiritual Energy Approach*, Abhinav, 2000.

Alimuddin Khan: *Basics of Energy*, Anmol, 2009.

Index

A

Acceleration 170
Alternative Oxidase 294, 299, 303
Anoxygenic 216
Antenna System 217, 223, 227, 235
Apoenzyme 168, 171, 172
Arabidopsis 292, 296
Aspartic Acid 63
Atmospheric 212, 219
ATP Synthase 230, 231
Auxins 246, 247, 248, 252, 254, 256

B

Backbone Atoms 53
Bacterial Photosynthesis 237, 239, 244
Bacteriochlorophylls 240, 241, 243
Bacteriophage 59, 60, 63, 64, 65, 68
Brown Species 242

C

Calvin Benson cycle 245
Carbohydrates 232
Carbohydrates 212, 213, 216, 219
Carboxylic Acid 239
Carboxypeptidase 47, 48
Carotenoids 242
Carotenoids 217, 221, 222, 237, 238, 240, 241, 243, 244
Chemical Potential 2, 10, 17, 218, 219, 230
Chlorobiaceae 236, 240, 241, 242, 245
Chloroflexaceae 240, 245
Choice of Material 290
Chymotrypsin 49
Coenzymes 169, 172, 173
Cyanogen Bromide 49, 50
Cytochrome c 78

D

Dehydrogenases 172
Desulfovibrio 244
Dinucleotide 172
Disruption of the Tissue 290

E

Efficiency 233
Electric Potentials 14
Electrochemical 12, 13, 230
Electrochemical Potential 13, 230
Electron Transport 228, 235
Energy Conversion 233
Enzymes 6, 219, 233, 234

F

Ferredoxin 301
Filtration 291

Flavoproteins 78, 80
Fractions 294

G

Gliding Bacteria 216
Glycogen 78, 79
Glycoproteins 78
Glycosidases 172

H

Hemoproteins 78, 80
Homogenization Medium 290
Hydrophobic 220, 225, 228

I

Inhibitors 176, 183, 184, 185, 186
Intermediate 50, 59
Intramolecular 49
Isolation of Mitochondria 290
Isomerization 168

L

Light Absorption 221
Lipoproteins 78, 79

M

Macromolecule 71
Membrane Potential 14
Metabolites 74, 75
Metalloproteins 78, 79, 80
Mitochondria 290, 291, 292, 297, 299
Mitochondrial Activities 295
Multiple Energy 297

N

Nucleoproteins 79

O

Oligopeptides 52, 60
Ovalbumin 71, 75
Oxidative Phosphorylation 297
Oxidative Phosphorylation 240, 299
Oxidoreductase 168, 172
Oxygen Evolution 214
Oxygenic Photosynthesis 234

P

Pentose Pathway 238
Phosphoproteins 78, 79
Plant Mitochondria 297, 299
Potential Factor 4
Primary Acceptor 236, 238
Primary Photochemistry 223
Prochlorophyta 240, 244
Purification of Mitochondria 291
Purple Bacteria 224, 237, 239, 240, 242

Q

Quantum Yield 233

R

Reaction Centre 214, 215, 216, 217, 218
Redox Reactions 11
Respiration 243
Ribonuclease 64, 65, 66, 71, 77
Ribozymes 167
Rubisco 219, 232, 233

S

Separation Methods 81
Somatotropin 71, 73
Special Features 299
Spontaneous 5, 6, 8
spontaneous 5, 6, 7, 8
Sulphur Bacteria 237, 240, 241, 242

T

Thiazolinone 46
Thyrotropin 71, 73

W

Water Vapour 11